ESSENTIALS OF
INTENTIONAL INTERVIEWING

ESSENTIALS OF INTENTIONAL INTERVIEWING

Counseling in a Multicultural World

SECOND EDITION

Allen E. Ivey
Courtesy Professor, University of South Florida, Tampa

Mary Bradford Ivey
Vice President, Microtraining Associates

Carlos P. Zalaquett
University of South Florida, Tampa

with
Kathryn Quirk
Screening for Mental Health

BROOKS/COLE
CENGAGE Learning

Australia • Brazil • Japan • Korea • Mexico • Singapore • Spain • United Kingdom • United States

BROOKS/COLE
CENGAGE Learning™

**Essentials of Intentional Interviewing:
Counseling in a Multicultural World,
Second Edition**
Allen E. Ivey, Mary Bradford Ivey,
Carlos P. Zalaquett, with Kathryn Quirk

Acquisitions Editor: Seth Dobrin

Assistant Editor: Alicia McLaughlin

Editorial Assistant: Suzanna Kincaid

Media Editor: Elizabeth Momb

Program Manager: Tami Strang

Content Project Manager: Rita Jaramillo

Design Director: Rob Hugel

Art Director: Caryl Gorska

Print Buyer: Rebecca Cross

Rights Acquisitions Specialist:
Tom McDonough

Production Service: Anne Draus,
Scratchgravel Publishing Services

Text Researcher: Pablo d'Stair

Copy Editor: Margaret C. Tropp

Cover Photo: © PhotoLink/Photodisc/
Getty Images

Compositor: MPS Limited, a Macmillan
Company

For product information and technology assistance, contact us at
Cengage Learning Customer & Sales Support, 1-800-354-9706.

For permission to use material from this text or product,
submit all requests online at **www.cengage.com/permissions.**
Further permissions questions can be e-mailed to
permissionrequest@cengage.com.

Library of Congress Control Number: 2011922246

Student Edition:

ISBN-13: 978-0-8400-3456-4

ISBN-10: 0-8400-3456-3

Brooks/Cole
20 Davis Drive
Belmont, CA 94002
USA

Cengage Learning is a leading provider of customized learning solutions with
office locations around the globe, including Singapore, the United Kingdom,
Australia, Mexico, Brazil, and Japan. Locate your local office at
www.cengage.com/global.

Cengage Learning products are represented in Canada by
Nelson Education, Ltd.

To learn more about Brooks/Cole, visit **www.cengage.com/brookscole.**

Purchase any of our products at your local college store or at our preferred
online store **www.cengagebrain.com**

Printed in the United States of America
4 5 6 7 8   19 18 17 16 15

Love is listening.

PAUL TILLICH

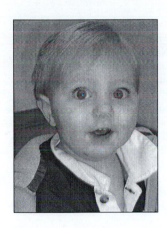

Dedicated to our latest grandson:
Tegan William Bradford Quirk

ALLEN AND MARY BRADFORD IVEY

Dedicated to my family:
Jenifer Zalaquett,
Andrea Zalaquett, and
Christine Zalaquett

CARLOS P. ZALAQUETT

About the Authors

ALLEN E. IVEY and **MARY BRADFORD IVEY** have been teaching and writing together for 30 years and have presented keynotes and workshop presentations throughout North America, the South Pacific, Asia, Europe, and the Middle East. Allen is Distinguished University Professor (emeritus) at the University of Massachusetts, Amherst, and Courtesy Professor of Counseling at the University of South Florida, Tampa. Mary is Courtesy Professor of Counseling at the University of South Florida, Tampa, and has served as visiting professor at the University of Massachusetts, Amherst; the University of Hawai'i, Manoa; and Flinders University, South Australia. Her primary career has been in schools, and her comprehensive elementary program was named one of the top ten in the nation at the Christa McAuliffe Conference.

The Iveys have received numerous honors but are particularly proud of their work in multicultural studies. Mary and Allen are Fellows of the American Counseling Association, and Allen was named a Multicultural Elder at the National Multicultural Conference and Summit. Allen has written more than 40 books and 200 articles and chapters, translated into 23 languages, and Mary has coauthored 15 of these, in addition to her own articles and chapters.

Allen's undergraduate work was in psychology at Stanford University, followed by a Fulbright Grant to study social work at the University of Copenhagen, Denmark. His doctorate is from Harvard University. Mary studied social work at Gustavus Adolphus College. She earned her masters in counseling at the University of Wisconsin and her doctorate in organizational development at the University of Massachusetts.

When they are not on the road, you'll find Allen and Mary in the boat pulling grandchildren on tubes on Lake Sunapee in New Hampshire or skiing on nearby Mt. Sunapee. Mary plays tennis while Allen bikes the hilly 24 miles around the lake. They also can be found worshipping the sun in Florida.

CARLOS P. ZALAQUETT is Associate Professor in the Department of Psychological and Social Foundations at the University of South Florida. He serves as the coordinator of the Mental Health Counseling Specialization and the Graduate Certificate in Mental Health Counseling. He is also the Director of the USF Successful Latina/o Student Recognition Awards Program and the Executive Secretary for United States of America and Canada of the Society of Interamerican Psychology. Carlos is the author or coauthor of more than 50 scholarly publications and four books, including the Spanish version of *Basic Attending Skills*. He has received several awards for his work on behalf of Latinas/os in education, such as the USF Latinos Association's Faculty of the Year Award and the Tampa Hispanic Heritage's Man of Education Award. He is an internationally recognized expert on mental health, diversity, and education. He has conducted workshops and lectures in seven countries on the topics of counseling, psychotherapy, and cyberbullying.

About the Contributor

KATHRYN M. QUIRK received her Master's in counseling from Cambridge College, Cambridge, Massachusetts. She completed two years of field work in Boston public schools in addition to working in a clinical outpatient setting. She has earned her initial professional license and certification as a school adjustment counselor. Her focus has been on inner-city youth and multicultural counseling. She also has 15 years of professional experience in public relations and communications. She now combines her counseling and past experience as Marketing and Communications Manager for Screening for Mental Health, Inc. (www.mentalhealthscreening.org). Kathryn has published articles about mental health education; this is her first contribution to a textbook.

C O N T E N T S

Essentials of Intentional Interviewing: Counseling in a Multicultural World, Second Edition, is based on the most researched framework on helping skills—microcounseling. These helping skills serve as the foundation of counseling, psychology, social work, and education. Students enjoy the focus on specific skills and identified competencies, the emphasis on diversity and wellness, and the many practice exercises. Many students develop an individual Portfolio of Competencies that can be shared as they move to practicum and internship settings. This book also provides the basics of interviewing in many other fields, ranging from nursing to business to communication studies.

This book was conceived because we wanted to write a competency-based brief text for courses in basic interviewing skills. This book contains the most important information from our larger book, *Intentional Interviewing and Counseling,* and can produce similar results. A key aspect of our vision is ensuring that students learn concepts that they can immediately take into direct practice in role-plays—and later into actual practice.

Through the step-by-step procedures of microcounseling, students will learn to predict the impact of diverse skills on client thinking, feeling, and behavior and will be able to intentionally flex and use another skill when the prediction does not hold. This is a results-oriented approach, thoroughly tested and researched in thousands of settings and in more than 450 data-based research articles.

At the same time, this book moves beyond being a skills text. As students become competent in skills, they are constantly encouraged to generalize the skills beyond the textbook into daily life and interviewing practice. Equally important, students see very specifically in the text how the microskills are played out in actual counseling and clinical practice through transcripts of interviews showing decisional counseling, crisis counseling, and brief counseling. Thus, students not only learn how to use some theoretical basics but also prepare for the next level of their training.

Microcounseling introduced diversity and multicultural issues to our field as important issues in 1974, and we have continually expanded that awareness over the years. The chapters in this book thoroughly infuse a broad definition of diversity. In "National and International Perspectives" boxes in each chapter, authorities around the world present their viewpoints on skills and strategies. We believe that multicultural factors enrich our understanding of individual uniqueness and that all interviewing involves multicultural issues. *All of us are cultural beings with unique individual differences that contribute to broader humanity.*

Every sentence and concept in this book have been thoroughly reviewed for relevance and clarity for beginning helpers. Discussions of research and more complex therapeutic applications are concise. This book focuses on the most essential ideas for a successful introduction to the key microskills of helping.

Chapter 15, Determining Personal Style, focuses on students' self-examination and evaluation of their progress during the term. Students integrate their Portfolios of Competencies and think through their favored approaches to the interview.

Some instructors will want to reorder chapters to meet their own instructional goals. We have tried to organize the chapters in such a way as to make alternative sequencing easy.

WHO IS THIS BOOK FOR?

This book is designed to meet the needs of both beginners and more advanced students. It will clarify interviewing and counseling skills for community college students, undergraduates, and graduate students in courses oriented toward the helping professions such as human services and all types of mental health work, counseling, nursing, psychology, and social work.

In addition, the microskills approach has been proven effective in training in many other groups, including communication students, nutrition counselors, AIDS counselors and refugees in Africa, and trainees in management and leadership. All of us can profit from gaining more skill in communication.

Some of you may be interested in examining and considering the larger book, *Intentional Interviewing and Counseling.* There you will find a full chapter on observational skills, details of motivational interviewing, more information on research background, and detailed material concerning the relationship of neuroscience to counseling and interviewing.

WHAT OUTCOMES CAN STUDENTS EXPECT?

Although this edition includes many innovations discussed below, the central core of *Essentials of Intentional Interviewing* remains the same. This book will enable students to:

- ▲ Draw out client stories and understand the importance of thoughts, feelings, behaviors, and meanings in those stories.
- ▲ Predict how clients will respond to the use of interviewing skills, and if this prediction does not hold true, students will be able to flex intentionally and meet client needs by selecting another approach.
- ▲ Complete a full interview using only listening skills by the time they are halfway though this book.
- ▲ Understand and apply three foundations of effective helping: ethics, multicultural competence, and a strength-based wellness approach.
- ▲ Develop beginning competence in four approaches to the interview: decisional counseling, person-centered counseling, brief counseling, and crisis counseling.
- ▲ Define their natural style of helping and their own integration of helping skills. They will be able to analyze their own interviewing behavior and its effectiveness with clients.

▲ Develop a Portfolio of Competencies, bringing together their learning and individual orientations to the helping fields. The ancillary DVD and web-based CourseMate programs bring the cognitive concepts to action with many practical exercises and videos.

WHAT IS NEW IN THIS SECOND EDITION OF *ESSENTIALS OF INTENTIONAL INTERVIEWING?*

There are many important new features in this edition. Key among them are a new interactive DVD and an interactive CourseMate website. Both of these offer video clips of skills, quizzes for assignments or self-study, case studies, and many interactive activities to supplement and support learning.

This edition of *Essentials* has been thoroughly revised and updated:

▲ *Relationship-story and strengths-goals-restory-action.* The five-stage structure of the interview remains, but with a new set of labels that clarify the meaning and intent of each stage. As students master the stages of the interview, they are better prepared to become competent in the many strategies of theories of our field.

▲ *Crisis counseling.* Given the pressures we face today, it has become imperative that students be introduced to crisis conseling. We show how even beginning students can learn the basics of working in difficult situations.

▲ *New practice exercises.* Practice exercises have been placed in each chapter with specifics of how to use the interactive DVD and CourseMate website. We believe that reading about the practice of counseling and therapy is only a beginning. The critical issue is intentional practice, aiming toward excellence. These activities, coupled with role-plays in specific skills, provide a solid background for students as they move on in their studies and eventually are placed in practicums and internships.

▲ *Increased emphasis on preparing for treatment and case management.* This section has been expanded, and we have added a checklist summarizing the many factors that need to be considered when preparing for an interview. Beginning students need to recall many crucial concepts, particularly in their first sessions.

▲ *Resilience.* This is a quality that we want both our clients and counselors to exhibit. The ability to bounce back from error or trauma is critical to life. This has been integrated with the strategy of reflection of meaning and focuses on life purpose as a key dimension of helping. Research evidence shows that people who have a sense of purpose and meaning are better able to survive and move on with their lives.

▲ *Other new content.* Psychoeducational skills, increased emphasis on advocacy and social justice, more information on Internet counseling, more research material on neuroscience and counseling, how to furnish a comfortable office, and how to prepare for working in the community are some of the additional content you will find in this edition.

THE MOST UP-TO-DATE ANCILLARY SYSTEM AVAILABLE

A robust set of student and faculty resources are available for *Essentials of Intentional Interviewing: Counseling in a Multicultural World,* Second Edition. Visit the CourseMate website at **www.cengagebrain.com** to learn more or to obtain access to the ancillary materials.

Student Resources.   Resources to facilitate student learning from this book include a DVD with video examples, learning activities, a Portfolio of Competencies, and more. The same material is now available online through the CourseMate website for those students who prefer to access the material on the web or for those taking an online class. The DVD or a printed access code card for the CourseMate website can be packaged with the book at the request of the instructor, or they can be purchased directly online at **www.cengagebrain.com.**

Instructor Resources.   Resources for instructors include an Instructor's Guide with a sample syllabus, PowerPoints® for flexible teaching, and an eBank Test Bank, with many possibilities for testing and evaluation.

Finally, both student and instructor ancillaries are available in a course cartridge that can be loaded into a course management system such as Blackboard. If you are doing online teaching, this package has just about everything that you'll need.

Phone and Online Help for Using These Materials Is Available.   Contact your local Cengage Learning representative for ordering options and for access to important resources. For technical support, visit Cengage Learning online at **www.cengage .com/support**. In addition, the authors are available and eager to be of assistance with any problems or issues that occur: Allen Ivey (allenivey@gmail.com), Mary Bradford Ivey (mary.b.ivey@gmail.com), and Carlos P. Zalaquett (carlosz@usf.edu).

ACKNOWLEDGMENTS

Thomas Daniels, Memorial University, Cornerbrook, has been central to the development of the microskills approach for many years. His summary of research on more than 450 data-based studies is available by request. We appreciate and thank Penny John, one of our favorite students, for permission to use her interview as an example in Chapter 14.

Weijun Zhang's writing and commentaries are central to this book. We also thank James Lanier, Courtland Lee, Robert Manthei, Mark Pope, Kathryn Quirk, Azara Santiago-Rivera, Sandra Rigazio-DiGilio, and Derald Wing Sue for their written contributions.

We appreciate Kristi Kanel joining us with her insights on crisis counseling and how they relate to the microskills framework. Discussions with Viktor Frankl helped clarify the presentation of reflection of meaning. William Matthews was especially helpful in formulating the five-stage model of the interview. Machiko Fukuhara, professor emeritus, Tokiwa University, and president of the Japanese Association of Microcounseling, has been our friend, colleague, and coauthor for many years. Her understanding and guidance have contributed in many direct ways to the clarity of our concepts and to our understanding of multicultural issues. We give special thanks and recognition to this wise partner.

David Rathman, chief executive officer of Aboriginal Affairs, South Australia, has constantly supported and challenged this book, and his influence shows in many ways. Matthew Rigney, also of Aboriginal Affairs, was instrumental in introducing us to new ways of thinking. These two people first showed us that traditional, individualistic ways of thinking are incomplete, and therefore they were critical in the development of the focusing skill with its emphasis on the cultural/environmental context. Lia and Zig Kapelis of Flinders University and Adelaide University are thanked for their support and participation while Allen and Mary served as visiting professors in South Australia.

The skills and concepts of this book rely on the work of many different individuals over the past 40 years, most notably Eugene Oetting, Dean Miller, Cheryl Normington, Richard Haase, Max Uhlemann, and Weston Morrill at Colorado State University, who were there at the inception of the microtraining framework.

Many of our students at the University of South Florida in Tampa, the University of Massachusetts in Amherst, the University of Hawai'i in Manoa, and Flinders University in South Australia also contributed in important ways through their reactions, questions, and suggestions.

Finally, it is always a pleasure to work with the group at Brooks/Cole, particularly our editors, Seth Dobrin and Julie Martinez. We appreciate the entire team that worked on this edition: Arwen Petty, Suzanna Kincaid, Elizabeth Momb, Tami Strang, Rita Jaramillo, Rob Hugel, Caryl Gorska, Rebecca Cross, Tom McDonough, Anne Draus, Pablo d'Stair, and Peggy Tropp.

We are grateful to the following reviewers for their valuable suggestions and comments: Susan A. Adams, Texas Woman's University; Kathleen Bieschke, Pennsylvania State University; Joyce Clohessy, Westmoreland County Community College; Lori Curtis, University of Missouri, St. Louis; Kenneth C. Hergenrather, The George Washington University; Nancy Jensen, University of Hartford; Ellen Marshall, Delaware Technical & Community College; Seth Olson, University of South Dakota; Jan A. Rodgers, Dominican University; and James M. Sam, Siena Heights University.

We would be happy to hear from readers with suggestions and ideas. Please contact us via email. We appreciate the time that you as a reader are willing to spend with us.

Allen E. Ivey (allenivey@gmail.com)
Mary Bradford Ivey (mary.b.ivey@gmail.com)
Carlos P. Zalaquett (carlosz@usf.edu)

ESSENTIALS OF
INTENTIONAL INTERVIEWING

LISTENING TO CLIENT STORIES ON A BASE OF ETHICS, MULTICULTURAL UNDERSTANDING, AND WELLNESS

"Tell me your story." This is the request that interviewers, counselors, or therapists often make to clients as the session begins. Although as helpers we may approach clients differently, we all want to learn the essence of who they are and how and why they arrived at the session.

Those who tell stories of their life concerns and challenges need empathic and caring listeners. The relationship between the storyteller and the listener is basic to establishing a working alliance.

Storytelling is a process that provides both information and meaning. Stories enable the listener to assess and understand events, experiences, beliefs, and emotional foundations. The listener can then help the teller "rewrite" these stories, find solutions, and initiate actions based on mutually agreed-upon goals.

The first three chapters of this book focus on the foundations of storytelling, listening, and relationship.

Chapter 1, "The Science and Art of Interviewing and Counseling," offers an overview and a road map of what this book can do for you. Most important, we ask you to reflect on what brings you here to the helping field. We also want to start by identifying your natural helping skills. You are not taking this course by chance; something has led you to this work.

Chapter 2, "Ethics, Multicultural Competence, and Wellness," presents three crucial aspects of all interviewing and counseling. Ethics—the professional standards that all major helping professions observe and practice—provides interviewers with guidelines on issues such as competence, informed consent, confidentiality, power, and social justice. Multicultural competence is about cultural awareness and sensitivity to the worldview of our clients. Positive psychology and wellness enable clients to identify their strengths and resources. A wellness approach helps resolve issues and problems with what clients "can do" rather than with what they "can't do."

Chapter 3, "Attending and Observation Skills: Basic to Communication," discusses the most essential dimension of listening. By encouraging clients to talk, you demonstrate that you truly want to hear their stories and concerns. Listening with full attention builds more solid relationships. This chapter also focuses on basics of nonverbal communication. You can respond more appropriately to the unique person before you through keen observation of visual and verbal cues such as a sudden change in topic, breaks in eye contact, a body shift, or a voice change (tone or volume). Keep in mind that clients who are culturally different from you may have different verbal and nonverbal patterns from your own.

THE SCIENCE AND ART OF INTERVIEWING AND COUNSELING

We humans are social beings. We come into the world as the result of others' actions. We survive here in dependence on others. Whether we like it or not, there is hardly a moment of our lives when we do not benefit from others' activities. For this reason it is hardly surprising that most of our happiness arises in the context of our relationships with others.

—*The Dalai Lama*

How are interviewing and counseling both a science and art? How can we benefit clients through the relationship that is called "helping"?

CHAPTER GOALS

Awareness, knowledge, and skills developed through the concepts of this chapter and this book will enable you to

▲ Explore the interview as both a science and art. In this process we ask you to begin reflecting on yourself as a potential helper.

▲ Define and discuss similarities and differences among interviewing, counseling, and psychotherapy, as well as the related fields of clinical counseling, clinical social work, psychology, and psychiatry.

▲ Gain knowledge of the microskills approach to the interview, a step-by-step approach that provides a flexible base on which to define your personal style and theory of counseling.

▲ Consider important factors related to the place you conduct your interviews, whether in an office, in the community, or online.

▲ Record and document your own natural interviewing style as the baseline of the expertise that you bring to this book.

1.1
YOU AS HELPER, YOUR GOALS, YOUR COMPETENCIES

KEY CONCEPT QUESTIONS

▲ **What is interviewing, counseling, and psychotherapy?**

▲ **Why do you want to be a helper?**

▲ **What are your natural skills?**

Here is a real first interview. Listening and relationship come first.

Sienna, a 16-year-old teen, is 8 months' pregnant with her first child. She says, "I wonder when I'll be able to see Freddy [baby's father] again. I mean I want him involved; he wants to be with me, and the baby. But my mom wants me home. His mom said she's looking for a three-bedroom apartment so we could possibly live there, but I know my mom will never go for it. She wants me to stay with her until I graduate high school and, well, to be honest, so that this never happens again [she points to her belly]."

I listen carefully to her story and then respond, "I'm glad to hear that Freddy wants to be involved in the care of his son and maintain a relationship with you. What are your goals? How do you feel about talking through this with your mom?"

"I don't know. We don't really talk much anymore," she says as she slumps down in her chair and picks away at her purple nail polish.

She then describes her life before Freddy, focused mainly on the crowd she hung around, a group of girls whom she says were wild, mean, and tough. Her mood is melancholy, and she seems anxious and discouraged. I say, "Well, it seems that there's a lot to talk about. How do you feel about continuing our conversation before sitting down with your mom?"

Surprisingly, she says, "No. Let's talk next week with her. The baby is coming and, well, it'll be harder then." As we close the session, I ask her, "How has this session been helpful for you?" Sienna responds, "Well, I guess you're going to help me talk about some important issues with my mom, and I didn't think I could do that."

This was the first step in a series of five interviews. As the story evolved, we invited Freddy for a session. He turned out to be employed and was anxious to meet his responsibilities, although finances remained a considerable challenge. This was followed by a meeting with both mothers in which a workable action plan for all families was generated. I helped Sienna find a school with a special program for pregnant teens.

This case exemplifies the reality of helping. We often face complex issues with no clear positive ending. If we can develop a relationship and listen to the story carefully, clearer goals develop and solutions usually follow.

▲ EXERCISE 1.1   **Love Is Listening**

Paul Tillich says "Love is listening." *Listening, love, caring,* and *relationship* are closely related. These four words could be said to be the center of the helping process. What relevance do these words have in the interview with Sienna? What are your reactions and thoughts about the centrality of these words?

This chapter introduces counseling as science and art. Interviewing and counseling now have a solid research and scientific base that enables us to identify many qualities and skills that make for effectiveness. Science has demonstrated that the specifics of listening skills are identifiable and they are central to competent helping. But a scientific approach by itself is not enough. You as interviewer or counselor are similar to an artist whose skills and knowledge produce beautiful paintings out of color, canvas, and personal experience. You are the listener who will provide color and meaning to the interpersonal relationship we call helping.

Another way to think about science and art is in the very different areas of music and sports. The violinist learns specific skills (fingering, mastering scales, music theory). The basketball player also develops specific skills over time (shooting, passing, ball handling). Both integrate these basics into the real world through the harmony of the orchestra and the flow of the basketball team.

In this book, you will be introduced to concepts, skills, and theory that are known through research to facilitate client growth. But you are the person, the artist, who will integrate these skills into the session, make the relationship work, and enable client growth and change.

Like the artist, the musician, or the basketball player, you bring a natural talent to share with others in "harmonious relationship." Before we delve into the systematic aspects of the helping relationship, we would like to ask you to reflect on what brings you into the helping field. What brought you to this moment? What natural talents do you bring with you? What do you need to learn to grow further?

 OPTIONAL DVD AND COURSEMATE WEBSITE AT WWW.CENGAGEBRAIN.COM

For *Essentials of Intentional Interviewing,* Second Edition, we have created optional interactive resources to help you further understand major concepts and sharpen your interviewing competencies. Additional case studies, video clips of sessions, and interactive exercises to clarify what happens in the interview or session will help you master the skills discussed in the textbook. To allow the most flexibility, these learning tools are available on an interactive DVD and also on the CourseMate website at www.cengagebrain.com. Simply go to www.cengagebrain.com and use the search field to find *Essentials of Intentional Interviewing*, Second Edition, to locate the CourseMate website.

Throughout the book we will put an icon like this one nearby references to these digital resources. Again, the resources on the DVD and the CourseMate website are the same, so you do not need both to use all of the materials. We encourage you to use the resources presented at the beginning and end of each chapter. We designed them to make your learning more meaningful and to help ensure competence in interviewing and counseling practice.

If you did not receive a DVD or a Printed Access Card packaged with your book, you can still purchase access to these resources at www.cengagebrain.com. Check with your professor first to find out if the online resources are required.

 Assess your awareness, knowledge, and skills as you begin the chapter:

1. Self-Assessment Quiz: The chapter quiz will help you determine your current level of knowledge. You can take it before and after reading the chapter.
2. Portfolio of Competencies: Before you read the chapter, please fill out the downloadable Self-Evaluation Checklist to assess your existing knowledge and competence in attending and listening skills. Then, at the end of the chapter, complete the checklist again to summarize your competencies after study and practice.

 Complete Video Activity: Getting to Know the Authors
Complete Interactive Self-Assessment 1: Personal Strengths and Your Natural Style
Complete Interactive Self-Assessment 2: Self-Awareness and Emotional Understanding
Review Case Study: What Do You Say Next? Working With a Difficult Case

▲ EXERCISE 1.2   **Personal Reflection**

Why do *you* want to be a helper?
What natural skills do *you* bring?
What do others see as *your* talents?
What learning goals do *you* have?

1.2
INTERVIEWING, COUNSELING, AND RELATED FIELDS

KEY CONCEPT QUESTION

▲ How are interviewing, counseling, and psychotherapy defined?

The terms *counseling* and *interviewing* are used interchangeably in this text. The overlap is considerable (see Figure 1.1), and at times interviewing will touch briefly on counseling and psychotherapy. Both counselors and psychotherapists typically draw

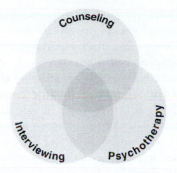

FIGURE 1.1 The interrelationship of interviewing, counseling, and psychotherapy.

on the interview in the early phases of their work. You cannot become a successful counselor or therapist unless you have solid interviewing skills.

Interviewing is the more basic process used for gathering data, helping clients resolve their issues, and providing information and advice to clients. Interviewers may be found in many settings, including employment offices, schools, hospitals, businesses, law offices, and a wide variety of helping professions.

Counseling is a more intensive and personal process than interviewing. Although information gathering is important, counseling is more about listening to and understanding a client's life challenges and, with the client, developing strategies for change and growth. Counseling is most often associated with the professional fields of counseling, human relations, counseling and clinical psychology, pastoral counseling, and social work; it is also part of the role of medical personnel and psychiatrists.

Psychotherapy focuses on more deep-seated personality or behavioral difficulties. Psychotherapists interview clients to obtain basic facts and information as they begin working with them, and often use counseling as well. The skills and concepts of intentional interviewing are equally important for the successful conduct of longer-term psychotherapy.

PROFESSIONAL FIELDS AND THE MULTIPLE APPLICATIONS OF THE INTERVIEW

Figure 1.1 illustrates the considerable overlap among the three major forms of helping. For example, a personnel manager may interview a candidate for a job but in the next hour may counsel an employee who is deciding whether to accept a promotion that requires a move to another city. A school guidance counselor may interview a student to check on course selection but may counsel the next student about college choice or a conflict with a friend. A psychologist may interview one person to obtain research data and in the next hour counsel another concerned with an impending divorce. In the course of a single contact, a social worker may interview a client to obtain financial data and then move on to counsel the same client about personal relationships.

Examples of overlap among the three approaches can be seen in the duties of a variety of professional helpers. For example, **school counselors** primarily focus on the counseling function, but constantly interview parents and students as well. Often school counselors find themselves doing limited therapy as they are the only helpers available for the child or adolescent. **Human service workers and professionals** also vary widely in their responsibilities. Although interviewing is basic, they may often counsel clients on specific issues. At times human service workers may find themselves as first line responders in a crisis and then engage in brief supportive therapeutic interventions.

Clinical mental health counseling (CACREP, 2009) is a new term gaining prominence. **Clinical mental health counselors** are skilled in both diagnostics and psychotherapeutic treatment, in addition to having excellent interviewing and counseling skills. They have extensive training in the art of therapy, somewhat similar to psychologists and psychiatrists. Clinical counselors are trained in all areas of mental health and in working with more severe client cases.

Social work is a broad field with many types of practice, and interviewing and counseling are typically basic to this work. **Clinical social workers** focus on psychotherapy, but they also involve themselves in many interviewing and counseling activities. They have extensive training in therapy, somewhat similar to mental health counselors and psychiatrists. In addition, they are fully trained in social work principles and ethics and thus are particularly aware of contextual issues such as socioeconomics, the impact of community, and social justice concerns.

Psychologists evaluate, diagnose, and treat a wide variety of human issues; their role may range from working with the severely distressed to providing typical counseling services, sometimes with a major emphasis on education and prevention. This group includes clinical psychologists, counseling psychologists, community psychologists, school psychologists, and many types of research psychologists.

Psychiatrists are medical professionals who have a traditional M.D. degree plus many years of additional study, internships, and residencies in hospitals. Psychiatrists once typically engaged in psychoanalytically oriented therapy, but they now tend to emphasize diagnosis and treatment with medication. Despite this general trend, many conduct thriving individual and group therapeutic practices using many of the methods and theories discussed in this book.

This list does not exhaust all the possibilities. Among others are professional marriage counselors, group counselors, coaches, rehabilitation counselors, crisis counselors, and substance abuse counselors. It is also possible to be certified in many situations with a bachelor's degree.

Mentoring and supervision are two additional roles that need to be considered, although they are not professions. **Mentoring** is often described as a process in which an experienced person passes on knowledge and helps a beginning professional open doors to opportunities. But effective interviewers and counselors often take on a mentoring role when they provide longer-term support such as college and career planning (Zalaquett & Lopez, 2006). **Supervision** is an educational and monitoring role in which the supervisor simultaneously teaches and supports beginning professionals. But supervision should not end, even with a master's, Ph.D., or M.D. All of us can learn from others; being an effective helper requires a lifetime commitment to learning and improvement.

All these fields demand competence in interviewing and counseling skills, plus some training in diagnosis and treatment with more severely stressed clients so that helpers can refer appropriately. These professionals work with children, adolescents, and adults; with individuals, couples, families, groups, and organizations. They provide a continuum of help that ranges from personal growth to concerns with issues of daily living to severe distress.

▲ EXERCISE 1.3   **Fields of Helping**

Although all the helping fields discussed here are related, each one has a central core. Which professional area appeals to you most?

Review Website: Interviewing, Counseling, and Psychotherapy
Complete Weblink Critique Exercise: Crisis and Trauma

1.3
THE MICROSKILLS APPROACH

KEY CONCEPT QUESTIONS

▲ **What are microskills, and how do they lead to effective intentional interviewing, counseling, and psychotherapy?**

▲ **What is the overview of this book?**

As with any profession, your natural talent and style will be enhanced greatly by careful study and practice of certain skills. **Microskills** are communication skill units that help you to interact more intentionally with a client. The microskills model was developed through years of study and analysis of hundreds of interviews. It is designed to show you how to use skills to master the art of communicating effectively and intentionally with a client.

As you master each microskill, you are prepared for learning the next important skill. This approach is similar to that employed by professionals of all kinds—from Olympic athletes and musicians to clothing designers, journalists, and craftspeople. World-famous golf pros and tennis stars, as well as accomplished musicians and dancers, all began with and continue to practice basic skills.

Like the athlete, musician, or artist, you begin with natural talent, but talent needs to be amplified by careful study and practice of specific skills. Extensive feedback from experts and experienced trainers also facilitates enhanced performance and true mastery. The microskills of the interview are critical dimensions of effective interviewing and counseling, just as the specifics of tennis are to Serena Williams, rounding a corner for a NASCAR driver, or harmony and style for a new pop singer, group, or band. All take persistence and practice to become truly expert. Knowing how is not enough; doing it well with expertise is very different. The microskills hierarchy (see Figure 1.2) summarizes the successive skill steps as you will discover them in this text. Interview skills rest on a base of *ethics, multicultural competence, and wellness,* which will be the focus of Chapter 2. Interviewing and counseling professions rest on a solid base of ethical principles, including confidentiality, competence, informed consent, power, and social justice. The RESPECTFUL model, which you will encounter very soon in this chapter, is integral to the development of multicultural competence. And clients will establish a better relationship with you and change their behavior and thinking more readily with a positive wellness approach.

Resting on this broad base is the first microskill, *attending behavior,* or culturally and individually appropriate listening. This includes culturally appropriate patt

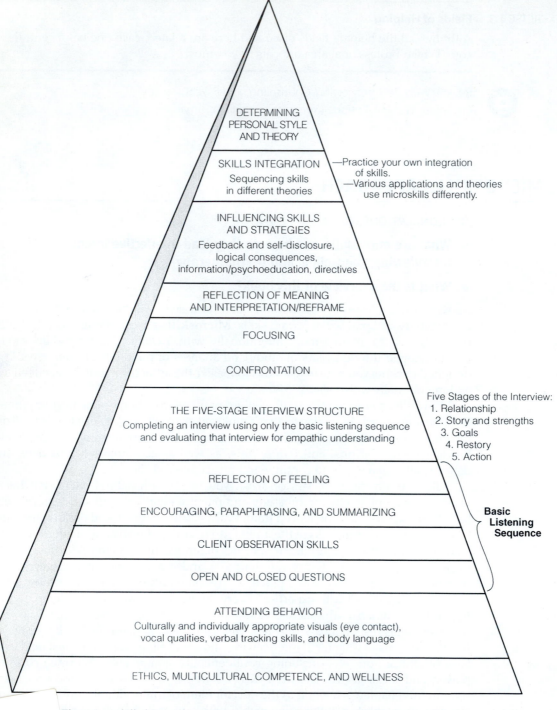

DETERMINING
PERSONAL STYLE
AND THEORY

SKILLS INTEGRATION

Sequencing skills
in different theories

—Practice your own integration
 of skills.
—Various applications and theories
 use microskills differently.

INFLUENCING SKILLS
AND STRATEGIES

Feedback and self-disclosure,
logical consequences,
information/psychoeducation, directives

REFLECTION OF MEANING
AND INTERPRETATION/REFRAME

FOCUSING

CONFRONTATION

THE FIVE-STAGE INTERVIEW STRUCTURE

Completing an interview using only the basic listening sequence
and evaluating that interview for empathic understanding

Five Stages of the Interview:
1. Relationship
2. Story and strengths
3. Goals
4. Restory
5. Action

REFLECTION OF FEELING

ENCOURAGING, PARAPHRASING, AND SUMMARIZING

**Basic
Listening
Sequence**

CLIENT OBSERVATION SKILLS

OPEN AND CLOSED QUESTIONS

ATTENDING BEHAVIOR

Culturally and individually appropriate visuals (eye contact),
vocal qualities, verbal tracking skills, and body language

ETHICS, MULTICULTURAL COMPETENCE, AND WELLNESS

The microskills hierarchy: A pyramid for building cultural intentionality. (Copyright © 1982, llen E. Ivey. Reprinted by permission.)

of verbal tracking, visual eye contact, vocal qualities, and body language ("3 V's + B"). These skills provide a critical foundation for demonstrating that you are a listening and caring person. Coupled with attending are your skills in observation, both of yourself and of your client (Chapter 3).

The microskills of *questioning, paraphrasing, summarizing,* and *reflection of feeling* are found in Chapters 4, 5, and 6. These skills elaborate attending behavior and attention; they enable you to demonstrate your ability to listen to the specifics of client concerns. You will communicate your understanding through drawing out their stories and concerns. You will paraphrase and summarize the main ideas expressed and enter their emotional word through accurate reflection of feelings.

With a solid background in these most important listening skills, you will learn in Chapter 7 how to *structure a well-formed interview*. You will want to include a focus on what clients *can do* as part of your standard interviewing plan. The *relationship— story and strengths—goals—restory—action* model summarizes a basic strength-based framework for human change that you can use to help ensure effective interviewing and counseling.

Relationship. No one wants to tell a story to someone who is not interested or who is not warm and welcoming. Unless you can develop rapport and trust with your client, expect little to happen. Your attending and empathic listening skills are key to maintaining the relationship throughout the entire interview. This is also called the working alliance.

Story and Strengths. Through effective listening, we learn the stories of clients' lives, their problems, challenges, and issues. But we also need to search for and listen to their strengths—what can they do, rather than what can't they do. Listen for stories in which clients describe times they have overcome obstacles. Listen for and be "curious about their competencies—the heroic stories that reflect their part in surmounting obstacles, initiating action, and maintaining positive change" (Duncan, Miller, & Sparks, 2004, p. 53).

Goals. How would you and the client like the story to develop? What is an appropriate ending? Remember, *"If you don't know where you are going, you may end up somewhere else."* Too many interviews wander and never have a focus. It is important to establish mutually determined goals.

Restory. Once you understand clients' stories and strengths and establish agreed-upon goals, you can help them restory—generate new ways to talk, think, and feel about themselves. One important strategy for restorying is using only listening skills to conduct a full interview. But restorying will most often also involve the influencing skills of the second half of this book.

Action. Through the use of microskills, you can enable clients to transform their new ideas and stories into concrete action in the "real world." Something must change in terms of thoughts, feelings, and behaviors outside the interview for interviewing and counseling to be effective.

We then move to the second half of the book, which focuses on influencing skills, or active change skills. The first influencing skill, *confrontation* (Chapter 8), is considered basic to the developmental change process. Here we examine how to identify the often mixed and confusing messages that clients bring to us ("I love my wife, but

I want to meet that person on the Internet . . .") Through identifying and confronting conflict and incongruence, we can help clients synthesize, resolve issues, and set clearer goals for change.

The microskill of *focusing* (Chapter 9) shows you how to engage the person in the interview while maintaining awareness of the effects of family and environmental/contextual factors. Focusing can help clients discover further complexities and issues that they maybe missed before. This skill is a key to effective work in diversity and multicultural interviewing.

The influencing skills of *reflection of meaning, interpretation/reframing, feedback, directives,* and others (Chapters 10, 11, 12) help clients restory and think about their issues in new ways. These skills enable clients to generate new stories and ideas, which in turn lead to new thoughts, feelings, and behaviors.

Chapter 13, *Decisional Counseling, Skill Integration, Treatment Plans, and Case Management,* is designed to integrate the earlier chapters of this book while also addressing case management and treatment planning for future sessions. Decisional counseling—an important, practical, and widely used interviewing theory—will be illustrated and demonstrated, including a detailed transcript analysis.

Crisis counseling and brief counseling, two useful applications of the microskills model, are presented in Chapter 14. If you have developed mastery in the skills of this book, you will be able to develop beginning competence, an important step toward professional performance.

At the apex of the microskills hierarchy is *determining personal style and theory.* Chapter 15 asks you to review your work with this book and the competencies you have developed. At this point, you will have a further opportunity to write your own narrative, your own personal story about interviewing and counseling. This chapter is designed to help you "put it together" in your own way.

Box 1.1 summarizes relevant research findings regarding the microskills.

Complete Interactive Exercise 1: The Microskills Hierarchy
Complete Group Practice Exercise 1: The Microskills Hierarchy

1.4
OUR MULTICULTURAL WORLD

KEY CONCEPT QUESTION

▲ How is all interviewing and counseling multicultural?

Multiculturalism is now defined quite broadly. Once it referred only to the major racial groups, but now the definition has expanded in multiple ways. The story is that we are all multicultural. If you are White, male, heterosexual, from Alabama, an Episcopalian, and able bodied, you have a distinct cultural background. Just change Alabama to Connecticut or California, and the client is very different culturally. Similarly, change the color, gender, sexual orientation, religion, or physical ability, and the client's cultural background changes significantly. Multiculturalism means just that—many cultures. Culture is like air: We breathe it without thinking about it, but it is essential for our being. Culture is not "out there"; it is inside you as a markedly cultural being.

BOX 1.1   Microskills Research

More than 450 microskills research studies have been conducted (Daniels, 2010; Daniels & Ivey, 2006). The model has been tested nationally and internationally in more than 1,000 clinical and teaching programs. You can view a comprehensive summary of this research on the DVD that accompanies this book. Microcounseling was the first systematic video-based counseling model to identify specific observable interviewing skills. It was also the first skills training program that emphasized multicultural issues. Some of the most important research findings include the following:

▲ *You can expect results from microskills training.* Several critical reviews have found microtraining an effective framework for teaching skills to a wide variety of people, ranging from beginning interviewers to experienced professionals, who need to relate to patients and clients more effectively. Teaching your clients many of the microskills will facilitate their personal growth and ability to communicate with their families or coworkers.

▲ *Practice is essential.* Practice the skills to mastery if the skills are to be maintained and used after training. *Use it or lose it*! Complete practice exercises and generalize what you learn to real life. Whenever possible, audio or video record your practice sessions.

▲ *Multicultural differences are real.* People from different cultural groups (e.g., ethnicity/race, gender) have different patterns of skill usage. Learn about people different from you, and use skills in a culturally appropriate manner.

▲ *Different applications and counseling theories have varying patterns of skill usage.* Expect person-centered counselors to focus almost exclusively on listening skills whereas cognitive behaviorists use more influencing skills. Microskills expertise will help you define your own theory and integrate it with your natural style.

▲ *If you use a specific microskill, then you can expect a client to respond in predictable ways.* You can predict how the client will respond to your use of each microskill, but each client is unique and predictability is not perfect. Cultural intentionality prepares you for the unexpected and teaches you to flex with another microskill.

Recent findings in neuroscience back up much of what is said in this book. Attention and attending behavior lead to your brain and your client's brain firing in measurable ways. Without attention, no learning or change occurs. The critical concept of empathy relates to mirror neurons in the forebrain. Mirror neurons enable you to understand and even experience the emotions of other people. Important in this book is a positive wellness approach to counseling and therapy. Recent studies have shown that meditation increases gray matter in brain regions involved in learning and memory processes, emotion regulation, and perspective taking (Hölzel et al., 2011). Functional magnetic imaging (fMRI) has revealed that negative emotions, located primarily in the amygdala, can be modified and brought under control through a positive approach to counseling and therapy (Likhtik, Popa, Apergis-Schoute, Fidacaro, & Paré, 2008). Some key resources for further study in this area may be found in Chapter 15.

The central point is that *all counseling is multicultural and has a cultural/contextual background that changes the meaning of the interview and the way clients and counselors behave.*

Multicultural competence is imperative in the interview process. We live in a multicultural world where every client that you encounter will be different from the last and different from you. Without a basic understanding of and sensitivity to a client's uniqueness, the interviewer will fail to establish a relationship and true grasp of a client's issues. Throughout, this book will examine the multicultural issues and opportunities we all experience.

Race and ethnicity present special circumstances in the multicultural context.

Cultural and social influences are not the only influences on mental health service and delivery, but they have been historically underestimated—and they do count. Cultural differences must be accounted for to ensure that minorities, like all Americans, receive mental health care tailored to their needs.

This quotation is from a U.S. Surgeon General's Report titled "Mental Health, Culture, Race and Ethnicity." The report also acknowledged that African Americans, Asian/Pacific Islander Americans, Hispanic Americans, and Native Americans have diminished access to mental health services and are unlikely to receive the same quality care as others. We encourage you to visit the website for the full report: www.surgeongeneral.gov/library/mentalhealth/cre.

Professional associations in counseling and psychology have developed guidelines for multicultural awareness and proficiency, often with special attention to underserved groups and racial minorities. Remember that the Euro-White population will become a minority around the year 2040, certainly within the time that you will be a practicing helper.

The American Medical Association and the National Medical Association have joined forces to form the Commission to End Health Care Disparities. This organization is designed to foster cultural competence among physicians through educational and training programs. In March 2005, New Jersey became the first state to mandate cultural competency courses, tying cultural competency training to licensure or relicensure for physicians. Needless to say, all these efforts have been met with controversy, but multiculturalism continues to take a central place in society.

It is critical that you examine yourself and seek to understand yourself more as a multicultural being. The following discussion provides that opportunity.

RESPECTFUL INTERVIEWING AND COUNSELING

The RESPECTFUL model (D'Andrea & Daniels, 2001) points out that all of us are multicultural beings. Multiculturalism refers to far more than race and ethnicity. As you review the list below, first identify your own multicultural dimensions. Then examine your beliefs and attitudes toward those who are similar to and multiculturally different from you on each issue below.

R   Religion/spirituality. What is your religious and spiritual orientation? How does this affect you as an interviewer or counselor?

E   Economic/class background. How will you work with those whose financial and social background differs from yours?

S   Sexual identity. How effective will you be with those whose gender or sexual orientation differs from yours?

P   Personal style and education. How will your personal style and educational level affect your interviewing practice?

E   Ethnic/racial identity. The color of our skin is one of the first things we notice. What is your reaction to different races and ethnicities?

C   Chronological/lifespan challenges. Children, adolescents, young adults, mature adults, and older persons all face different issues and challenges. Where are you in the developmental lifespan?

T   Trauma. It is estimated that 90% or more of all people experience serious trauma(s) in their lives. Trauma underlies the issues faced by many of your clients. War, flood, rape, and assault are powerful examples, but divorce, loss of a parent, or being raised in an alcoholic family are more common sources of trauma. The constant repetition of racist, sexist, or heterosexist actions and comments can also be traumatic. What is your experience with life trauma?

F   Family background. We learn culture in our families. The old model of two parents with two children is challenged by the reality of single parents, gay families, and varying family structures. How has your life experience been influenced by your family history (both your immediate family and your intergenerational history)?

U   Unique physical characteristics. Become aware of disabilities, special challenges, and false cultural standards of beauty. Help clients think about themselves as physical beings and the importance of nutrition and exercise. How well do you understand the importance of the body in the interview, and how will you work with others different from you?

L   Location of residence and language differences. Whether in the United States, Great Britain, Canada, or Australia, there are marked differences between the south and north, the east and west, urban and rural. Moreover, many of you reading this book, and certainly many of your clients, will come from widely varying nations. The small town of Amherst, Massachusetts, has 23 different languages represented in their student body. Large cities will often have 100 or more different languages and nationalities represented. Remember that a person who is bilingual is advantaged and more skilled, not disadvantaged. What languages do you know, and what is your attitude toward those who use a different language from you?

As you review this list, perhaps you can see more clearly that all interviewing and counseling is multicultural. Broaden your definition of diversity beyond race and ethnicity to include gender, lifespan, and the other factors in the RESPECTFUL model. All your sessions will involve these dimensions (and more).

Intersections among several multicultural factors are also critical. For example, consider the biracial family (e.g., Chinese and White, African descent and Latino). Both children and parents are deeply affected, and categorizing an individual into just one multicultural category is inappropriate. Or think of the Catholic lesbian woman who may be economically advantaged (or disadvantaged). Or the South Asian male with a Ph.D. who is gay. For many clients, sorting out the impact of their multiculturality may be a major issue in counseling.

▲ EXERCISE 1.4   **Culture Counts! You as a Multicultural Being**

Review the RESPECTFUL list and identify your own multicultural identities and the impact of intersections among them. Then examine your beliefs and attitudes toward those who are similar to and multiculturally different from you on each of these issues. Finally, you may want to pair up with some of your classmates and compare your views.

The American Counseling Association has stated that members should "recognize diversity in our society and embrace a cross-cultural approach in support of the worth, dignity, potential, and uniqueness of each person." You can use this self-assessment exercise to determine how able are you to work with people who may be culturally different from you.

Complete Interactive Exercise 2: We Are All Multicultural Beings
Complete Group Practice Exercise 2: Examine Your Multicultural Background

1.5
CULTURAL INTENTIONALITY:
Developing Multiple Responses

KEY CONCEPT QUESTION

▲ **Why do we want an interviewer to have many responses as possible available for any situation?**

Your present communication style and social skills are valuable natural tools on which you can build your unique approach to helping others. Imagine that you are interviewing a client. What would you say in response to the following?

▲ EXERCISE 1.5   **Multiple Responses**

ROBERTO: (talking about a conflict on the job) I just don't know what to do about my new boss. It seems he's always on me, blaming me even when I do a good job. He's new on the job, I know. Perhaps he doesn't have much experience as a supervisor. But he's got me all jumpy. I'm so nervous I can't sleep at night, and yesterday I even threw up in the bathroom. My family isn't doing well, either. I've been arguing with Farah and she doesn't seem to understand what's going on and is upset. Even the kids aren't doing well in school. What do you suggest I do?

How would you respond to Roberto? How would you respond if the client were a woman going through the same issues with her husband?

Compare your responses with others'. What do you learn from their ideas? What is the "correct" response in this case?

Of course, there are many potentially useful responses in any interviewing situation. The absolutely "correct" response likely does not exist. Asking an open question ("Could you tell me more?") may be particularly useful so that you can understand issues more fully. Or reflecting feelings may be helpful ("Looks like you are upset over the situation with the new boss"). Self-disclosure and direct advice may be what is needed for some clients ("My experience with such situations is . . . and I suggest you try . . .").

Beginning interviewers can be eager to find the "right" answer for the client. In fact, some often give quick, inappropriate, patch-up advice that can lead in wrong directions and make the problem worse. How ideal it would be to find the perfect empathic response to unlock the door and free the client for more creative living, but we need to be patient and flexible as we work with amazingly different and unique clients.

BOX 1.2   National and International Perspectives on Counseling Skills

Problems, Concerns, Issues, and Challenges—How Shall We Talk About the Story?
JAMES LANIER, UNIVERSITY OF ILLINOIS, SPRINGFIELD

Counseling and therapy historically have tended to focus on client problems. The word *problem* implies difficulty and the necessity of eliminating or solving the problem. Problem may imply deficit. Traditional diagnosis such as that found in the *Diagnostic and Statistical Manual of Mental Disorders-IV-TR* (American Psychiatric Association, 2000) carries the idea of problem a bit further, using the word *disorder* with such terms as *panic disorder*, *conduct disorder*, *obsessive-compulsive disorder*, and many other highly specific *disorders*. The way we use these words often defines how clients see themselves. The forthcoming DSM-5 carries the idea of pathology even further with many new categories of "disorder" (www.dsm5.org/Pages/Default.aspx).

I'm not fond of problem-oriented language, particularly that word "disorder." I often work with African American youth. If I asked them, "What's your problem?" they likely would reply, "I don't have a problem, but I do have a concern." The word *concern* suggests something we all have all the time. The word also suggests that we can deal with it—often from a more positive standpoint. Defining *concerns* as *problems* or *disorders* leads to placing the blame and responsibility for resolution almost solely on the individual.

Finding a more positive way to discuss client concerns is relevant to all your clients, regardless of their background. *Issue* is another term that can be used instead of *problem*. This further removes the pathology from the person and tends to put the person in a situational context. It may be a more empowering word for some clients. Carrying this idea further, *challenge* may be defined as a call to our strengths. Some might even talk about *an opening for change*.

Beyond that, the concepts of the wellness and positive asset search make good sense for the youth with whom I have worked. Change is most easily made from a position of strength—criticism and problem-oriented language can weaken. However, don't be afraid to challenge people to grow. Confrontation can help your clients develop in positive ways.

Cultural intentionality is a core goal of effective interviewing. Culturally intentional interviewing is concerned with how many potential responses may be helpful. We can define it as follows:

> Cultural intentionality is acting with a sense of capability and deciding from among a range of alternative actions. The intentional individual can generate alternatives in a given situation and approach a problem from different vantage points, using a variety of skills and personal qualities, adapting styles to suit different individuals and cultures.

The culturally intentional interviewer remembers a basic rule of helping: *If a helping lead or skill doesn't work—try another approach!* A critical issue in interviewing is that the same comment may have different effects on individuals who have different personal life experiences and multicultural backgrounds. Intentional interviewing requires awareness that cultural groups each have their own patterns of communication. For example, in European and North American cultures, middle-class patterns call for rather direct eye contact, but in some cultural groups direct eye contact is considered rude and intrusive. Many Spanish-speaking groups have more varied vocal tones and a more rapid speech rate than do English-speaking people.

See Box 1.2 for a discussion of how you can help clients talk about their issues from a more positive perspective.

INTENTIONAL PREDICTION

This text is action- and results-oriented. It is founded on research revealing that you can anticipate relatively specific results when you use any given microskill or strategy in the interview. If you work intentionally in the interview, you can anticipate what the client is likely to say next. And even if the expected does not happen, you can intentionally flex and come up with a helpful alternative comment.

Let us illustrate intentional prediction with an important skill discussed in later chapters—reflection of feeling. Central to clarifying client emotions, this skill facilitates more meaningful and rapid resolution of clients' issues, concerns, and problems. If you reflect feelings, you can *predict* that clients will talk more about emotions and do so in a clearer fashion. Note below the brief definition of this skill and the prediction that you can make when you use the skill intentionally.

But intentionality also requires you to be flexible. Anticipations and predictions do not always hold true. This is where cultural intentionality really comes in. It is important that you have an alternative response ready when what you predict does not happen or you make an error. Our plumber once said, "It's not the errors I make; it's my ability to fix them that counts." Don't expect perfection in any interview; intentionality is what is most important.

Reflection of Feeling	*Predicted Result*
The interviewer identifies a client's key emotions and feeds them back to clarify affective experience. With some clients, the brief acknowledgment of feeling may be more appropriate.	Clients will be able to experience their emotional states more clearly. They also may correct the interviewer's reflection with a more accurate descriptor.

A summary of all brief interviewing skill definitions and the predicted results from their use can be found in Appendix I. Again, it is important to stress that predictability and the ability to anticipate results of your interventions will never reach 100%. Remember that if your first use of a skill is ineffective, you have others prepared to keep the session flowing.

If something you say or do doesn't work the first (or the second) time, don't try it again! Intentionality asks you to be flexible and be ready, as no client is the same as any other.

   Review Web Document: Intentional Counseling and Interviewing (ICI)

1.6

OFFICE, COMMUNITY, PHONE, AND INTERNET: Where Do We Meet Clients?

KEY CONCEPT QUESTIONS

▲ **Where do interviews take place?**

▲ **What type of environment should be considered for the interview?**

▲ How can I maintain an office and/or physical presence in a respectful fashion for my clients?

First, let's recognize that interviewing and counseling occur in many places other than a formal office. There are street counselors, who work with youth organizations, homeless shelters, and the schools, as well as those who work for the courts, who go out into the community and get to know groups of clients. Counseling, interviewing, and therapy can be very informal, taking place in the clients' home, a neighborhood coffee shop or nearby park, and while they play basketball or just hang out on the street corner. The "office" may not exist, or it may be merely a cubicle in a public agency where the counselor can make phone calls, receive mail, and work at a computer, but not necessarily a place where he or she will meet and talk with clients. The office is really a metaphor for your physical bearing and dress—smiling, culturally appropriate eye contact, a relaxed and friendly body style.

Mary Bradford Ivey, as a school counselor, learned early on that if she wanted to counsel recent immigrant Cambodian families, home visits were essential. She sat on the floor as the family did. She attended cultural events, ate and cooked Cambodian food, and attended weddings. She brought the Cambodian priest into the school to bless the opening ceremonies. She provided translators for the parents so that they could communicate with the teachers. She worked with school and community officials to advocate for the special needs of these immigrants. The place of counseling and developing your reputation as a helper varies widely. Maintaining a pleasant office is important, but not enough.

COUNSELING IN AN OFFICE

However, in schools, colleges, and some agencies, the office is where you will see most of your clients. When preparing to meet with a client, here are a few things to think about:

Desk and Chairs.   Having the desk facing the door and forcing clients to sit across from you does not establish a friendly setting. Generally speaking, seek to have a variety of chair situations and distances so that clients and you can choose freely what is most appropriate and comfortable.

However, few of us have the luxury of much space. Often you will have a desk against the wall and you can turn to talk to clients. Sometimes you simply have to sit behind a desk. Encourage the client to sit at the corner of the desk nearest you, although some will choose to sit directly across from you or in the corner of your room.

Warmth and Comfort.   Regardless of physical setting, you as a person can light up the room. Smiling and a warm, friendly voice make up for difficult office situations.

How you decorate your office with pictures and objects is important and will be noticed by your clients. Generally speaking, keep personal objects and photos to a minimum. One acceptable approach is to select noncontroversial pictures and photos such as flowers and landscapes. Blandness is "safe," but even there some clients will see it as a sign of coldness and distance.

Another approach, used by Mary, is to consider the clientele likely to come to your setting. Working in a school setting, she sought to have objects and artwork representing various races and ethnicities. The brightness of the artwork worked with well with children, and many parents commented favorably on seeing their culture represented in her office. But most important, make sure that nothing in your office can be considered objectionable by any of those whom you serve.

Games and Other Possibilities.   Games, books, playhouses, and play materials are examples of what is needed in the office of anyone who works extensively with children. We are increasingly discovering that adult clients also benefit from the use of many of these materials. Sand play and similar activities can be the best way to reach some clients.

Children and Adolescents.   Children like to play with small objects and toys while talking with you. Have several on the desk and around the office. If you work with younger children, have small chairs available and use one yourself. Sit at the children's level rather than forcing them to look up to you. Taking children or adolescents to the play area or gym during counseling can be useful, but be careful to maintain confidentiality.

Adolescents and adults often find it helpful to have manipulative pieces to touch while talking. These should be provided more commonly, as they bring the possibility for sensorimotor experience and the use of metaphor. Always have tissues available for those times when clients cry. Hand the box to them gently with support.

The Waiting Room.   The waiting room, if you have one, should be pleasant, with the basic characteristics described above. Typically, however, it will be bland and "officelike." The key privacy and ethical policies of the school or agency should be prominently displayed. A smiling and supportive staff here makes a real difference in how you will be perceived.

INTERNET COUNSELING

Insert "online counseling" into your search engine, and a number of services will appear. Following is a composite result from Allen Ivey's visits to several online services:

Meeting life's challenges is difficult

[Internet counseling center] enables you to talk with real-life professional counselors 24/7 in full confidentiality.

Choose a counselor.

First session is free.

Easy payments arranged.

We expect that more and more of these services will appear on the Internet. Some may be quite helpful to clients at a reasonable cost, but others may be risky. Now search for "coaching services," and you will find an amazing array of possibilities. However, view all these sites with some attention as to how ethical and professional standards can be met on the Internet.

It is possible that you are reading this book as part of an online course. If so, you understand the potential of Internet counseling. Now, with the increasing use of video cameras in laptop computers, face-to-face counseling and therapy via Skype and other services are possible. With video communication, observing nonverbal communication becomes possible, although only to a limited extent.

Distance Credentialed Counselor (DCC) is a national credential currently offered by the Center for Credentialing and Education (CCE). Holders of this credential adhere to the National Board for Certified Counselors' Code of Ethics and the Ethical Requirements for the Practice of Internet Counseling. These professionals adapt their counseling services for delivery to clients via technology-assisted methods, including telecounseling (telephone), secure e-mail communication, chat, videoconferencing, or other appropriate software (CCE, n.d.). Professional associations are still working on these new approaches to counseling delivery. As with all types of counseling and therapy, ethics is the first concern. In Chapter 2 we present some beginning issues in ethics. You will also find Internet links to the professional ethics of the main national helping organizations.

Review Website: Bureau of Labor. Counselors
Review Website: Bureau of Labor. Psychologists
Review Website: Bureau of Labor. Social Workers

1.7
YOUR NATURAL STYLE AND BEGINNING EXPERTISE:
An Important Audio or Video Exercise

KEY CONCEPT QUESTION

▲ **How will you document your natural personal style and current skill level before you begin systematic microskills training?**

On page 16 you were asked to give your response to Roberto, a client who came to you to discuss multiple issues. We asked how you would organize your response. This is an indication of your natural style and expertise.

The microskills learned through this text will provide you with additional alternatives for intentional responses for the client. However, these responses must be genuinely your own. If you use a skill or strategy simply because it is recommended, it is likely to be ineffective for both you and your client. Not all parts of the microskills framework are appropriate for everyone. You have a natural style of communicating, and these concepts must enhance your natural style, not detract from it.

You will need varying patterns of helping skills with the clients with whom you work. Couple your natural style with awareness and knowledge of individual and multicultural differences. How will clients respond to your natural style? You will need to be able to "flex" and be intentional as you encounter diversity among clients.

You may work more effectively with some clients than with others. Developing trust and a working alliance takes more time with some clients than with others. Many clients lack trust with interviewers who come from a cultural background different from their own. You may be less comfortable with teenagers than you are with children or adults. Some have difficulty with elders.

You are about to engage in a systematic study of the interviewing process. By the end of the book, you will have experienced many ideas for analyzing your interviewing style and skill usage. Along the way, it will be helpful to have a record of where you were before you began this training.

It is invaluable to identify your personal style and current skill level before you begin systematic training. Eventually, it is YOU who will integrate these ideas into your own practice. Let's start with your own work. Please read Box 1.3 and make plans to record your natural style *before* continuing further in this text.

Complete Interactive Exercise 3: Your Natural Helping Style: An Important Audio or Video Exercise
Complete Video Activity: We Hope You Were Able to Videotape Your First Interview

---

BOX 1.3   Discovering Your Natural Style of Interviewing

Many of you now have cameras with video and sound capability. Cell phones and computers often have video capabilities readily available. All these provide the opportunity for effective feedback. There is nothing like seeing yourself with a client as it "really happened." You will want to make transcripts of some of your practice sessions. With a sound recorder, you can back up and check what was said more easily, but all technical equipment is becoming more flexible and efficient. Backing up and reviewing key parts of an interview several times is immensely valuable.

Guidelines for Audio or Video Recording
1. Find a volunteer client willing to role-play a concern, problem, opportunity, or issue.
2. Interview the volunteer client for at least 15 minutes. Seek to avoid sensitive topics.
3. Use your own natural communication style.
4. Ask the volunteer client, "May I record this interview?"
5. Inform the volunteer client that the tape recorder can be turned off any time he or she wishes.

6. Select a topic. You and the client may choose interpersonal conflict, a specific issue selected by the client, or one of the elements from the RESPECTFUL model.
7. Follow the ethical guidelines discussed in Chapter 2. Common sense demands ethical practice and respect for the client.
8. Obtain feedback. You will find it very helpful to get immediate feedback from your client. As you practice the microskills, use the Client Feedback Form (Box 1.4). You may even find it helpful to continue using this form, or some adaptation of it, in your work as an interviewing professional.
9. Compare this baseline with subsequent recordings of your work throughout this text.

It is important to conduct this exercise as soon as possible. Your choice of audio- or videotape will document an accurate baseline of your natural style and skill level. As you progress through this text, compare this session with your later work.

BOX 1.4   Client Feedback Form

_____ (Date)

_____   _____
(Name of Interviewer)   (Name of Person Completing Form)

Instructions: Rate each statement on a 7-point scale where 1 = strongly agree, 7 = strongly disagree, and 4 = neutral. You and your instructor may wish to change and adapt this form to meet the needs of varying clients, agencies, and situations.

	Strongly Agree		Neutral		Strongly Disagree	
1. (Awareness) The session helped you understand the issue, opportunity, or problem more fully.	1	2	3	4	5	6	7
2. (Awareness) The interviewer listened to you. You felt heard.	1	2	3	4	5	6	7
3. (Knowledge) You gained a better understanding of yourself today.	1	2	3	4	5	6	7
4. (Knowledge) You learned about different ways to address your issue, opportunity, or problem.	1	2	3	4	5	6	7
5. (Skills) This interview helped you identify specific strengths and resources you have to help you work through your concerns and issues.	1	2	3	4	5	6	7
6. (Skills) You will take action and do something in terms of changing your thinking, feeling, or behavior after this session.	1	2	3	4	5	6	7

What did you find helpful? What did the interviewer do that was right? Be specific. For example, not "You did great," but rather, "You listened to me carefully when I talked about _____."

What, if anything, did the interviewer miss that you would have liked to explore today or in another session? What might you have liked to have happen that didn't?

Use this space or the other side for additional comments or suggestions.

© 2009 Carlos Zalaquett and Allen Ivey. Permission granted to users of this text to reprint this form.

CHAPTER SUMMARY

Key Points of "The Science and Art of Interviewing and Counseling"	CourseMate and DVD Activities to Build Interviewing Competence

1.1 You As Helper, Your Goals, Your Competencies

▲ Counseling is a science and an art.

▲ Interviewing and counseling have a solid research and scientific base.

▲ The identified listening skills are central to competent helping, but are not enough.

▲ You are the listener who will provide meaning to the interpersonal relationship we call helping.

▲ You can integrate these skills into the session to make the relationship work and enable client growth and change.

▲ Ask yourself: What brought you here? What natural talents do you bring? What do you need to learn to grow further?

1. Video Activity: Getting to Know the Authors
2. Interactive Self-Assessment 1: Personal Strengths and Your Natural Style.
3. Interactive Self-Assessment 2: Self-Awareness and Emotional Understanding.
4. Case Study: What Do You Say Next? Working With a Difficult Case. Reflect on how you would react and what you would say to Todd.

1.2 Interviewing, Counseling, and Related Fields

▲ Interviewing is a basic process designed for gathering data, solving problems, and providing information and advice.

▲ Counseling is more comprehensive and is generally concerned with helping people cope with life challenges and develop new opportunities for further growth, whereas psychotherapy is focused on more deep-seated personality or behavioral difficulties.

▲ Multiple fields that use the skills of this text range from business interviewing through in-depth work in clinical social work, counseling or clinical psychology, and psychiatry.

▲ All use the microskills presented in this text, but with varying emphasis and objectives.

1. Website: Interviewing, Counseling, and Psychotherapy
2. Weblink Critique Exercise: Crisis and Trauma. What do you think of these websites related to crises, disasters, or trauma? What did you learn?

1.3 The Microskills Approach

▲ Microskills are communication skill units that help you develop the ability to interact more intentionally with a client.

▲ The microskills hierarchy provides a visual picture of the skills and strategies.

▲ Natural talent in many activities is enhanced by the study and practice of single skills.

1. Interactive Exercise 1: The Microskills Hierarchy
2. Group Practice Exercise 1: The Microskills Hierarchy

Key Points of "The Science and Art of Interviewing and Counseling"

CourseMate and DVD Activities to Build Interviewing Competence

▲ Applications in counseling and interviewing differ in how they use the microskills.

▲ More than 450 data-based studies have tested the microskill model, which is used in more than 1,000 settings throughout the world.

1.4 Our Multicultural World

▲ Culture counts! Discussion of multicultural differences early in the session is often important, but common sense also dictates that such discussion is not appropriate for all situations.

▲ Race and ethnicity are especially important multicultural issues, and we need to be aware that being White, being male, and being economically advantaged often puts a person in a privileged group.

▲ The RESPECTFUL model lists ten key multicultural dimensions, thus showing that cultural issues will inevitably be part of the interviewing and client relationship.

▲ Developing multicultural awareness, knowledge, and skills is a lifelong process of continuing learning.

1. Interactive Exercise 2: We Are All Multicultural Beings

1.5 Cultural Intentionality: Developing Multiple Responses

▲ Cultural intentionality is defined as acting with a sense of capability and choosing from among a range of alternative actions.

▲ When you use a specific microskill or strategy, you can predict how the client will respond. Be flexible; if one skill isn't working, try another.

▲ The use of microskills will increase your cultural intentionality and flexibility by showing you many ways to respond to a single client statement or concern.

1. Web document: Intentional Counseling and Interviewing (ICI). A good example demonstrating that intentional counseling and the microskills are taught around the world.

1.6 Office, Community, Phone, and Internet: Where Do We Meet Clients?

▲ Interviewing and counseling occur in many places other than a formal office, and interviewers and counselors may spend much of their time outside the formal office.

▲ When working in an office, keep comfort and safety in mind and provide a sense of warmth.

▲ Use decorations that embrace diversity and games, objects, and books that are developmentally appropriate.

▲ Online is another setting where counseling is being provided. As with all type of counseling and therapy, ethics is the first concern.

1. Website: Bureau of Labor: Counselors. Counselors work in diverse community settings designed to provide a variety of counseling, rehabilitation, and support services.

2. Website: Bureau of Labor: Psychologists.

3. Website: Bureau of Labor: Social Workers.

Key Points of "The Science and Art of Interviewing and Counseling"	**CourseMate and DVD Activities to Build Interviewing Competence**

1.7 Your Natural Style and Beginning Expertise: An Important Audio or Video Exercise

▲ The microskills are meant to add to and enhance, not detract from, your natural style of communicating.

▲ Your ability to vary the use of microskill patterns allows you to flex as you encounter client diversity.

▲ It is critical to document a baseline of your natural style and skill level before you begin the systematic study of the microskills.

▲ Using the resources and portfolio of competencies can help you become an effective listening and helper.

1. Interactive Exercise 3: Your Natural Helping Style: An Important Audio or Video Exercise.

2. Video Activity: We Hope You Were Able to Video-tape Your First Interview. Of course, audiotape is fine, but if you videotape, then this exercise is for you.

 Assess your awareness, knowledge, and skills as you conclude the chapter:

1. **Flashcards:** Use the flashcards to check your understanding of key concepts and facilitate memorization of key information.

2. **Self-Assessment Quiz:** The quiz will help you assess your current knowledge and prepare for course examinations.

3. **Portfolio of Competencies:** Evaluate your present level of competence in attending and listening skills using the downloadable Self-Evaluation Checklist. Self-assessment of your attending skills competence demonstrates what you can do in the real world.

ETHICS, MULTICULTURAL COMPETENCE, AND WELLNESS

ETHICS, MULTICULTURAL COMPETENCE, AND WELLNESS

I am (and you are also)
Derived from family
Embedded in a community
Not isolated from prevailing values
Though having unique experiences
In certain roles and statuses
Taught, socialized, gendered, and sanctioned
Yet with freedom to change myself and society.

—*Ruth Jacobs**

---

*R. Jacobs, *Be an Outrageous Older Woman*, 1991, p. 37. Reprinted by permission of Knowledge, Trends, and Ideas, Manchester, CT.

How can this chapter help you and your clients?

CHAPTER GOALS   Effective interviews build on professional ethics, multicultural sensitivity, and a positive wellness approach. Specifics are stressed that apply to use in the interview.

Awareness, knowledge, and skills developed through the concepts of this chapter will enable you to

▲ Apply key ethical principles in interviewing and counseling.
▲ Develop your own informed consent form.
▲ Define multicultural competence, including key aspects of awareness, knowledge, and skills.
▲ Apply wellness and positive psychology in an assessment interview.

Assess your awareness, knowledge, and skills as you begin the chapter:

1. Self-Assessment Quiz: The chapter quiz will help you determine your current level of knowledge. You can take it before and after reading the chapter.
2. Portfolio of Competencies: Before you read the chapter, fill out the downloadable Self-Evaluation Checklist to assess your existing knowledge and competence in attending skills. Then, at the end of the chapter, complete the checklist again to summarize your competencies after study and practice.

2.1
ETHICS IN INTERVIEWING AND COUNSELING

KEY CONCEPT QUESTIONS

▲ **How competent are you to work with various issues presented by clients, and when should you refer?**

▲ **How do you inform clients about key issues as they begin interviewing and counseling?**

▲ **What are major guidelines for confidentiality?**

▲ **How is power defined in relation to interviewing practice?**

▲ **What is your responsibility to work with issues of social justice?**

If you behave ethically and intentionally, you can anticipate that the relationship will proceed more smoothly and your client will be protected. A sense of ethics is part of trust building. Note the following brief description of ethical behavior and its implications for the session.

Ethics	Predicted Result
Observe and follow professional standards, and practice ethically. Particularly important issues for beginning interviewers are *competence, informed consent, confidentiality, power,* and *social justice.*	Client trust and understanding of the interviewing process will increase. Clients will feel empowered in a more egalitarian session. When you work toward social justice, you contribute to problem prevention in addition to healing work in the interview.

Complete Interactive Exercise 1: Ethical Self-Assessment

All major helping professions throughout the world have codes for ethical practice. The codes promote empowerment for both interviewers and their clients. Ethical codes aid the helping process by (a) teaching and promoting the basics of ethical and appropriate practice, (b) protecting clients by providing accountability, and (c) serving as a mechanism to improve practice (Corey, Corey, & Callanan, 2010). Ethical codes can be summarized with the following statement: "Do no harm to your clients; treat them responsibly with full awareness of the social context of helping." As interviewers and counselors, we are responsible for our clients and for society as well. At times these responsibilities conflict, and you may need to seek detailed guidance from documented ethical codes, your supervisor, or other professionals.

Box 2.1 lists websites of some key ethical codes in English-speaking areas of the globe. All codes provide guidelines on competence, informed consent, confidentiality, and diversity. Issues of power and social justice are explicit in social work and human services and implicit in other codes.

▲ EXERCISE 2.1   **Examining Ethical Codes**

As you work through this chapter, please select the ethical code from Box 2.1 that is most relevant to your interests and examine it in detail. Also select the code of another helping profession, and look for similarities and differences among the two codes and the ethical presentation here.

COMPETENCE

The American Counseling Association's (2005) statement on professional competence includes diversity. Note the emphasis on continued learning and expanding one's qualifications over time.

> C.2.a. ***Boundaries of Competence.*** Counselors practice only within the boundaries of their competence, based on their education, training, supervised experience, state and national professional credentials, and appropriate professional experience. Counselors will gain knowledge, personal awareness, sensitivity, and skills pertinent to working with a diverse client population.

Regardless of the human services profession with which you identify, competence is key.

In working with a client you need to constantly monitor whether you are competent to counsel the individual on each issue presented. For example, you may be able to help the client work out difficulties occurring at work, but you discover a more complex problem that requires family counseling. You may need to refer the client to another counselor for family counseling while continuing to work with the job issues. If a client demonstrates severe distress or presents an issue with which you are uncomfortable, seek supervision.

CONFIDENTIALITY

The American Counseling Association's (2005) ethical code states:

> Section B: Introduction. Counselors recognize that trust is the cornerstone of the counseling relationship. Counselors aspire to earn the trust of clients by creating an

BOX 2.1   Professional Ethical Codes With Websites

Listed below are some important ethical codes. Website addresses are correct at the time of printing but can change. For a keyword web search, use the name of the professional association and the words *ethics* or *ethical code.*

American Academy of Child and Adolescent Psychiatry (AACAP)	http://www.aacap.org
American Association for Marriage and Family Therapy (AAMFT) Code of Ethics	http://www.aamft.org
American Counseling Association (ACA) Code of Ethics	http://www.counseling.org
American Psychological Association (APA) Ethical Principles of Psychologists and Code of Conduct	http://www.apa.org
American School Counselor Association (ASCA)	http://www.schoolcounselor.org
Australian Psychological Society (APS) Code of Ethics	http://www.psychology.org.au
British Association for Counselling and Psychotherapy (BACP) Ethical Framework	http://www.bacp.co.uk/ethical_framework/
Canadian Counselling Association (CCA) Codes of Ethics	http://www.ccacc.ca
Commission on Rehabilitation Counselor Certification (CRCC) Code of Professional Ethics for Rehabilitation Counselors	http://www.crccertification.com/pages/crc_ccrc_code_of_ethics/10.php
International Union of Psychological Science (IUPsyS) Universal Declaration of Ethical Principles for Psychologists	http://www.am.org/iupsys/resources/ethics/univdecl2008.html
National Association of School Nurses (NASN)	http://www.nasn.org
National Association of School Psychologists (NASP)	http://www.naspweb.org
National Association of Social Workers (NASW) Code of Ethics	http://www.naswdc.org
National Career Development Association (NCDA)	http://www.ncda.org
New Zealand Association of Counsellors (NZAC) Code of Ethics	http://www.nzac.org.nz
School Social Work Association of America (SSWAA)	http://www.sswaa.org
Ethics Updates provides updates on current literature, both popular and professional, that relate to ethics.	http://ethics.sandiego.edu
Ethical codes of Latin American countries can be found at the Society of Interamerican Psychology's Grupo de Trabajo de Ética y Deontología Profesional webpage under "Informacion del Grupo y Documentos."	http://www.sipsych.org/grupoetica

ongoing partnership, establishing and upholding appropriate boundaries, and maintaining **confidentiality**. Counselors communicate the parameters of confidentiality in a culturally competent manner.

As a student taking this course, you are a beginning professional; you usually do not have legal confidentiality. Nonetheless, you need to keep to yourself what you hear in class role-plays or practice sessions. Trust is built on your ability to keep confidences. Be aware that state laws on confidentiality vary.

Professionals encounter many challenges to confidentiality. Some states require you to inform parents before counseling a child, and information from interviews must be shared with them if they ask. If issues of abuse should appear, you must report this to the authorities. If the client is a danger to self or others, then rules of confidentiality change; the issue of reporting such information needs to be discussed with your supervisor. As a beginning interviewer, you will likely have limited, if any, legal protection, so limits to confidentiality must be included in your approach to informed consent.

HIPAA PRIVACY

The Health Insurance Portability and Accountability Act (HIPAA) took effect in 1996. It is included here because, among other functions, it requires the protection and confidential handling of protected health information.

Following is a summary of some key elements of the Privacy Rule, including who is covered, what information is protected, and how protected health information may be used and disclosed. For a complete outline of HIPPA requirements, visit the website of the U.S. Department of Health and Human Services, Office for Civil Rights, at http://www.hhs.gov/ocr/privacy/hipaa/understanding/summary/index.htm.

1. *Protected Health Information.* The Privacy Rule defines "protected health information (PHI)" as all individually identifiable health information held or transmitted by a covered entity or its business associate, in any form or media, whether electronic, paper, or oral. "Individually identifiable health information" is information, including demographic data, that identifies the individual, or could reasonably be used to identify the individual, and that relates to:

   ▲ The individual's past, present, or future physical or mental health or condition
   ▲ The provision of health care to the individual
   ▲ The past, present, or future payment for the provision of health care to the individual

   Individually identifiable health information includes many common identifiers such as name, address, birth date, and Social Security number.

   The Privacy Rule excludes from protected health information employment records that a covered entity maintains in its capacity as an employer as well as education and certain other records subject to, or defined in, the Family Educational Rights and Privacy Act, 20 U.S.C. §1232g.

2. *De-identified Health Information.* There are no restrictions on the use or disclosure of de-identified health information. This is information that makes it impossible for others to identify a client. There are two ways to de-identify

information: (1) a formal determination by a qualified statistician; or (2) the removal of specified identifiers of the individual and of the individual's relatives, household members, and employers. De-identification is adequate only if the covered entity has no actual knowledge that the remaining information could be used to identify the individual.

When you visit a physician, you are asked to sign a version of the privacy statement. Mental health agencies make their privacy statements clearly available to clients and often post them in the office.

▲ EXERCISE 2.2   **Ethical Case Scenario**

The case study of Michael is designed to provide an overview of many issues discussed so far in the chapter. Michael's instructor has asked each student to interview a person from a different cultural background from his or her own and write a two-page summary. In addition, to help expand cultural knowledge, the professor has asked each student to review another's paper before class and write comments. (Needless to say, all names and towns have been changed here, including Michael's.)

The following is a brief edited summary of what Michael wrote. Note that he has not disguised the name of the person interviewed. This is ethical violation #1.

ASSIGNMENT

The Diverse Experience of Culturally Diverse People: An Investigation Project

My interview of a person from a culturally/ethnically diverse group different from my own was of a gentleman named Carlos Patricio Montenegro. Mr. Montenegro is a 27-year-old man who is originally from Ixtapa, Mexico. He has been residing in the United States for 11 years and owns his own plumbing company called "Montenegro Can Fix It." He currently lives with his wife, Maria Gonzales Montenegro, and 4-month-old daughter Margarita in Tampa, Florida. Mr. Montenegro discussed the differences between American life and the life he lived in Mexico.

Mr. Montenegro explained that the most important values and beliefs of his culture are family and respect. Because he understands that this is a confidential interview, he shared with me that he has been unfaithful to his wife twice since living in this country and is ashamed of himself and is terrified that anybody would ever find out.

He also explained that if one could make a living for himself by a self-made business and provide a home for his family, life in Mexico was great. He suggested the reason he moved to the USA was for the opportunity this country had to offer. Mr. Montenegro taught me that Cinco De Mayo is not a celebration of independence; September 15th is the true Independence Day for Mexico. [The report continues in some detail.]

What are some of the ethical issues involved in this report? Did Michael inform the client about what was to happen in the interview? In terms of multicultural issues, do you think Michael is ready and competent to counsel Mexican Americans? Does the session have a wellness and positive psychology tone?

What would you say to Michael? Imagine you are a student friend or the instructor. What actions are possible in this situation?

INFORMED CONSENT

Counseling is an international profession. The Canadian Counselling Association (2007) approach to informed consent is particularly clear:

> *B4. Client's Rights and Informed Consent.* When counselling is initiated, and through-out the counselling process as necessary, counsellors inform clients of the purposes, goals, techniques, procedures, limitations, potential risks and benefits of services to be performed, and other such pertinent information. Counsellors make sure that clients understand the implications of diagnosis, fees and fee collection arrangements, record-keeping, and limits of confidentiality. Clients have the right to participate in the ongoing counselling plans, to refuse any recommended services, and to be advised of the consequences of such refusal.

Complete Interactive Exercise 2: Informed Consent

The American Psychological Association (2002) stresses that psychologists should inform clients if the interview is to be supervised:

> *Standard 10.01* When the therapist is a trainee and the legal responsibility for the treatment provided resides with the supervisor, the client/patient, as part of the informed consent procedure, is informed that the therapist is in training and is being supervised and is given the name of the supervisor.

In addition, the APA specifies:

> *Standard 4.03* Before recording the voices or images of individuals to whom they provide services, psychologists obtain permission from all such persons or their legal representatives.

When you work with children, the ethical issues around informed consent become especially important. Depending on state laws and practices, it is often necessary to obtain written parental permission before interviewing a child or before sharing information about the interview with others. The child and family should know exactly how the information is to be shared, and interviewing records should be available to them for their comments and evaluation. An important part of informed consent is stating that both child and parents have the right to withdraw their permission at any point. Needless to say, these same principles apply to all clients—the main difference is parental awareness and consent.

When you enter into role-plays and practice sessions, it is important that you inform your volunteer "clients" about their rights, your own competence, and what clients can expect from the session. For example, you might say,

> I'm taking an interviewing course, and I appreciate your being willing to help me. I am a beginner, so only talk about things that you want to talk about. I would like to audiotape (or videotape) the interview, but I'll turn it off immediately if you become uncomfortable and erase it as soon as possible. I may share the tape in a practicum class or I may produce a written transcript of this session, removing anything that

could identify you personally. I'll share any written material with you before passing it in to the instructor. Remember, we will stop any time you wish. Do you have any questions?

You can use this statement as an ethical starting point and eventually develop your own approach to this critical issue. The sample practice contract in Box 2.2 may be helpful as you begin.

BOX 2.2   Sample Practice Contract

The following is a sample contract for you to adapt for practice sessions with volunteer clients. (If you are working with a minor, add that form must be signed by a parent as appropriate under HIPPA standards.)

Dear Friend,

I am a student in interviewing skills at [insert name of class and college/university]. I am required to practice counseling skills with volunteers. I appreciate your willingness to work with me on my class assignments.

You may choose to talk about topics of real concern to you, or you may prefer to role-play an issue that does not necessarily relate to you. Please let me know before we start whether you are talking about yourself or role-playing.

Here are some important dimensions of our work together:

Confidentiality. As a student, I cannot offer any form of legal confidentiality. However, anything you say to me in the practice session will remain confidential, except for certain important exceptions that state law requires me to report. Even as a student, I must report (1) a serious issue of harm to yourself; (2) indications of child abuse or neglect; (3) other special conditions as required by our state [insert as appropriate].

Audio- and/or Videotaping. I will be recording our sessions for my personal listening and learning. If you become uncomfortable at any time, we can turn off the recorder. The tape(s) may be shared with my supervisor [insert name and phone number of professor or supervisor] and/or students in my class. You'll find that recording does not affect our practice session so long as you and I are comfortable. Without additional permission, recordings and any written transcripts are destroyed at the end of the course.

Boundaries of Competence. I am an inexperienced interviewer; I cannot do formal counseling. This practice session helps me learn interview skills. I need feedback from you about my performance and what you find helpful. I may give you a form that asks you to evaluate how helpful I was.

_____   _____

Volunteer Client   Interviewer

Date _____

▲ EXERCISE 2.3   **Informed Consent**

Develop an Informed Consent Form. Box 2.2 presents a sample informed consent form, or practice contract. In a small group, develop your own informed consent form that is appropriate for your practice sessions, school, or agency.

Complete Group Practice Exercise 1: Develop an Informed Consent Form

POWER

The National Organization for Human Services (NOHS, 1996) comments on **power,** an important ethical issue that often receives insufficient attention:

> *Statement 6.* Human service professionals are aware that in their relationships with clients power and status are unequal. Therefore, they recognize that dual or multiple relationships may increase the risk of harm to, or exploitation of clients, and may impair professional judgment.

Power differentials occur in society. The very act of helping has power implications. The client may begin counseling with perceived lesser power than the interviewer. Awareness of and openness to talking about these issues help you work toward a more egalitarian relationship with the client. If your gender is opposite that of your client, it can be helpful to bring up the gender difference. For example, "How comfortable are you discussing this issue with a man?" If your client is uncomfortable, it is wise to discuss this issue further. Referral may be necessary.

You will encounter many situations in which institutional or cultural **oppression** becomes part of the counseling relationship, even though you personally may not have been involved in that oppression. For example, a woman may have had bad experiences with men. An African American being counseled by a European American may perceive the interviewer as potentially prejudiced, and a gay person may not feel safe with a heterosexual counselor. Those who have a disability may expect to be treated with a lack of real understanding by those more physically able. In each of these cases, discussion of differences in background and culture can be helpful early in the session.

Dual relationships occur when you have more than one relationship with a client. Another way to think of this is the concept of *conflict of interest*. If your client is a classmate or friend, you are engaged in a dual relationship. These situations may also occur when you counsel a member of your church or school community. Personal, economic, and other privacy issues can become complex issues. You can examine statements on dual relationships in more detail in the ethical codes.

In the past, the ethical ideal was to avoid all dual relationships; however, the term itself has multiple meanings. For this reason, some current codes of ethics (e.g., American Counseling Association, 2005) do not mention "dual relationship" but recognize three types of roles and relationships with clients: sexual/romantic relationships, nonprofessional relationships, and professional role change. The first type is banned because of its damaging effect on the clients. The other two may allow counselors to interact with a client in a nonprofessional activity as long as the interaction is potentially beneficial to the client and is not of a romantic or sexual nature. Of course you should always use caution.

SOCIAL JUSTICE

The National Association of Social Workers (1999) suggests that action beyond the interview may be needed to address **social justice** issues. The code includes a major statement on social justice.

> *Ethical Principle: Social workers challenge social injustice.*
>
> Social workers pursue social change, particularly with and on behalf of vulnerable and oppressed individuals and groups of people. Social workers' social change efforts are focused primarily on issues of poverty, unemployment, discrimination, and other forms of social injustice. These activities seek to promote sensitivity to and knowledge about oppression and cultural and ethnic diversity. Social workers strive to ensure access to needed information, services, and resources; equality of opportunity; and meaningful participation in decision making for all people.
>
> There are two major types of social justice action. The first and most commonly discussed is the importance of action in the community to work against the destructive influences of poverty, racism, and all forms of discrimination. These preventive strategies are now considered an important dimension of the "complete" counselor or therapist. Getting out of the office and understanding societies' influence on client issues is central.

▲ EXERCISE 2.4   **Social Justice**

We now know that childhood poverty, adversity, and stress produce lifelong damage to the brain. These changes are visible in cells and neurons and include permanent changes in DNA (Marshall, 2010). Therefore, just treating children of poverty through supportive counseling is not enough. For significant change to occur, prevention and social justice action are critical.

The incidence of hypertension among African Americans is reported to be 4 to 7 times that of Whites, and the long-term impact of racism may be an important contributor (Hall, 2007). Here again social justice action has both mental and physical implications.

Many professional associations focus on social justice issues. Among them are Counselors for Social Justice (http://counselorsforsocialjustice.com) and the International Association for the Advancement of Social Work with Groups (http://www.aaswg.org).

The second type of social justice action occurs in the interview. When a female client discusses mistreatment and harassment by her supervisor, the issue of oppression of women should be named as such. The social justice perspective requires you to help her understand that the problem is not caused by her behavior or how she dresses. By naming the problem as sexism and harassment, you often free the client from self-blame and empower her for action. You can also support her in efforts to effect change in the workplace. On a broader scale, you can work in the larger community outside the interview to promote fairer treatment for women in the workplace. Helping clients work through issues in the interview may not be enough. You also have a responsibility to promote community change through social action.

These same points hold true for any form of oppression that you encounter in the session, whether racism, ableism, heterosexism, classism, or other forms of prejudice. We need to remember that our clients live in relationship to the world. The microskill of focusing discussed later in this book provides specifics for bringing the cultural/environmental/social context into the interview (Chapter 9).

▲ EXERCISE 2.5   **Issues of Oppression**

Helping clients discover how issues of oppression have an impact on their personal issues is considered controversial by some. What are your thoughts?

2.2
DIVERSITY AND MULTICULTURAL COMPETENCE

KEY CONCEPT QUESTIONS

▲ **What are basic standards for working cross-culturally?**

▲ **How is dealing with diversity an ethical imperative?**

Following are some predictions that you can make when you take the broad array of multicultural issues into consideration. Remember the concept of intentionality. As you learn more about this area, you will increase your ability to respond appropriately to more and more clients who may be different from you. Be ready to flex intentionally as you learn more.

Multicultural Issues	*Predicted Result*
Base interviewer behavior on an ethical approach with an awareness of the many issues of diversity. Include the multiple dimensions described in this chapter. All of us have many intersecting multicultural identities.	Anticipate that both you and your clients will appreciate, gain respect, and learn from increasing knowledge in intersecting identities and multicultural competence. You, the interviewer, will have a solid foundation for a lifetime of personal and professional growth.

Complete Interactive Exercise 3: Multicultural Assessment

DIVERSITY AND ETHICS

The American Counseling Association (2005) focuses the Preamble to its Code of Ethics on diversity as a central ethical issue:

> The American Counseling Association is an educational, scientific, and professional organization whose members work in a variety of settings and serve in multiple capacities. ACA members are dedicated to the enhancement of human development throughout the life-span. Association members recognize diversity and embrace a cross-cultural approach in support of the worth, dignity, potential, and uniqueness of each individual within their social and cultural contexts.

The Ethical Standards for Human Service Professionals (NOHS, 1996) include the following three assertions:

> *Statement 17.* Human service professionals provide services without discrimination or preference based on age, ethnicity, culture, race, disability, gender, religion, sexual orientation, or socioeconomic status.
>
> *Statement 18.* Human service professionals are knowledgeable about the cultures and communities within which they practice. They are aware of multiculturalism in society and its impact on the community as well as individuals within the community. They respect individuals and groups, their cultures and beliefs.
>
> *Statement 19.* Human service professionals are aware of their own cultural backgrounds, beliefs, and values, recognizing the potential for impact and their relationships with others.

Diversity and multiculturalism have become central to the helping professions throughout the world. If a client's needs are rooted in a multicultural issue with which you are not competent, you may need to refer the client. Over the long term, however, referral is inadequate. You also have the responsibility to build your multicultural competence through constant study and supervision and to minimize your need for referral.

 Complete Interactive Exercise 4: Assessing Your Multicultural Identity

MULTICULTURAL COMPETENCE

Multicultural guidelines and specific competencies for practice have been developed by the American Counseling Association and the American Psychological Association (APA, 2002; Roysircar, Arredondo, Fuertes, Ponterotto, & Toporek, 2003; Sue & Sue, 2007). In these statements the words *multiculturalism* and *diversity* are defined broadly to include many dimensions.

The multicultural competencies include awareness, knowledge, and skills. You need to become aware of specific issues, develop knowledge about multicultural issues, and master skills for the interview and daily practice in our multicultural world. Expect the issue of multicultural competence to become increasingly important to your professional helping career. Cultural competency training is now a requirement for medical licensure in New Jersey, and at least four other states have similar legislation pending (Adams, 2005). You can anticipate that you will be increasingly required to be multiculturally competent. Developing this competence will take a lifetime of learning, as there is endless information to absorb.

Let us examine the multicultural guidelines and competencies in more detail.

AWARENESS: BE AWARE OF YOUR OWN ASSUMPTIONS, VALUES, AND BIASES

Awareness of yourself as a cultural being is a vital beginning. Unless you see yourself as a cultural being, you will have difficulty in developing awareness of others. It is important that you understand your own multicultural background and the differences that may exist between you and those who come from other backgrounds. Learn about groups different from yours, and recognize your limitations and the need for referral of clients when necessary.

The guidelines also speak of how contextual issues beyond a person's control affect the way a person discusses issues and problems. Oppression and discrimination, sexism, racism, and failure to recognize and take disability into account may deeply affect clients without their conscious awareness. Is the problem "in the individual" or "in the environment"? For example, you may need to help clients become aware that issues such as tension, headaches, and high blood pressure may result from the stress caused by harassment and oppression. Many issues are not just client problems but rather problems of a larger society.

Privilege is power given to people through cultural assumptions and stereotypes. McIntosh (1988) comments on the "invisibility of Whiteness." European Americans tend to be unaware of the advantages they have because of the color of their skin. The idea of special privilege has been extended to include men, those of middle- or upper-class economic status, and others in our society who have power and privilege.

Whites, males, heterosexuals, middle-class people, and others enjoy the convenience of not being aware of their privileged state. The physically able see themselves as "normal" with little awareness that they are only "temporarily able" until old age or a trauma occurs. Out of privilege comes stereotyping of the less dominant group, thus further reinforcing the privileged status.

You, the interviewer, face challenges. For example, if you are a middle-class European American heterosexual male and the client is a working-class female of a different race, she is less likely to trust you and rapport may be more difficult to establish. You must improve your awareness, knowledge, and skills to work with clients culturally different from you.

In summary, you first need to learn about yourself and whether you have a privileged status. Then your lifetime task is to avoid stereotyping any group or individual and to constantly learn about various cultural groups. Individual differences within a cultural grouping often have greater impact than the cultural "label." Your client is a unique human being. Although diversity factors influence development, always recognize the person before you as special and different from all others. Awareness of multicultural issues and diversity actually enhances individual differences and the ways each client is unique.

KNOWLEDGE: UNDERSTAND THE WORLDVIEW OF THE CULTURALLY DIFFERENT CLIENT

Worldview is formally defined as the way you and your client interpret humanity and the world. Because of varying multicultural backgrounds, we all view people and the larger world differently. Multicultural competence stresses the importance of being aware of our negative emotional reactions and biases toward those who are different from us. If you have learned to view certain groups through inaccurate stereotypes, you especially need to listen and learn respect for the worldview of the client; be careful not to impose your own ideas.

All of us need to develop knowledge about various cultural groups, their history, and their present concerns. If you work with Spanish-speaking groups, it is critical to learn the different history and issues faced by those from Mexico, Puerto Rico, the Caribbean, and Central and South America. What is the role of immigration? How do the experiences of Latinas/Latinos who have been in Colorado for several centuries differ from those of newly arrived immigrants? Note that diversity is endemic to the broad group we often term "Hispanic."

The same holds true for European Americans and all other races and ethnicities. Old-time New England Yankees in Hadley, Massachusetts, once chained Polish immigrants in barns to keep hired hands from running away. The tables are now turned and it is the Poles who control the town, but quiet tensions between the two groups still remain. Older gay males who once hid their identity are very different from young activists. Whether it is race or religion, ability or disability, we will constantly be required to learn more about our widely diverse populations.

Traditional approaches to counseling theory and skills may be inappropriate and/or ineffective with some groups. We also need to give special attention to how socioeconomic factors, racism, sexism, heterosexism, and other oppressive forces may influence a client's worldview.

Understanding various worldviews often comes first through academic study and reading. Another important approach is to become actively involved in the client's community, attending community events, social and political functions, celebrations and festivals, and—most important—getting to know on a personal basis those who are culturally different from you.

Race, gender, sexual orientation, ability, and other multicultural dimensions do matter. You will need to continually study, learn, and experience more about this complex area throughout your career. Despite the election of President Barack Obama, racial disparities remain. Racial minorities still have more school dropouts and at all levels tend to be more dissatisfied with the educational system. While college attendance has nearly doubled, recent court decisions have resulted in fewer minorities at "top" state universities. Beyond the schools, minorities encounter more poverty and violence, income disparities, and a variety of discriminatory situations. Just getting a taxi in downtown New York is very difficult for an African American, even in business attire.

These large and small insults and slights, termed *microaggressions*, result in many physical and mental health problems (Sue, 2010). As you work with minority clients, it is vital that you remain aware of the impact that these external system pressures and oppression may be having on their concerns. One useful intervention in such situations is encouraging and helping these clients to examine how racism, sexism, heterosexism, and other forms of oppression may relate to their present headaches, stomach upsets, high blood pressure, and an array of psychological stressors.

In short, do not enter the helping field without awareness and knowledge that multicultural differences are always present. Furthermore, clients from many backgrounds experience the harassment of insults and slights, which can be ultimately damaging. Be prepared to make this awareness part of your practice.

A final critical element in multicultural competence is to seek supervision and increase your own awareness, knowledge, and skills when you recognize that you are uncomfortable and perhaps even deficient in knowledge and skills.

SKILLS: DEVELOP APPROPRIATE INTERVENTION STRATEGIES AND TECHNIQUES

A classic study found that 50% of minority clients did not return to counseling after the first session (cited in Sue & Sue, 2007). This book seeks to address cultural intentionality by providing you with ideas for multiple responses to your clients. *If your first response doesn't work, be ready with another.* Attend and use listening skills to understand and learn the worldview of others as they tell you their stories (Chapters 3–7). Focusing (Chapter 9) can help clients, who may be blaming themselves for a problem with a classmate or instructor, determine whether their issues are actually related to discrimination.

BOX 2.3   National and International Perspectives on Counseling Skills

Multiculturalism Belongs to All of Us
MARK POPE, CHEROKEE NATION AND PAST PRESIDENT OF THE AMERICAN COUNSELING ASSOCIATION

Multiculturalism is a movement that has changed the soul of our profession. It represents a reintegration of our social work roots with our interests and work in individual psychology.

Now, I know that there are some of you out there who are tired of culture and discussions about culture. You are the more conservative elements of us, and you have just had it with multicultural this and multicultural that. And, further, you don't want to hear about the "truth" one more time.

There is another group of you that can't get enough of all this talk about culture, context, and environmental influences. You are part of the more progressive and liberal elements of the profession. You may be a member of a "minority group" or you have become a committed ally. You may see the world in terms of oppressor and oppressed.

Perhaps now you are saying, "good analysis" or alternatively, "he's pathetic" (especially if you disagree with me). I'll admit it is more complex than these brief paragraphs allow, but I think you get my point.

Here are some things that perhaps can join us together for the future:

1. We are all committed to the helping professions and the dignity and value of each individual.
2. The more we understand that we are part of *multiple cultures*, the more we can understand the multicultural frame of reference and enhance individuality.
3. *Multicultural* means just that—many cultures. Racial and ethnic issues have tended to predominate, but diversity also includes gender, sexual orientation, age, geographic location, physical ability, religion/spirituality, socioeconomic status, and other factors.
4. Each of us is a multicultural being and thus all interviewing and counseling involve multicultural issues. It is not a competition as to which multicultural dimension is the most important. It is time to think of a "win/win" approach.
5. We need to address our own issues of prejudice—racism, sexism, ageism, heterosexism, ableism, classism, and others. Without looking at yourself, you cannot see and appreciate the multicultural differences you will encounter.
6. That said, we must always remember that the race issue in Western society is central. Yes, I know that we have made "great progress," but each progressive step we make reminds me how very far we have to go.

All of us have a legacy of prejudice that we need to work against for the liberation of all, including ourselves. This requires constantly examining yourself, honestly and painfully. You are going to make mistakes as you grow multiculturally; but see these errors as an opportunity to grow further.

Avoid saying, "Oh, I'm not prejudiced." We need a little discomfort to move on. If we realize that we have a joint goal in facilitating client development and continue to grow, our lifetime work will make a significant difference in the world.

Traditional counseling strategies are being adapted for use in a more culturally respectful manner (Ivey, D'Andrea, & Ivey, 2012). It is important to be mindful of the history of cultural bias in assessment and testing instruments and the impact of discrimination on clients. Over time, you will expand your knowledge and skills with traditional strategies and also with newer methods designed to be more sensitive to diversity. Box 2.3 depicts the ongoing process of becoming multiculturally aware.

Review Case Study: Minority Experience in Counseling Training
Complete Group Practice Exercise 2: Examine the Multicultural Guidelines and Competencies

▲ EXERCISE 2.6   **Multiple Responses for Multicultural Issues**

Differences in gender can become a major multicultural issue. Imagine that you are a male counselor working with Susan and you sense some timidity and wonder if she is frightened and not truly comfortable with you. What would you do?

SUSAN: (hesitatingly) Yes, I am scared and really don't [hesitates again] know what to do. [Sits back in her chair with a frightened look on her face.]

Choose the interviewer response that you think will be most helpful:

1. I really sense you are frightened and scared about what you're facing.
2. I sense how scary this is for you. It is hard for you even to share it with me. I'm a male, and I wonder if you would be more comfortable talking with a woman. On the other hand, you've talked fairly freely so far. How is it for you talking with me?
3. Susan, your issues are very complex, and I think it may be best to refer you to another counselor, most likely a woman, at this point. What do you think?

Response #1: This is a fairly good reflection of feeling and certainly catches Susan's major emotions at this moment. On the other hand, her nonverbal hesitation suggests that something else is going on and needs to be explored. (Go back and try again.)

Response #2: This response catches the feelings in the moment and summarizes the observation that the session has gone well so far. Yet it also recognizes that Susan faces real challenges. It offers Susan the possibility of a referral if she wants to work with a female helper. (We'd recommend this at this early stage.) The open question at the end encourages her to share what is going on in the here and now of the session. It is also an attempt at moving toward a more egalitarian relationship.

Response #3: Referral is a possibility, but this sudden suggestion would break the flow of the interview. It is another way an interviewer can maintain power. The implication here is "If you are not satisfied with me, then I'll send you to someone else." This may be OK later, but now it is too early and referral can be seen as a rejection.

POLITICAL CORRECTNESS: OR IS THE ISSUE RESPECT FOR DIFFERENCES?

Political correctness (also **politically correct**, **P.C.**, or **PC**) is a term used to describe language that is calculated to provide a minimum of offense, particularly to the racial, cultural, or other identity groups being described. . . . The existence of PC has been alleged and denounced by conservative, liberal, and other commentators. The term itself and its usage are hotly contested. (http://en.wikipedia.org/wiki/Political_correctness)

Given this controversy, what is the appropriate way to name and discuss cultural diversity? We argue that interviewers and counselors should use language empathically and we urge that you use terms that the client prefers. *Let the client define the name that is to be used.* Respect is the issue here. The client's point of view is what counts in the issue of naming. A woman is unlikely to enjoy being called a girl or a lady, but you may find some who use these terms. Some people in their 70s resent being called elderly or old, whereas others embrace and prefer this language. At the same time, you may find that the client is using language in a way that is self-deprecating.

A woman struggling for her identity may use the word *girl* in a way that indicates a lack of self-confidence. The older person may benefit from a more positive view of the language of aging. A person struggling with sexual identity may find the word *gay* or *lesbian* difficult to deal with at first. You can help clients by exploring names and social identifiers in a more positive fashion.

Race and ethnicity present particularly important issues. *African American* is considered the preferred term, but some clients prefer *Afro-Canadian* or *Black*. Others may feel more comfortable being called Haitian, Puerto Rican, or Nigerian. A person from a Hispanic background may well prefer Chicano, Mexican, Mexican American, Cuban, Puerto Rican, Chilean, or Salvadorean. Some American Indians prefer Native American, but most prefer to be called by the name of their tribe or nation—Lakota, Navajo, Swinomish. Some Caucasians would rather be called British Australians, Irish Americans, Ukrainian Canadians, or Pakistani English. These people are racially White but also have an ethnic background.

The language of nationalism and region is also important. American, Irish, Brazilian, New Zealander, or the nickname "Kiwi" may be the most salient self-identification. *Yankee* is a word of pride to those from New England and a word of derision from many Southerners. Midwesterners, those in Outback Australia, and Scots, Cornish, and Welsh in Great Britain often identify more with their region than with their nationality. Many in Great Britain resent the more powerful region called the Home Counties. And you must recognize that the Canadian culture of Alberta is very different from the cultures of Ontario, Quebec, and the Maritime provinces.

Capitalization is an important issue. The *New York Times* style manual does not capitalize Black and several other multicultural terms, but the capital has become a standard in counseling and psychology. White is also not capitalized by the *Times*, but it is helpful for White people to discover that they, too, have a general racial cultural identity. Capitalization of the major cultural groupings of Black and White is becoming more and more the standardized usage in counseling, human relations, psychology, and social work.

▲ EXERCISE 2.7   **Political Correctness**

What occurs for you when you explore the importance of naming? What are your thoughts about capitalization?

2.3
MAKING POSITIVE PSYCHOLOGY WORK
THROUGH A WELLNESS APPROACH

KEY CONCEPT QUESTIONS

▲ **What is the history and importance of a wellness approach and positive psychology?**

▲ **How do we identify wellness and client strengths?**

▲ **How can we plan for our own wellness and that of our clients?**

If you help clients recognize their strengths, you can expect them to use this recognition in positive ways. But too heavy an emphasis may lead to their ignoring what you say. We do need to work through client difficulties and life challenges. Be intentional and have multiple possibilities to encourage your clients to adopt a lifelong positive orientation and wellness style.

Wellness	Predicted Result
Help clients discover and rediscover their strengths through wellness assessment. Find strengths and positive assets in the clients and in their support system. Identify multiple dimensions of wellness.	Clients who are aware of their strengths and resources can face their difficulties and discuss problem resolution from a positive foundation.

 Complete Personal Wellness Form

▲ EXERCISE 2.8   **Multiple Responses for Wellness Issues**

The following exchange occurs just after a good relationship has been established and Ashar starts telling his story.

ASHAR: It's really difficult. I'm getting comments all the time about being a Muslim. Yes, I'm from Lebanon, but I'm a Maronite Christian. Coming here to the university this week, I was stopped for half an hour before they would let me on the plane. Its getting worse and worse. My family's coming to visit soon from Beirut, and I dread what they are going to face with customs at the airport.

Here are three possible interviewer responses:

1. Yes, I understand that airport security is very rough now. Could you tell me more?
2. I'm impressed with your ability to get through an experience like that, and I appreciate your coming to talk about it. Part of our work will be to explore strengths and positives that your family has. Please continue with your story.
3. Ashar, you've done well to get through that. I can understand your concerns. I see that you're angry and frustrated, but also worried about your parents. Where would you like to go next with this?

Response #1: None of these responses is truly wrong. The first one is a very typical counselor response in that it focuses immediately on the problem with no emphasis on strengths and ability at this point. It is not wrong, particularly because the counselor wisely encourages Ashar to tell his story.

Response #2: This response is the most useful of the three. There is an emphasis on wellness strengths and positives, but perhaps it is a bit too early. We encourage a positive wellness approach, and this response, too, is clearly not wrong.

Response #3: This response uses the client's name, thus personalizing the interview. The interviewer comments first on client strengths and then on the client's body language (anger, frustration, and worry are often easy to observe). This observation is likely appropriate, but if the relationship had not been established earlier, it might have been inappropriate. Like the other two responses, this one also encourages continuation of the story as Ashar would like to tell it.

The wellness approach does not deny human problems and difficulties. Rather, it provides a positive foundation from which issues can be addressed more effectively. When clients discuss their concerns in a positive atmosphere of strength and wellness, this enhances their chances of working through complex issues successfully. If you start by listening only to client stories of problems and failures, clients' positive assets and strengths can be overlooked.

POSITIVE PSYCHOLOGY: THE SEARCH FOR STRENGTHS

Recently, the field of counseling has developed an extensive body of knowledge and research supporting the importance of **positive psychology,** a strength-based approach. Psychology has overemphasized the disease model. Seligman (2009, p. 1) states, "We've become too preoccupied with repairing damage when our focus should be on building strength and resilience." Positive psychology brings together a long tradition of emphasis on positives within counseling, human services, psychology, and social work.

Clients come to us to discuss their problems, their issues, and their concerns. They are talking with us about what is *wrong* with their lives and may even want us to *fix* things for them. There is no question that our role is to enable clients to live their lives more effectively and meaningfully. An important part of this problem-solving process is helping clients discover their strengths.

Leona Tyler (1961), one of the first women to serve as president of the American Psychological Association, developed a system of counseling based on human strengths:

> The initial stages . . . include a process that might be called exploration of resources. The counselor pays little attention to personality weaknesses . . . [and] is most persistent in trying to locate . . . ways of coping with anxiety and stress, already existing resources that may be enlarged and strengthened once their existence is recognized.

Tyler's positive ideas have been central to the microskills framework since its inception (Ivey & Gluckstern, 1974; Ivey, Gluckstern, & Ivey, 2006). The strength- and resource-oriented model, *relationship—story and strengths—goals—restory—action,* is an elaboration of Tyler's original ideas. Chapter 14 discusses applications of the story model, extending the ideas to crisis counseling and brief counseling.

Work with severely distressed clients will operate most effectively if it incorporates wellness concepts as a foundation for therapy. For a positive, multiculturally sensitive approach to clinical and therapeutic work, we suggest that you consult *Theories of Counseling and Psychotherapy* (Ivey et al., 2007). There you will find many specific skill-oriented strategies for therapy with a positive wellness orientation.

WELLNESS ASSESSMENT: DEFINITIONS AND QUESTIONS

The wellness orientation to interviewing and counseling has been most clearly defined and thoroughly researched by Jane Myers and Thomas Sweeney (2004, 2005). Their Indivisible Self model incorporates five key dimensions that they have identified. Each dimension has practical subcategories for assessing clients and facilitating their growth and development.

The Individual Self holistic model stresses the importance of context. As appropriate to the individual client before you, it may be helpful to explore the multiple contexts of human development. For example, what is going on locally (family, neighborhood, community)? Problems here obviously affect the individual; even more important, strengths and wellness assets can be found here as well.

The Indivisible Self model points out that change in any part of the wellness system can be beneficial throughout the whole person—or it may damage many of the 17 dimensions of wellness. A problem or a positive change in one part of the system affects all the other parts. For example, a person may have all dimensions of wellness operating effectively but then encounter a difficult contextual issue such as parents divorcing, a major flood or hurricane, or a major personal trauma. However, the individual may use wellness assets to surmount these challenges and come out of them stronger.

Other contextual issues that may be important to your clients' wellness include the institutions that define so much of their experience, such as education, religion, government, and business/industry. At an even broader level, politics, culture, environmental changes, global events, and the media can deeply affect clients. A change in social services, global warming, or a call-up for military service are three examples of contextual issues that can affect the individual.

A final contextual issue is lifespan development. Issues for a child entering the teenage years are very different from those for a teen entering the military, work, or college. Marriage or selection of a life partner, raising a family, and older maturity all present different contextual issues that need to be considered.

Again, the Indivisible Self concept reminds us that the individual is totally connected with the social context and with all parts of his or her developing personhood all the time.

The following exercise provides brief definitions of the 17 personal dimensions and some beginning wellness questions for you to explore with your clients. For a more detailed presentation of wellness research and the actual listing of factor analytic structures, we recommend consulting Myers and Sweeney (2004, 2005).

▲ EXERCISE 2.9   **Your Own Wellness Assessment**

Unless you do it, you won't remember it! This is a vital exercise that will illustrate the importance and value of a wellness approach. You seldom will do this with all your clients, but including wellness as a central part of each session can be critical for success and effectiveness.

Again, this takes more than reading—it also takes thought and action. First focus on yourself to develop a solid understanding of the several wellness dimensions. Identify concrete examples and specifics available to you in each area. Write down your wellness strengths and personal assessment so that you are familiar with the process.

Then find a student colleague, friend, or family member and work through the wellness assessment with that person.

Dimension 1: The Essential Self

Four aspects of the core self serve as a foundation for personal exploration of wellness. Each of these areas can provide important resources and strengths for positive growth.

Spirituality.   There is considerable evidence that those who have a spiritual or religious orientation have more positive attitudes and better mental health than those lacking such supports. Define spirituality and religion broadly, as a thoughtful agnostic or atheist often has many of the characteristics of a highly religious person. At times, the words *values* or *meaning* should be substituted for *spirituality* and *religion*.

▲ What strengths and supports do you gain from your spiritual/religious orientation? Be as specific as possible.
▲ How could you draw on this resource when faced with life challenges?
▲ Can you give a specific example of how spirituality has helped you in the past?

Gender Identity.   This area has two dimensions—gender and sexual orientation. Identifying positive men and women as role models and finding other positives about your own gender may help you develop unique strengths. Sexual orientation relates to one's identity as a heterosexual, gay, lesbian, bisexual, transsexual, transgender, or questioning. You will find some clients who are unaware that heterosexuality is a sexual orientation. This lack of knowledge can lead to heterosexism. Again, seeking positive models and personal strengths can be a helpful route to wellness.

▲ What strengths can you draw on as a female or male?
▲ Who are some positive gender role models you have looked to during your life?
▲ What strengths do you draw from your sexual orientation—as a heterosexual, a gay, a lesbian, a bisexual, or a transgendered person?
▲ Who in your community supports your sexual orientation?
▲ Can you provide concrete examples of how your gender and sexual orientation have been important to you and your development?

Cultural Identity—Race and Ethnicity.   Research reveals that a positive attitude toward one's race and ethnicity is part of mental health and wellness. Being aware of the strengths of your race/ethnicity can be helpful in establishing who you are and your cultural history. Getting in touch with positive aspects of our ancestry, whether Aboriginal, African American, Italian, Korean, Maori, Navajo, or Swedish, can help us build strengths from our traditions and our families.

▲ What strengths do you draw from your race? Your ethnicity?
▲ Do you have family or role models that suggest ways of living effectively?
▲ Can you provide concrete examples of how your race and ethnicity have been important to you and your development?

Self-Care.   Part of wellness is how well people take care of themselves. Cleanliness, avoidance of drugs, health maintenance, and safety habits (such as wearing seatbelts) are all examples that lead to a longer life. Those clients who do not engage in self-care may be depressed or have other issues. How well versed are you in substance abuse issues and other dimensions of health?

▲ How well do you care for yourself? Do you avoid drugs and alcohol?
▲ Do you attend to your health and personal hygiene?
▲ How careful or safety conscious are you in work and play situations?
▲ Do you take normal precautions and avoid risky or harmful behavior?
▲ Can you provide concrete examples of how you take care of yourself?

Dimension 2: The Social Self

Connection with others is essential for wellness. We are selves-in-relation, and closeness to others is a central aspect of wellness. Two major components of the social self are identified here.

Friendship. We are people in connection, not meant to be alone. It takes time to nurture relationships. This component focuses on your ability to be a friend and to have friends in healthy long-term relationships.

- ▲ Tell about your friends and what strengths they provide for you.
- ▲ Do you have a special friend, one with whom you have had a long-term relationship? What does that mean to you?
- ▲ Can you tell something specific about yourself as a friend, and what you have done to be a good friend to others?

Love. Caring for special people, such as family members or a loved one, results in intimacy, trust, and mutual sharing. Sexual intimacy and sharing with a close partner are key areas of wellness.

- ▲ Describe some positive family stories. What are some positive memories about grandparents, parents, siblings, or your extended family?
- ▲ How does your family value you? As a grandparent/parent? Brother/sister? Child?
- ▲ If your immediate family relationships are not close, please share your experiences with your equivalent of family. (Examples: church/mosque/synagogue, community, cultural group, friendship groups)
- ▲ Give an example of a positive love relationship and what this means to you.

Dimension 3: The Coping Self

To live effectively, we need to be able to cope with the situations around us, and four basic elements to help us have been identified. Each of these is related to different issues in interviewing and counseling, and often different theoretical approaches will be useful with the various elements.

Leisure. People who take time to enjoy themselves daily are better equipped to return to work or school the next day with more energy and less stress. This area is all too often forgotten in counseling's problem-solving approach. When you have time for fun, often it is much easier to solve problems.

- ▲ What leisure time activities do you enjoy?
- ▲ Equally important, do you take time to do them?
- ▲ When was the last time you did something fun, and how did it feel?
- ▲ Can you tell about a specific time when having fun and taking leisure time really benefited you?

Stress Management. Our approach to life, coupled with multiple commitments to family, career, religion, community, and even leisure activities, provides us with endless opportunities to be "stressed out." Data are accumulating that stress is perhaps the central issue in producing mental ill-health and that stress due to either short- or long-term trauma produces bodily changes and affects brain development. Can you help yourself build stress management resources?

▲ What do you do when you encounter stress?
▲ What specific skills and strategies do you use to cope with stress, and do you remember to use these strategies?
▲ Give at least one example of when you managed stress well.
▲ Exercise alleviates stress. Can you tell of a time when you exercised or did something else to help you calm down and relax?

Self-Worth. Self-esteem and feeling good about oneself are required for personal comfort and effective living. We need to accept our imperfections as well as acknowledge our strengths. This part of wellness is obviously especially important; unless we feel positive about ourselves, the aspects of wellness will be weak at best. It also illustrates the holistic and relatedness qualities found in the Indivisible Self model.

▲ What gives you a sense of self-worth and self-esteem?
▲ Can you tell about some specific times that you did something kind or helpful for others that you feel especially good about?
▲ How do you value your life contribution?
▲ What would you like to contribute to others and the world in the future?

Realistic Beliefs. Life is obviously not all positives. We also need a clear grasp of reality, the ability to examine our beliefs and those of others. We can get stuck with negative beliefs about ourselves and the world that undermine effective problem solving. You will find that cognitive-behavioral theory and some of the strategies presented in the influencing skills section of this book are especially helpful here.

▲ How able are you to face up to difficult situations and see things as they really are?
▲ Do you have realistic beliefs and expectations about yourself and your abilities?
▲ Do you have realistic beliefs and expectations about others and their abilities?
▲ What has gone well for you in the past? The present? What positive anticipations do you have for the future?
▲ Is there a specific time you participated in a realistic assessment of yourself or others?

Dimension 4: The Creative Self

Research reveals five creative ways to have a positive impact on the world. Each of these can serve as a springboard for a wellness approach.

Thinking. This element includes the thoughts and thinking patterns that guide your life. Effective problem solving can lead to better personal adjustment. Important in this process is avoiding negative thoughts about yourself and others. An optimistic view is clearly helpful.

▲ What is the nature of your "inner speech"—words and ideas that you say to yourself "inside your head"? Are you encouraging to yourself? To others?
▲ How are you at problem solving? Tell about a time when you effectively resolved a difficult issue.
▲ Can you give an example or two of when positive thinking and optimism worked for you?

Emotions.   Coupled with our thoughts are our feelings (e.g., glad, sad, mad, scared). The ability to experience emotion appropriate to the situation is vital to a healthy lifestyle.

▲ When have you felt and expressed emotion with a good result? Negative emotion? Positive emotion?
▲ Can you understand and support another's emotional experience and become attuned to the way this person experiences the world?
▲ How do you accept emotional support from others?

Control.   People who feel in control of their lives see themselves as making a difference; they are in charge of their own "space." They do not seek to control others. It is a subjective feeling that you know what is happening, what is going to happen, and that you can control present and future events.

▲ When have you been able to control difficult situations in a positive way?
▲ When have you had a positive sense of self-control? In relation to self? In relation to others?
▲ Provide specific examples of how you are in control of your own destiny. Again, provide concrete positive examples.

Work.   We need work to sustain ourselves; it is an activity that takes as much of our time as sleep—or more. Much of our self-worth comes from our ability to contribute to the world through the work we do.

▲ What jobs have you most enjoyed or been most proud of?
▲ What kind of volunteer work do you do?
▲ What do you see as your major contributions or most supportive habits on the job? Write specific examples.

Positive Humor.   Laughing works! It opens your mind and refreshes the body. Humor is part of creativity and enjoying the moment. People with a sense of humor can often find something positive in the midst of real problems.

▲ What makes you laugh?
▲ Tell about your sense of humor.
▲ Is there some specific time when a sense of humor or laughing helped you deal with a difficult situation?

Dimension 5: The Physical Self

The last two aspects of the research on wellness reveal an area that needs far more attention in interviewing and counseling. If a person is not doing well physically, even the best self-concept, ability to handle emotions, or ability to relate effectively with others is not enough. We strongly suggest that you study and bring this dimension into your own self and into interviewing and counseling practice.

Nutrition.   Eating a good diet is part of a wellness program. If a person is failing to eat well, referral to dietary counseling may be helpful. But focus here on strengths.

▲ How aware are you of the standards of good nutrition?
▲ How well do your present weight and eating habits reflect good nutritional standards?

▲ Can you provide concrete examples of how you have taken care of yourself in terms of nutrition in the past and present?

Exercise. New research appears almost daily on the values of exercise and keeping the body moving. For example, recent evidence indicates that general health, memory, and cognitive functioning are all supported by regular exercise. Help your clients keep their bodies moving. One useful treatment for clients who may be depressed is exercise and relaxation training. Self-evaluate your exercise. Also, make evaluation of exercise part of your interviews and help your clients plan for the future.

▲ What do you do for exercise?
▲ What types of exercise do you like best?
▲ How often do you exercise?
▲ Can you provide concrete examples of how exercise has been beneficial for you?
▲ How can you start a program of exercise?

Completing this exercise will familiarize you with the wellness assessment. Later, when you work with a volunteer or real clients, you will be more knowledgeable and competent. As part of the assessment, you will note weaknesses that can be addressed in counseling and through a wellness plan. But the focus is on finding strengths and positive assets for problem solving in the future.

▲ EXERCISE 2.10   **Ideas for Personal Wellness Plan**

Now that you have completed the wellness assessment, briefly describe your ideas for a personal wellness plan for yourself.

 Complete Group Practice Exercise 3: Practice Wellness Interview

INTENTIONAL WELLNESS PLAN

Sweeney and Myers (2005) suggest that counselors need to develop an **intentional wellness plan** with their clients. The first step is a concrete assessment of wellness strengths. The second step is an honest appraisal of areas for improvement. It is particularly important not to overwhelm the client with too many immediate improvements for overall wellness—you could easily lose a discouraged client. Keep it simple. With the client, select one or two items from the wellness assessment and negotiate a contract for action. Check with your client regularly to see how the plan is working. As the client grows and develops, you can move to other dimensions.

We suggest developing an informal wellness plan as part of one session or as a dimension of a longer-term treatment plan. Clients work more effectively on their issues and challenges with a positive wellness approach. The growing interest in positive psychology supports wellness practices in a variety of settings and with clients of all ages. You may ask the client to complete a full wellness assessment as a homework assignment. This will give you a good picture of client strengths that you can draw on during difficult sessions.

CHAPTER SUMMARY

Key Points of "Ethics, Multicultural Competence, and Wellness"	CourseMate and DVD Activities to Build Interviewing Competence

2.1 Ethics in Interviewing and Counseling

▲ Ethical codes can be summarized as follows: "Do no harm to your clients; treat them responsibly with full awareness of the social context of helping."

▲ Counselors must practice within boundaries of their competence, based on education, training, supervised experience, state and national professional credentials, and appropriate professional experience.

▲ Informed consent requires us to tell clients of their rights. When taping sessions, we need permission from the client.

▲ Interviewers and counselors have more perceived power than clients. Efforts need to be made to equalize power in the relationship.

▲ Helping professionals are asked to work outside the interview to improve society and are called upon to act on social justice issues.

1. Interactive Exercise 1: Ethical Self-Assessment. This will provide important personal background as you read this chapter.
2. Interactive Exercise 2: Informed Consent. Study the Informed Consent Form and develop your own informed consent form.
3. Group Practice Exercise 1: Develop an Informed Consent Form.

2.2 Diversity and Multicultural Competence

▲ It is an ethical imperative that interviewers and counselors be multiculturally competent and continually increase their awareness, knowledge, and skills in multicultural areas.

▲ The interviewer needs to be client centered rather than directed by "politically correct" terminology. This is an issue of respect, and clients need to say what is comfortable and appropriate for them.

1. Interactive Exercise 3: Multicultural Self-Assessment. This will help you assess how able are you to work with people who may be culturally different from you.
2. Interactive Exercise 4: Assessing Your Multicultural Identity. This will help you assess your race/cultural identity development.
3. Case Study: Minority Experience in Counseling Training.

2.3 Making Positive Psychology Work Through a Wellness Approach

▲ Most student interviewers can benefit from a wellness assessment and wellness plan. An effective wellness plan can improve the quality of your life and work.

▲ Most clients can benefit from a wellness assessment and wellness plan. You can use a wellness assessment and wellness planning in every interview if you work through the areas of wellness step by step.

1. Complete Personal Wellness Form. Completing this exercise will help you further assess your personal wellness strengths.
2. Group Practice Exercise 3: Practice Wellness Interview.

Assess your awareness, knowledge, and skills as you conclude the chapter:

1. Self-Assessment Quiz: The quiz will help you assess your current knowledge and prepare for course examinations.
2. Portfolio of Competencies: Evaluate your present level of competence in attending and listening skills using the downloadable Self-Evaluation Checklist. Self-assessment of your attending skills competence demonstrates what you can do in the real world.

ATTENDING AND OBSERVATION SKILLS
Basic to Communication

ATTENDING BEHAVIOR

ETHICS, MULTICULTURAL COMPETENCE, AND WELLNESS

Your first task in a helping relationship is to listen. The second is to listen and observe with individual and multicultural sensitivity.

—*Allen Ivey*

How can attending and observation skills be used to help your clients?

CHAPTER
GOALS

Attending behavior encourages client talk. You will want to use attending behavior to help clients tell their stories and to reduce your own talk-time. Through selective attention and sometimes silence, you can help clients refocus their stories in more positive ways.

Observation skills help you understand what is going on between you and your clients both verbally and nonverbally. This understanding can be vital to establishing a helping relationship. Observation skills will help you respond appropriately to both individual and multicultural differences.

Awareness, knowledge, and skills developed through the concepts in this chapter will enable you to

▲ Communicate to the client that you are interested in what he or she is saying by your individually and culturally appropriate attending behavior.
▲ Become aware of verbal and nonverbal attending behavior styles in the interview—including the styles of both you and your client.
▲ Modify your patterns of attending to establish rapport with each individual.
▲ Note varying individual and cultural styles of listening and talking.

Assess your awareness, knowledge, and skills as you begin the chapter:

1. Self-Assessment Quiz: The chapter quiz will help you determine your current level of knowledge. You can take it before and after reading the chapter.
2. Portfolio of Competencies: Before you read the chapter, fill out the download-able Self-Evaluation Checklist to assess your existing knowledge and competence in attending skills. Then, at the end of the chapter, complete the checklist again to summarize your competencies after study and practice.

3.1
DEFINING ATTENDING BEHAVIOR

KEY CONCEPT QUESTION

▲ What are the behavioral skills of listening?

Listening is the core of developing a relationship and making real contact with our clients. How can we define effective listening more precisely? The following exercise may help you to identify listening in terms of clearly observable behaviors.

▲ EXERCISE 3.1   **Identify and Define Listening Skills**

One of the best ways to identify and define listening skills is to experience the opposite—poor listening. Think of a time when someone failed to listen to you. Perhaps a family member or friend failed to hear your concerns, a teacher or employer misunderstood your actions and treated you unfairly, or you had that all-too-familiar experience of calling a computer helpline and never getting someone who listened to your problem. These situations illustrate the importance of being heard and the frustration you feel when someone does not listen to you.

Find a partner to role-play an interview in which one of you plays the part of a poor listener. The poor listener should feel free to exaggerate in order to identify concrete behaviors of the ineffective interviewer. If no partner is available, think of a specific time when you felt that you were not heard. What feelings and thoughts occur to you when you recall that someone important did not listen to you?

List the specific behaviors of poor listening that you identified. Later, compare your thoughts with the ideas presented in this chapter.

An exaggerated role-play is often humorous. However, on reflection, your strongest memory of poor listening may be disappointment and even anger at not being heard. Examples of poor listening and other ineffective interviewing behaviors are numerous—and instructive. If you are to be effective and competent, do the opposite of the ineffective interviewer: Attend and listen!

This exercise demonstrates clearly that attending and listening behaviors make a significant difference to the interview. Neuroscience brain imaging has proven the importance of attending in another way. When a person attends to a stimulus such as the client's story, many areas of the brain of both interviewer and client become involved (Posner, 2004). Specific areas of the brain show activity. In effect, attending and listening "light up" the brain in effective interviewing, counseling, and psychotherapy. Without attention, nothing will happen.

Now let us turn to further discussion of how skills and competence can "light up" the interview.

ATTENDING BEHAVIOR: THE SKILLS OF LISTENING

Attending behavior is defined as supporting your client with individually and culturally appropriate verbal following, visuals, vocal quality, and body language. Attending behavior will have predictable results in client conversation. When you use each of the microskills, you can anticipate what the client is likely to do. These predictions are never 100% perfect, but research has shown what you can usually expect (Daniels, 2010). And if your first attempt at listening is not received, you can intentionally flex and change the focus of your attention or try another approach to show that you are hearing the client.

Attending Behavior	Predicted Result
Support your client with individually and culturally appropriate visuals, vocal quality, verbal tracking, and body language.	Clients will talk more freely and respond openly, particularly about topics to which attention is given. Depending on the individual client and culture, anticipate fewer breaks in eye contact, a smoother vocal tone, a more complete story (with fewer topic jumps), and a more comfortable body language.

Obviously, you can't learn all the possible qualities and skills of effective listening immediately. It is best to learn important behaviors step by step. Attending behavior is the first critical skill of listening, and it is a necessary part of all interviewing and counseling. To communicate that you are indeed listening or attending to the client, you need the following "three V's + B":*

1. **Visual/eye contact.** Look at people when you speak to them.
2. **Vocal qualities.** Communicate warmth and interest with your voice. Think of how many ways you can say, "I am really interested in what you have to say," just by altering your vocal tone and speech rate.
3. **Verbal tracking.** Track the client's story. Don't change the subject; stay with the client's topic.
4. **Body language.** Be yourself—authenticity is essential to building trust. To show interest, face clients squarely, lean slightly forward with an expressive face, and use encouraging gestures. Especially critical, smile to show warmth and interest in the client.

The three V's + B reduce interviewer talk-time and provide clients with an opportunity to tell their stories with as much detail as needed. You will find it helpful to observe your clients' verbal and nonverbal behavior. Note the topics to which your clients attend and those that they avoid. You will find that clients who are culturally different from you may have differing patterns of attending and listening. Use client observation skills and adapt your style to meet the needs of the unique person before you.

HOW DOES ATTENDING BEHAVIOR RELATE TO MULTICULTURAL ISSUES?

Each person you work with will have a unique style of communicating. Furthermore, the multicultural background of each client may modify her or his communication style both verbally and nonverbally. People with varying disabilities also represent a cultural group with whom you may need to vary your attending style, as illustrated in Box 3.1. Use your observation skills and be ready to change your style of attending when necessary.

Review Case Study: Is Attending Enough?
Complete Video Activity 1: Sharpening Your Observation Skills
Complete Video Activity 2: A Negative and a Positive Example of Attending

3.2
EXAMPLE INTERVIEWS: Do I Want to Become a Counselor?

KEY CONCEPT QUESTION

▲ **How does attending behavior vary in ineffective and effective interviews?**

In the example interviews that follow, the client, Jared, is a first-semester sophomore exploring career choice. Like many reading this book, he is considering the helping

---

*We thank Norma Gluckstern Packard for the three V's acronym.

BOX 3.1   Attending Behavior and People With Disabilities

Attending behaviors may require modification if you are working with people with disabilities. It is your role to learn clients' unique ways of thinking and being and how they deal with important issues. Focus on the person, not the disability. For example, think of a person with hearing loss rather than "hearing impaired," a person with AIDS rather than "AIDS victim," a person with a physical disability rather than "physically handicapped." So-called handicaps are often societal and environmental rather than personal.

People who are blind or have limited vision	Clients who are blind or partially sighted may not look at you when they speak. Expect clients with limited vision to be more aware of and sensitive to your vocal tone. People who are blind from birth may have unique patterns of body language. It may be helpful to teach clients attending skills, such as orienting their face and body to other people, that enable them to communicate more easily with the sighted.
People who are deaf or have hearing loss	Many people who are deaf do not consider themselves impaired in any way. People who were born deaf have their own language (signing) and their own culture that often excludes the hearing. You are unlikely to work with this type of client unless you are skilled in sign language and are trusted among the deaf community.
	You may be skilled in sign language or you may counsel a deaf person through an interpreter. To be effective, you will need specific training in the use of an interpreter and a basic understanding of deaf culture. Eye contact is vital in counseling a deaf client or while using an interpreter. It will isolate the client when you speak to the interpreter instead of to the client or use phrases such as "Tell him. . . ."
	For those with moderate to severe hearing loss, speak in a natural way, but not fast. Extensively paraphrase the client's words. Speaking loudly is often ineffective, as ear mechanisms may not equalize for loud sounds. In turn, teaching those with hearing loss to paraphrase what others say to them can help them to communicate with others.
People with physical disabilities	We cannot place people with physical disabilities in any one group; each person is unique. Consider a person who uses a wheelchair, an individual with cerebral palsy or Parkinson's disease, someone who may have lost a limb, or a person who is physically disfigured by a serious burn. They all have the common problem of physical disability, but their body language and speaking style will vary. You must attend to each individual as a complete person from her or his unique perspective.
People who are temporarily able	We suggest that you consider yourself one of the many who are *temporarily able*. Age and life experience will bring most of us some variation of the challenges previously described. For older individuals, the issues discussed here may become the norm rather than the exception. Approach disabled clients with humility and respect.

The National Council on Disability has a searchable website for further details: http://www.ncd.gov.

professions as an alternative. Jared has already stated that he is considering psychology as a major and counseling as a career field. The first example is designed to illustrate ineffective interviewing; it provides a sharp contrast with the second, more positive example.

In both cases, the interviewer, Jerome, is reviewing a form outlining key aspects of career choice. In the first example, Jerome does several things wrong. As you read this session, notice specifics that can disrupt the interview. What do you observe?

NEGATIVE EXAMPLE

Interviewer and Client Conversation	Process Comments
1. **Jerome:** The next thing on my questionnaire is your job history. Tell me a little bit about it, will ya?	The vocal tone is overly casual, almost uninterested. The interviewer looks at the form, not at the client. He is slouching in a chair.
2. **Jared:** Well, I guess the job that, uh . . .	Client appears a bit hesitant and unsure as to what to say or do. There is a slight stammer.
3. **Jerome:** Hold it! There's the phone. [Long pause while Jerome talks to a colleague.] Uh, okay, okay, where were we?	The interview is for the client. Avoid phone and other interruptions during the session. Such behavior shows disrespect for Jared. And if a break is necessary, remember what was occurring just before you were called away.
4. **Jared:** The job that really comes to mind is my work as a camp counselor during my senior year of high school. I really liked counseling kids and . . .	The client's eyes brighten and he leans forward as he starts to talk about something he likes.
5. **Jerome:** [interrupts] Oh, yeah, I did a camp counselor job myself. It was at Camp Itasca in Minnesota. I wasn't that crazy about it, though. But I did manage to have fun. What else have you been doing?	When he is interrupted by the topic jump, Jared looks up with surprise, then casts his eyes downward, as if he realizes he is no longer the focal point of attention. It would have been better if Jerome had asked Jared for specifics about what he liked in his work as a camp counselor.
6. **Jared:** Ah . . . well, I . . . ah . . . wanted to tell you a little bit more about this counseling job, I . . .	Jared is interested in this topic and tries again, but notice the speech hesitations.
7. **Jerome:** [interrupts] I got that down on the form already, so tell me about something else. A lot of counseling types have done peer counseling in school. Have you?	The leading closed question puts Jared on the spot.
8. **Jared:** Ah . . . no, I didn't.	The topic jump now has Jared at a complete standstill. He looks puzzled and confused.
9. **Jerome:** Many effective counselors were peer helpers in school . . .	Jerome is now working on his agenda. One senses that he is not fond of his job and would rather be doing something else.

Interviewer and Client Conversation	Process Comments
10. **Jared:** [interrupts] No, I lived in a small town. I never learned about peer counseling until I got to college.	Jared is now interrupting and sits forward and talks a little more loudly. His anger is beginning to show.
11. **Jerome:** Too bad, it might have helped.	Jerome looks away and dismisses Jared's thoughts and feelings.
12. **Jared:** I only had 75 kids in my graduating class. One person that I always admired was my school counselor. He really got along with the kids. . . .	Jared is valiantly trying to focus the conversation on himself and his interest in counseling. He starts to look a bit more enthusiastic, but again is interrupted.
13. **Jerome:** [interrupts] Hmmm, that sounds like something I ought to write down. You came from a small school. My school counselor wasn't so great. I think it was just a job for him.	Jerome does remember he has a job to do. But he is looking intently at his form and still does not look at Jared. He then returns the focus to himself and the client sits back in discouragement and disarray.

This interview is extreme, but not so rare. It illustrates the many ineffective things an interviewer can do. Jerome clearly had poor visuals, vocals, verbal tracking, and body language.

Now let's give Jerome another chance. What differences do you note in the way Jerome handles Jared in the second example?

POSITIVE EXAMPLE

Interviewer and Client Conversation	Process Comments
1. **Jerome:** Jared, so far, I've heard that you are interested in counseling as a career. You've liked your psychology courses and you find friends come to you to talk about their problems. In this next phase of the session, I'd like to review a form with you that may help us plan together. The next item concerns your job history. Could you tell me a little bit about the jobs you've had in the past?	Jerome leans forward slightly, maintains good eye contact, and his vocal tone is warm and friendly. He personalizes the interview by using the client's name. He summarizes what he has heard and tells the client what is going to happen next. He asks an open question to obtain information from Jared's point of view.
2. **Jared:** Sure, the job that comes to mind is camp counselor at the YMCA during my senior year of high school and my first two years of college.	Jared's vocal tone is confident and relaxed. He seems eager to explore this area.
3. **Jerome:** Sounds like you really liked it. Tell me more.	Jerome observes Jared's enthusiastic words and encourages him to say more.
4. **Jared:** Well, . . . when I was a kid people always asked me what I wanted to be when I grew up. I had no idea, but when I worked at this camp, I realized that I truly liked working with kids, and I was good at it.	Jared smiles and continues.

Interviewer and Client Conversation	Process Comments
5. **Jerome:** Uh-huh.	Jerome continues good eye contact and encourages Jared to go on.
6. **Jared:** I like to work with people—maybe that's why I'm thinking of majoring in psychology or maybe human services or social work.	Jerome begins to see the relationship between past jobs Jared has held and possible future majors in college and career choices.
7. **Jerome:** I see. So you really liked helping kids in the camp. You think it might be a good career direction. Could you give me a specific example of what you particularly liked at the camp?	Jerome summarizes Jared's comments. Asking for specific examples moves client talk from general to specific. Jerome uses attending and verbal tracking and gives the lead to his client.
8. **Jared:** Well, there were kids from lots of different racial groups with differing amounts of money in their backgrounds. Yet I seemed to be able to organize them in a way that built on everybody's strengths. I think I helped them feel pretty good about themselves, and we seemed to avoid conflicts that way.	Jerome is identifying client wellness strengths, which can serve as a basis for both decision making and problem solving.
9. **Jerome:** Sounds like you really feel good about your work there.	Jerome focuses on Jared's underlying feeling tone.
10. **Jared:** I feel real good because I helped them and we had lots of fun at the same time. I learned how to play the ukulele, and we had some great singing sessions. I thought, you know, that this could be a profession for me.	Jared continues with enthusiasm and verifies good feelings. If feeling tones are correctly identified, the client will often acknowledge these feelings and continue discussion in more depth. Whether a positive or negative feeling, you can anticipate more discussion on that topic.
11. **Jerome:** So, considering counseling as a profession may be important. What are some other things along that line you might want to do?	Jerome pinpoints Jared's enthusiasm, paraphrases Jared's last important words, and then asks an open question to encourage more exploration on the topic.
12. **Jared:** Well, kids came to me a lot to talk about their issues with their families and friends. I might even want to be a camp director. I also thought I might want to become a counselor to help people with real personal problems. I just know I want to work with people, but I'm not exactly sure how.	Throughout all this discussion Jerome and Jared have a comfortable relaxed relationship with a solid demonstration of the three V's + B. We now have a far better interviewer and client relationship and a basis for further discussion of majors and possible careers.

The focus in this second session is on attending to Jared and drawing out some early dimensions of what his career story has been so far. We now have some positive ideas of where to go with this session. This time Jerome effectively demonstrates visuals, vocal qualities, verbal tracking, and culturally appropriate body language.

Complete Interactive Exercise 1: Identification and Classification

3.3

INDIVIDUAL AND MULTICULTURAL ISSUES IN ATTENDING BEHAVIOR

KEY CONCEPT QUESTION

▲ **What are some of the many individual and multicultural implications of attending behavior?**

Listen before you leap! A frequent tendency of the beginning counselor or interviewer is to try to solve the client's difficulties in the first 5 minutes. It is critical that you slow down, relax, and attend to client narratives. Think about it—the client most likely developed her or his concern over a period of time. Attending and giving clients talk-time demonstrates that you truly want to hear their story and major concerns. In addition, clients have vary-ing attending and interacting styles. By observing their patterns of conversation with you, you are also learning about how they interact with others outside of the session.

VISUAL/EYE CONTACT

Direct eye contact is considered a sign of interest in European–North American middle-class culture, with more eye contact while listening and less while talking. However, research indicates that some African Americans in the United States may have reverse patterns—looking more when talking and slightly less when listen-ing. Among some traditional Native American and Latin groups, eye contact by the young is a sign of disrespect. Imagine the problems this may cause when the teacher or counselor contradicts basic cultural values by saying, "Look at me!" Some tradi-tional Native American, Inuit, and Aboriginal Australian groups generally avoid eye contact, especially when talking about serious subjects. Cultural differences in eye contact abound, but we must recall that individual differences are often even greater and avoid stereotyping any client or group with so-called "normative" patterns.

You want to look at clients and notice breaks in eye contact. Clients often tend to look away when thinking about a complex issue or discussing topics that particularly distress them. It may be wise to avoid direct and solid eye contact when the client is uncomfortable. Also, think about your own breaks in eye contact. You may find yourself avoiding eye contact while discussing certain topics. There are counselors who say their clients talk about "nothing but sex" and others who say their clients never bring up the topic. Both types of counselors indicate to their clients whether the topic is appropriate through their style of eye contact. If you find that your clients are avoiding a topic, look at your own behavior, not just that of the client. Be sure that your behavior is not encouraging clients to avoid certain subjects.

VOCAL QUALITIES

Your voice is an instrument that communicates much of the feeling you have toward another person or situation. Changes in its pitch, volume, or speech rate convey specific meaning, just as changes in eye contact or body language do.

Keep in mind that different people are likely to respond to your voice differently. Try the following exercise with a group of three or more people.*

▲ EXERCISE 3.2   **Getting Feedback on Your Speech**

Ask the group to close their eyes and note your vocal qualities as you speak. Talk for 2 or 3 minutes in your normal tone of voice on any subject of interest to you. How do they react to your tone, your volume, your speech rate, and perhaps even your regional or ethnic accent? Ask the group for feedback. What does this feedback say to you?

This exercise reveals a central point of attending: People differ in their reactions to the same stimulus. Some people find one voice interesting; others may find that same voice boring or even threatening. People differ, and what is successful with one person or client may not work with another.

Accent is a particularly good example of how different people will react differently to the same voice. Obviously we need to avoid stereotyping people because their accent is different from ours. How do you react to the following accents—Australian, BBC English, Canadian, French, South Asian, New England U.S., and Southern U.S.? Are you aware that the person who speaks two or more languages is advantaged? How many languages can you speak?

As you consider the way you tell a story, you will find yourself giving louder volume and increased vocal emphasis to certain words and phrases; this is known as **verbal underlining.** Clients, of course, do the same. The key words a person underlines via volume and emphasis are often concepts of particular importance.

Awareness of your voice and observation of the changes in others' vocal qualities will enhance your attending skills. Speech hesitations and breaks and the timing of vocal changes can signal distress, anxiety, or discomfort. Clearing one's throat may indicate that words are not coming easily.

VERBAL TRACKING

Verbal tracking is staying with your client's topic to encourage full elaboration of the narrative. Just as people make sudden shifts in nonverbal communication, they change topics when they aren't comfortable. In middle-class U.S. communication, direct tracking is appropriate, but in some Asian cultures such direct verbal follow-up may be considered rude and intrusive.

Verbal tracking is especially helpful to both the beginning interviewer and the experienced interviewer who is lost or puzzled about what to say next in response to a client. *Relax;* you don't need to introduce a new topic. Ask a question or make a brief comment regarding whatever the client has said in the immediate or near past. Build on the client's topics, and you will come to know the client very well over time.

Observe yourself and use **selective attention.** Clients tend to talk about what interviewers are willing to hear. A famous training film (Shostrum, 1966) shows three eminent counselors (Albert Ellis, Fritz Perls, and Carl Rogers) all counseling the

---

*This exercise was developed by Robert Marx, School of Management, University of Massachusetts, Amherst.

same client, Gloria. Gloria changes the way she talks and responds very differently as she works with each counselor. Research on verbal behavior in the film revealed that Gloria tended to match the language of the three different counselors (Meara, Pepinsky, Shannon, & Murray, 1981; Meara, Shannon, & Pepinsky, 1979). Each expert indicated, by his nonverbal and verbal behavior, what he wanted Gloria to talk about!

Should clients match the language of the interviewer, or should you, the interviewer, learn to match your language and style to that of the client? Most likely, both approaches are relevant, but in the beginning, you want to draw out client stories from their own language perspective, not yours. What do you consider most important in the interview? Are there topics with which you are less comfortable? Some interviewers are excellent at helping clients talk about vocational issues but shy away from interpersonal conflict and sexuality. Others may find their clients constantly talking about interpersonal issues, excluding critical practical issues such as getting a job.

Consider the following example:

CLIENT: [speaks slowly, seems to be sad and depressed] I'm so fouled up right now. The first term went well and I passed all my courses. But this term, I am really having trouble with chemistry. It's hard to get around the lab in my wheelchair and I still don't have a textbook yet. [An angry spark appears in her eyes, and she clenches her fist.] By the time I got to the bookstore, they were all gone. It takes a long time to get to that class because the elevator is on the wrong side of the building for me. [looks down at floor] Almost as bad, my car broke down and I missed two days of school because I couldn't get there. [The sad look returns to her eyes.] In high school, I had lots of friends, but somehow I just don't fit in here. It seems that I just sit and study, sit and study. Some days it just doesn't seem worth the effort.

There are several different directions an interviewer could follow from this statement. Where would you go, given the multiple possible directions? List at least three possibilities for follow-up to this client statement.

One way to respond is to reflect the main theme of the client's story. For example, "You must feel like you're being hit from all directions. Which would you like to talk about first?" Different interviewers place emphasis on different issues. Some interviewers consistently listen attentively to only a few key topics while ignoring other possibilities. Be alert to your own potential patterning of responses. It is important that no issue gets lost, but it is equally important to avoid confusion by not seeking to solve everything at once.

Professor Howard Busby of Gallaudet College has commented on this case in a personal email to Allen and Mary. Dr. Busby, who is deaf, points out that the client's problem could be related to the disability, the issues at school, or both.

> The wheelchair might be the problem as much as how this client is dealing with it. The client obviously has a disability, but it does not have to be disabling unless the client makes it so. Although I am categorized as having a disability due to deafness, I have never allowed it to be disabling.
>
> My interpretation of the client is that this depression, on the surface, could be the result of mobility restriction. However, there are other factors that might have caused the depression, even if there were no need for a wheelchair. It is easy for the client to blame problems on the disability and thus distract from personal issues. The issue could be a poor campus environment, learned helplessness on the part of the client, or a combination of these and other multiple factors.

All of us would do well to consider Dr. Busby's comments. He is pointing out to us the importance of taking a broad and comprehensive view of all cases. We must avoid stereotyping, but we must also be sensitive to individual and cultural differences.

ATTENTIVE AND AUTHENTIC BODY LANGUAGE

The anthropologist Edward Hall once examined film clips of Southwestern Native Americans and European North Americans and found more than 20 different variations in the way they walked. Just as cultural differences in eye contact exist, body language patterns also differ.

A comfortable conversational distance for many North Americans is slightly more than arm's length, and the English prefer even greater distances. Many Latin people prefer half that distance, and some people from the Middle East may talk practically eyeball to eyeball. As a result, the slightly forward lean we recommend for attending is not appropriate all the time.

What determines a comfortable interpersonal distance is influenced by multiple factors. Hargie, Dickson, and Tourish (2004, p. 45) point out the following:

Gender: Women tend to feel more comfortable with closer distances than men.

Personality: Introverts need more distance than extraverts.

Age: Children and the young tend to adopt closer distances.

Topic of conversation: Difficult topics such as sexual worries or personal misbehavior may lead a person to more distance.

Personal relationships: Harmonious friends or couples tend to be closer. When disagreements occur, observe how harmony disappears. (This is also a clue when you find a client suddenly crossing the arms, looking away, or fidgeting.)

A person may move forward when interested and away when bored or frightened. As you talk, notice people's movements in relation to you. How do you affect them? Note your own behavior patterns in the interview. When do you markedly change body posture? A natural, authentic, relaxed body style is likely to be most effective, but be prepared to adapt and be flexible according to the individual client.

Your authentic personhood is a vital presence in the helping relationship. Whether you use visuals, vocal qualities, verbal tracking, or attentive body language, be a real person in a real relationship. Practice the skills, be aware, and be respectful of individual and cultural differences. Box 3.2 demonstrates the impact of our attending behavior on people from different cultures.

HELPING CLIENTS MOVE ON FROM REPETITIOUS AND IRRELEVANT CONVERSATION

Skilled counselors and interviewers use attending skills to open and close client talk, making the most effective use of limited time in the interview. If a client talks insistently about the same topic over and over again, this can become a major problem. Simply paying attention and staying on the topic may be nonproductive. Intentional nonattending may be useful to shift clients away from repetitive or negative topics. Otherwise, the interview can become boring, and you may then miss important data, as well as lose the client.

BOX 3.2   National and International Perspectives on Counseling Skills

Use With Care: Culturally Incorrect Attending Can Be Rude
WEIJUN ZHANG

The visiting counselor from North America got his first exposure to cross-cultural counseling differences at one of the counseling centers in Shanghai. His client was a female college student. I was invited to serve as an interpreter. As the session went on, I noticed that the client seemed increasingly uncomfortable. What had happened? Since I was translating, I took the liberty of modifying what was said to fit each other's culture, and I had confidence in my ability to do so. I could not figure out what was wrong until the session was over and I reviewed the videotape with the counselor and some of my colleagues. The counselor had noticed the same problem and wanted to understand what was going on. What we found amazed us all.

First, the counselor's way of looking at the client—his eye contact—was improper. When two Chinese talk to one another, we use much less eye contact, especially when it is with a person of the opposite sex. The counselor's gaze at the Chinese woman could have been considered rude or seductive in Chinese culture.

Although his nods were acceptable, they were too frequent by Chinese standards. The student client, probably believing one good nod deserved another, nodded in harmony with the counselor. That unusual head bobbing must have contributed to the student's discomfort.

The counselor would mutter "uh-huh" when there was a pause in the woman's speech. While "uh-huh" is a good minimal encouragement in North America, it happens to convey a kind of arrogance in China. A self-respecting Chinese would say *er* (oh), or *shi* (yes) to show he or she is listening. How could the woman feel comfortable when she thought she was being slighted?

He shook her hand and touched her shoulder. I told our respected visiting counselor afterward, "If you don't care about the details, simply remember this rule of thumb: in China, a man is not supposed to touch any part of a woman's body unless she seems to be above 65 years old and displays difficulty in moving around."

"Though I have worked in the field for more than 20 years, I am still a lay person here in a different culture," the counselor commented as we finished our discussion.

The first and usually the most effective way to help a client move on further with her or his issues is to break in and paraphrase or summarize what has been said, thus indicating that the client has been heard. For example:

> Sondra, may I break in here? I'd like to summarize what I've heard you say and I want to make sure I have a full understanding. What I heard you say was . . . [here you continue with your summary].

If the client has told you the same story three times, at a minimum your repetition makes it four!

Clients need to tell their stories, and if the story is traumatic, extensive storytelling and repetition may be necessary. Be patient and continue to listen carefully in those cases. With children, the stories also can go on and on, and the same principles apply. Attending to their stories and repeating back to them key elements will enable them to feel heard and move on.

Assuming that the client has not been traumatized, after you summarize what you've heard, you can ask a question referring to another topic, usually one earlier in the interview. Or, if you sense important specifics in the longer story, turn attention to those issues—"I heard . . . and that seems important. Could you tell me more about . . ." If you find yourself ineffective in helping the client move on, look for your own failure to maintain eye contact, your subtle shifts in body posture, and your vocal tone. The client, even though talking too much, is noticing this type of interviewer behavior.

In short, be patient, be kind, and listen fully!

THE USEFULNESS OF SILENCE

Sometimes the most useful thing you can do as a helper is to support your client silently. As a counselor, particularly as a beginner, you may find it hard to sit and wait for clients to think through what they want to say. Your client may be in tears, and you may want to give support through your words. However, sometimes the best support may be simply being with the person and not saying anything. Consider offering a tissue, as even this small gesture shows you care without any need to say anything. In general, it's always good to have a box or two of tissues for clients to take even without asking or being offered.

For a beginning interviewer, silence can be frightening. After all, doesn't counseling mean talking about issues and solving problems verbally? When you feel uncomfortable with silence, look at your client. If the client appears comfortable, draw from her or his ease and join in the silence. If the client seems disquieted by the silence, rely on your attending skills. Ask a question or make a comment about something relevant mentioned earlier in the session.

Finally, remember the obvious: *Clients can't talk while you do.* Review your interviews for talk-time. Who talks more, you or your client? With most adult clients, the percentage of client talk-time should generally be more than that of the interviewer. With less verbal clients or young children, however, the interviewer may need to talk slightly more than they do or tell stories to help clients verbalize. A 7-year-old child dealing with parental divorce may not say a word initially. But when you read a children's book on feelings about divorce, he or she may start to ask questions and talk more freely.

 Complete Interactive Exercise 2: Basic Competence: Shera Looks at Careers

3.4
OBSERVATION SKILLS

KEY CONCEPT QUESTION

▲ **What additional verbal and nonverbal behaviors are important to observe in both interviewer and client?**

Some authorities say that 85% or more of communication is nonverbal. Observing verbal and nonverbal behavior is a critical dimension of effective interviewing. Be self-aware and simultaneously aware of client actions and underlying emotional tone, which is often conveyed through nonverbals. How something is said can sometimes overrule the actual words used by you or your client.

Observation skills help you understand how the client likely is responding to you. If a client breaks eye contact, "something" is happening. That prediction is very likely to be accurate, but defining the reasons the client looked away will vary. The client may find the topic uncomfortable, or may not like what you just said, or perhaps just happened to hear a noise outside the window and looked to see what was happening. Your ability to observe will help you anticipate and understand what is happening with your client, but be careful to watch for individual and cultural differences.

Client Observation Skills	Predicted Result
Observe your own and the client's verbal and nonverbal behavior. Anticipate individual and multicultural differences in nonverbal and verbal behavior. Carefully and selectively feed back observations to the client as topics for discussion.	Observations provide specific data validating or invalidating what is happening in the session. Also, they provide guidance for the use of various microskills and strategies. The smoothly flowing interview will often demonstrate movement symmetry or complementarity. Movement dissynchrony provides a clear clue that you are not "in tune" with the client.

OBSERVE ATTENDING PATTERNS OF CLIENTS

Clients may break eye contact, shift their bodies, and change vocal qualities as their comfort level changes when they talk about various topics. You may observe clients crossing their arms or legs when they want to close off a topic, using rapid alterations of eye contact during periods of confusion, or exhibiting increased stammering or speech hesitations when topics are difficult. And if you watch yourself carefully on tape, you, as interviewer, will exhibit many of these same behaviors at times.

OBSERVE BODY LANGUAGE

Jiggling legs, making complete body shifts, or suddenly closing one's arms most often indicates discomfort. Hand and arm gestures may give an indication of how you and the client are organizing things. Random, discrepant gestures may indicate confusion, whereas a person seeking to control or organize things may move hands and arms in straight lines and point fingers authoritatively. Smooth, flowing gestures, particularly those in harmony with the gestures of others, such as family members, friends, or the interviewer, may suggest openness.

Often people who are communicating well "mirror" each other's body language. They may unconsciously sit in identical positions and make complex hand movements together as if in a ballet. This is termed **movement synchrony. Movement complementarity** is paired movements that may not be identical but are still harmonious. For instance, one person talks and the other nods in agreement. You may observe a hand movement at the end of one person's statement that is answered by a related hand movement as the other takes the conversational "ball" and starts talking.

You may observe discrepancies in nonverbal behavior. **Movement dissynchrony** occurs when a client is talking casually about a friend, for example, with one hand tightly clenched in a fist and the other relaxed and open, possibly indicating mixed feelings toward the friend. Lack of harmony in movement is common between people who disagree markedly or even between those who may not be aware they have subtle conflicts. You have likely seen this type of behavior in couples that you know have problems in communicating.

Some expert counselors and therapists deliberately "mirror" their clients. Experience shows that matching body language, breathing rates, and key words of the client can heighten interviewer understanding of how the client perceives and experiences the world. But be careful with deliberate mirroring. A practicum student reported difficulty with a client, noting that the client's nonverbal behavior seemed especially unusual. Near the end of the session, the client reported, "I know you guys; you try to mirror my nonverbal behavior. So I keep moving to make it difficult for you." You can expect that some clients will know as much about observation skills and nonverbal behavior as you do. What should you do in such situations? Use the skills and concepts in this book with *honesty* and *authenticity*. And talk with your clients about their observations of you without being defensive. Openness works!

INDIVIDUAL AND MULTICULTURAL ISSUES IN NONVERBAL BEHAVIOR

As you engage in observation, recall that each culture has a different style of nonverbal communication. For example, Russians say yes by shaking the head from side to side and no by moving the head up and down. Most Europeans do exactly the opposite.

A study was made of the average number of times in an hour friends of different cultural groups touched each other while talking in a coffee shop. The results showed that British friends in London did not touch each other at all, French friends touched 110 times, and Puerto Rican friends touched 180 times (cited in Asbell & Wynn, 1991).

Smiling is a sign of warmth in most cultures, but in some situations in Japan, smiling may indicate discomfort. Eye contact may be inappropriate for the traditional Navajo but highly appropriate and expected for a Navajo official who interacts commonly with European Americans.

Complete Interactive Exercise 3: Toward Intentional Competence
Complete Interactive Exercise 4: Naomi and Marcus Interview
Complete Interactive Exercise 5: Observation of Nonverbal Patterns

Be careful not to assign your own ideas about what is "standard" and appropriate nonverbal communication. It is important for the helping professional to begin a lifetime of studying nonverbal communication patterns and their variations. In terms of counseling sessions, you will find that changes in style may be as important as, or more important than, finding specific meanings in communication style. Edward Hall's *The Silent Language* (1959) remains a classic. Paul Ekman's work (2007) is the current standard reference for nonverbal communication. You can also visit several useful websites devoted to nonverbal communication; try http://nonverbal.ucsc.edu, or use a search engine with the key words "nonverbal communication." The visuals available on the Internet will provide clearer examples of nonverbal communication than we can provide through the written word.

3.5
MAGIC, SAMURAI, NEUROSCIENCE, AND THE IMPORTANCE OF PRACTICE TO MASTERY

KEY CONCEPT QUESTIONS

▲ **Are the skills of the samurai magical? How does their training relate to true competence?**

▲ **How does neuroscience relate to magic?**

Japanese masters of the sword learn their skills through a complex set of highly detailed training exercises. The process of masterful sword work is broken down into specific components that are studied carefully, one at a time. Extensive and intensive practice is basic to a samurai. In this process of mastery, the naturally skilled person often suffers and finds handling the sword awkward at times. The skilled individual may even find performance worsening during the practice of single skills. Being aware of what one is doing can interfere with coordination and smoothness in the early stages.

Once the individual skills have been practiced and learned to perfection, the samurai retire to a mountaintop to meditate. They deliberately forget what they have learned. When they return, they find the distinct skills have been naturally integrated into their style or way of being. The samurai then seldom have to think about skills at all; they have become samurai masters.

What is samurai magic, you may ask? Intentional practice!

Once upon a time, it was believed that giftedness was inherited. Thus, many of us have been taught that Mozart and Beethoven had a magical gift. Baseball fans still believe that Ted Williams and Joe DiMaggio "had it in their genes." It is a bit different from that. The "magic" of a solely genetic predisposition to giftedness is now recognized as a scientific error, but that error is still promoted in the popular media. Natural talent is there, but it needs to be developed and nurtured with careful practice. Expertise across all fields depends on persistence, practice, and the search for excellence (Ericsson, Charness, Feltovich, & Hoffman, 2006).

The neuroscience of "giftedness" has been detailed by David Shenk in his book *The Giftedness in All of Us* (2010). He finds that whether one is a master musician or a superstar athlete, natural talent may be there, but the real test is many hours and often years of detailed practice. We now know that Mozart, with many natural talents, was bathed in music by his demanding father, who was one of the first to focus on a detailed study of techniques and skills. From the age of 3, Mozart received intensive instruction, and his greatness magnified over time. Ted Williams carried his bat to school and practiced until dark.

Intentional practice is the magic! This means that you need to recognize and enhance your natural talents, but greatness only happens with extensive practice. Practice is the breakfast of champions. Skipping practice means mediocre performance.

Here is what Shenk (2010, pp. 53–54) found that relates directly to you and your commitment to excellence in interviewing and counseling:

1. *Practice changes your body.* Both the brain and body change with practice.
2. *Skills are specific.* Each skill must be practiced completely, if they are to be integrated in superior performance.
3. *The brain drives the brawn.* Changes in the brain are evident in scans. Areas of the brain relating to finger exercises or arm movements show brain growth in those areas. Expect the same in your brain as you truly master communication skills.
4. *Practice style is crucial.* One can understand attending behavior intellectually, but actually practicing the specific skills of attending is what will make the difference. One pass through is seldom enough.
5. *Short-term intensity cannot replace long-term commitment.* If Ted Williams did not continue to practice, his skills would have gradually been lost. You will want to take what you learn about counseling skills and use them regularly.
6. A *continuous feedback loop* is provided by practice, which leads to even more improvement. In addition, feedback from colleagues on your counseling style and skills is especially beneficial.

We are asking you to focus on your natural gifts in communication and then add to them through practice and sharpening of new skills. You may find a temporary and sometimes frustrating decrease in competence, just as can happen with samurai, athletes, and musicians. Some of you may experience some discomfort in practicing the skill of attending. Others may find attending so "easy" that you fail to become fully competent in this most basic of listenings skills (many experienced professionals still can't listen effectively to their clients).

Learn the skills of this book, but allow yourself time for integrating these ideas into your own natural authentic being. It does not take magic to make a superstar, but it does require systematic and intentional practice to achieve full competence in interviewing, counseling, and psychotherapy. Make your own magic.

Complete Interactive Exercise 6: Observation of Attending Patterns in Daily Life
Complete Group Practice Exercise: Group Practice With Attending Behavior
Complete Interactive Exercise 7: Examining Your Own Verbal and Nonverbal Styles

CHAPTER SUMMARY

Key Points of "Attending and Observation Skills"	CourseMate and DVD Activities to Build Interviewing Competence

3.1 Defining Attending Behavior

▲ Demonstrating poor listening skills through role-play is an effective way to identity the importance of listening and the specific skills of attending behavior.

▲ Attending behavior involves the 3V's + B: visuals, vocal qualities, verbal tracking, and body language.

▲ Be sensitive to clients' multicultural and individual characteristics. Never stereotype.

1. Case Study: Is Attending Enough? What would you do in a similar situation?

2. Video Activity 1: Sharpening Your Observation Skills. Use your observation skills to note the interviewer's behavior.

3. Video Activity 2: A Negative and a Positive Example of Attending. Introduction of the microskills and a demonstration of negative and positive attending.

3.2 Example Interviews: Do I Want to Become a Counselor?

▲ Be sensitive to clients' multicultural and individual characteristics. Never stereotype.

▲ When the interviewer demonstrates culturally appropriate attending skills, it shows interest in the client and promotes greater client talk-time.

▲ Expect individual and cultural differences on the four key dimensions of attending.

1. Interactive Exercise 1: Identification and Classification. Two interview excerpts will help you identify attending and nonattending responses and understand the importance of effective listening.

3.3 Individual and Multicultural Issues in Attending Behavior

▲ Vocal qualities express emotions, and each client may interpret your voice differently.

▲ Verbal tracking is the skill of focusing on client topics and identifying topics of concern.

▲ Respect multicultural and individual characteristics in the client's body language. Remain authentic to your own style.

▲ Nonattention may help clients shift from negative, nonproductive topics to more appropriate conversation. There are times when silence is the best approach.

1. Interactive Exercise 2: Basic Competence: Shera Looks at Careers. Select which of three possible interviewer responses to client comments seem most helpful. Compare your selections with ours.

Key Points of "Attending and Observation Skills"	**CourseMate and DVD Activities to Build Interviewing Competence**

3.4 Observation Skills

▲ Observing verbal and nonverbal behavior is critical to understanding the client. In addition, watch videos of your own patterns to assess your strengths and areas for improvement.

▲ Movement synchrony, movement complementarity, and mirroring may represent particularly effective moments and close relationships in the interview.

▲ Movement dissynchrony and noncommunicative body movements may indicate that there are issues not being discussed in the here and now of the interview, or that the client is uncomfortable with you or the session.

▲ Multicultural variations in verbal and nonverbal behavior will require a lifetime of learning on your part.

1. Interactive Exercise 3: Toward Intentional Competence: Practice Attending. Two different activities are offered to help you sharpen your observation and attending skills.
2. Interactive Exercise 4: Naomi and Marcus Interview. Here you will find five client comments, each of which is followed by three possible interviewer responses. Select the one that you think is most appropriate.
3. Interactive Exercise 5: Observation of Nonverbal Patterns. Use this exercise to sharpen your observation skills.

3.5 Magic, Samurai, Neuroscience, and the Importance of Practice to Mastery

▲ The samurai effect helps us understand the importance of practicing and assimilating each skill.

▲ Many highly demanding activities, including sports, driving, golf, dance, and music, are improved with single skills practice.

▲ Continued and intentional practice are needed to achieve mastery.

▲ The microskills approach breaks interviewer skills into single units. When learning single skills one by one, you may experience a temporary decrease in competence. However, continued intentional practice and experience develops competence and a natural integration of skills.

1. Interactive Exercise 6: Observation of Attending Patterns in Daily Life. Allow yourself to become an observer of what is occurring around you. Spend time noting how people interact.
2. Group Practice Exercise: Group Practice With Attending Behavior.
3. Interactive Exercise 7: Examining Your Own Verbal and Nonverbal Styles. Videotape yourself with another person in a real interview or conversation for at least 20 minutes. Then review your verbal and nonverbal behaviors.

 Assess your awareness, knowledge, and skills as you conclude the chapter:

1. Flashcards: Use the flashcards to check your understanding of key concepts and facilitate memorization of key information.
2. Self-Assessment Quiz: The quiz will help you assess your current knowledge and prepare for course examinations.
3. Portfolio of Competencies: Evaluate your present level of competence in attending and listening skills using the downloadable Self-Evaluation Checklist. Self-assessment of your attending skills competence demonstrates what you can do in the real world.

THE BASIC LISTENING SEQUENCE AND ORGANIZING AN EFFECTIVE INTERVIEW

II

SECTION

Without individually and culturally appropriate attending behavior, there can be no interviewing, counseling, or psychotherapy. Attending behavior is basic to all the communication skills of the Microskills Hierarchy.

This section, which includes Chapters 4–7, presents the **basic listening sequence (BLS)** that will enable you to draw out the major facts and feelings central to client concerns. Through the skills of questioning, encouraging, paraphrasing, reflecting feelings, and summarizing, you will learn how to draw out your clients' stories, including not only the basic facts but how they think about the situation and with what associated emotions. The basic listening sequence encompasses the skills of empathic communication.

The basic listening sequence begins with questioning skills (Chapter 4) followed by the clarifying skills of paraphrasing, encouraging, and summarizing (Chapter 5). Perhaps the most important skill for understanding clients empathically is reflecting feelings (Chapter 6). Once you become skilled in basic listening, you are prepared to conduct a complete interview using only listening skills (Chapter 7).

It is possible to have a very successful interview using only the skills of this second section of the book. Carl Rogers, who developed person-centered counseling, made it clear that effective interviewing and counseling can be conducted via listening only. Rogers carried this idea even further and rarely used questions, preferring to have clients carry as much responsibility for the interview as possible. We will present a brief summary of some of Rogers's key theoretical points.

Chapter 7 includes a transcript of a decisional counseling interview that uses only the basic listening sequence. Later, we will expand on decisional counseling as one of the basic tools of all interviewing.

With sufficient practice, you can reach the following performance objectives:

1. Master the basic listening sequence (using open questions, encouraging, paraphrasing, summarizing, and reflecting feelings) and draw out the thoughts, feelings, and behaviors relevant to clients' concerns.

2. Observe clients' reactions to your skill usage and modify your listening skills so that you are with them more fully in the here and now.
3. Conduct an interview using only listening and observing skills.

When you have accomplished these tasks, you may find that your clients have a surprising ability to resolve their own issues and challenges without further intervention on your part. You may also gain a sense of confidence in your own ability as an interviewer.

"When in doubt, listen!" This is the motto of this section and the entire microskills interviewing framework.

CHAPTER 4

QUESTIONS
Opening Communication

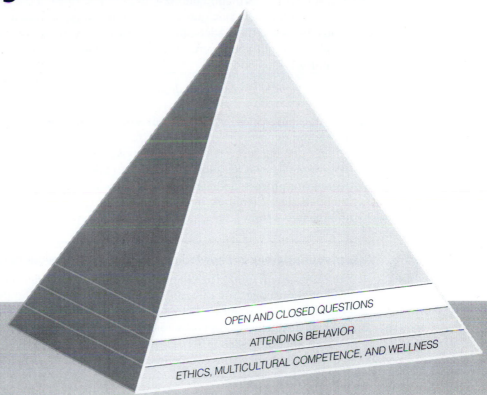

OPEN AND CLOSED QUESTIONS

ATTENDING BEHAVIOR

ETHICS, MULTICULTURAL COMPETENCE, AND WELLNESS

How you ask questions is very important in establishing a basis for effective communication. Effective questions open the door to knowledge and understanding. The art of questioning lies in knowing which questions to ask when. Address your first question to yourself: if you could press a magic button and get every piece of information you want, what would you want to know? The answer will immediately help you compose the right questions.

—*Robert Heller and Tim Hindle*

How can questions help you and your clients?

CHAPTER
GOALS

The basic listening sequence uses questions to draw out information and facilitate client goals. If you use open questions effectively, you can expect the client to talk more freely and openly. Closed questions will elicit shorter responses and may provide you with information and specifics.

Like attending behavior, questions can encourage or discourage client talk. With questions, however, the leadership comes mainly from the interviewer. The client is often talking within your frame of reference. Questions potentially can take away from client self-direction.

Awareness, knowledge, and skills developed through the concepts of this chapter will enable you to

▲ Enrich client stories by bringing out a more complete description, including important details.

▲ Choose the question style that is most likely to achieve a useful predicted result. For example, *what* questions often lead to talk about facts, *how* questions to feelings or process, and *why* questions to reasons.

▲ Open or close client talk, intentionally, according to the individual needs of the interview.

Assess your awareness, knowledge, and skills as you begin the chapter:

1. Self-Assessment Quiz: The chapter quiz will help you determine your current level of knowledge. You can take it before and after reading the chapter.
2. Portfolio of Competencies: Before you read the chapter, fill out the download-able Self-Evaluation Checklist to assess your existing knowledge and competence in attending skills. Then, at the end of the chapter, complete the checklist again to summarize your competencies after study and practice.

4.1
DEFINING QUESTIONS

KEY CONCEPT QUESTIONS

▲ **What are open and closed questions, and how do they affect client conversation?**

▲ **What are the potential problems of questions in the session?**

▲ **When are questions essential?**

▲ EXERCISE 4.1 **Identify and Define Listening Skills**

Benjamin is nearing completion of his junior year in high school. The school requires that each student be interviewed about plans after graduation—work, the armed forces, or college. You are the high school counselor and have called Benjamin in to check on his future plans. He is known as a "nice boy." His grades are average, placing him in the middle third of his class, and he is not particularly verbal or talkative.

What are some questions that you could use to draw him out and help him think ahead to the future? What would you anticipate achieving by using these questions? If you ask too many questions, what potential problems will you face?

You may want to compare your questions with our thoughts on page 90 at the end of this chapter.

Skilled attending behavior is the foundation of the Microskills Hierarchy; questioning provides a systematic framework for directing the interview. Questions help an interview begin and move along smoothly. They open up new areas for discussion, assist in pinpointing and clarifying issues, and aid in clients' self-exploration.

Questions are an essential component in many theories and styles of helping, particularly cognitive-behavioral counseling, interviewing, and much of career decision making. The employment counselor facilitating a job search, the social worker conducting an assessment interview, and the high school guidance counselor helping a student work on college admissions all need to use questions. Moreover, the diagnostic process, while not counseling, uses many questions.

This chapter focuses on two key styles of questioning—open and closed questions.

Open questions are those that can't be answered in a few words. They tend to facilitate deeper exploration of client issues. They encourage others to talk and provide you with maximum information. Typically, open questions begin with *what, how, why,* or *could.* For example, "Could you tell me what brings you here today?"

Closed questions enable you to obtain important specifics and can usually be answered in very few words. They may provide important information, but the burden of guiding the talk remains on the interviewer. Closed questions often begin with *is, are,* or *do.* For example, "Are you living with your family?"

 Complete Interactive Exercise 1: Open vs. Closed Questions

If you use open questions effectively, the client may talk more freely and openly. Closed questions elicit shorter responses and may provide you with information and specifics. Following are some of the results you can expect when using questions.

Intentional Prediction.   Effective use of questions enables you to anticipate what the client will do next. Clients, however, vary in their responses. Use your intentionality to draw on rewording your question or move to other skills. Falling back on basic attending behavior is often helpful.

Open and Closed Questions	*Predicted Result*
Open questions often begin with *who, what, when, where,* or *why.* Closed questions may start with *do, is,* or *are. Could, can,* or *would* questions are considered open but have the advantage of giving more power to the client, who can more easily say what he or she wants to respond.	Clients will give more detail and talk more in response to open questions. Closed questions may provide specific information but may close off client talk. Effective questions encourage more focused client conversations with more pertinent detail and less wandering. *Could, would,* and *can* questions are often the most open of all.

Complete Interactive Exercise 2: Basic Competence

FOUR KEY ISSUES AROUND QUESTIONS

EXCESSIVE USE OF QUESTIONS

An interviewing style focused primarily on questions puts the power and influence almost totally on the counselor. There are good reasons that some theories of helping state that virtually no questions should be used. Our task is to enter the client's world as he or she sees and experiences it. Excessive use of questions is often culturally inappropriate and could destroy a relationship. Your responsibility is to find a reasonable balance for using questions in the interview.

CLIENTS MAY HAVE NEGATIVE EXPERIENCE WITH QUESTIONS

Why do some people dislike questions? Take a minute to recall and explore some of your own experiences with questions. Perhaps you had a teacher or a parent who used questions in a manner that resulted in your feeling uncomfortable or even attacked— "Why did you *do* that?" What thoughts and feelings did this experience produce in you?

People often respond to this exercise by describing situations in which they were put on the spot or grilled by someone. They may associate questions with anger and guilt. Furthermore, questions may be used to direct and control client talk. School discipline and legal disputes typically use questions to control the person being interviewed. If your objective is to enable clients to find their own way, questions may inhibit your reaching that goal. It is for these reasons that some humanistically oriented helping professionals object to questions in the interview. Additionally, in many non-Western cultures, questions are inappropriate and may be considered offensive or overly intrusive.

Nevertheless, questions remain a fact of life in most cultures, and we encounter them everywhere. The physician or nurse, the salesperson, the government official, and many others find questioning clients basic to their profession. Many counseling theories frequently use questions. The issue, then, is how to question wisely and intentionally.

SOMETIMES QUESTIONS ARE ESSENTIAL—"WHAT ELSE?"

Clients do not always provide you with important information, and sometimes the only way to get at missing data is by asking questions. For example, the client may talk about being depressed and unable to act. As a helper, you could listen to the story carefully but still miss important underlying issues relating to the depression. You could ask an open question: "What important things are happening in your life right now or with your family?" The client's answer might tell you that a separation or divorce is impending, that a job has been lost, or that there is some other important dimension underlying the concern. What you first interpreted as depression becomes modified as a shorter-term issue, and the conversation takes a different direction.

An incident in Allen's life—when his father became blind after open heart surgery—illustrates the importance of questions. Was the blindness a result of the surgery? No; it was because the physicians failed to ask the basic open question "Is anything else happening physically or emotionally in your life at this time?" If that question had been asked, the physicians would have discovered that Allen's father had developed severe and unusual headaches the week before surgery was scheduled, and they could have diagnosed an eye infection that is easily treatable with medication.

In counseling, a client may speak of tension, anxiety, and sleeplessness. You listen carefully and believe the problem can be resolved by helping the client relax and plan changes in her work schedule. However, you ask the client, "What else is going on in your life?" In response the client shares a story of sexual harassment, and the goals of the session change.

Finally, at the close of any session, consider asking your client something like "What else should we have discussed today?" or "What have we missed today?" Another good question that will give you feedback on the session and some surprising answers is "What one thing stood out for you today from our conversation?" Often the answers to these questions open up new avenues that need to be considered in the future.

QUESTIONS CAN HELP CLIENTS SEARCH FOR POSITIVE ASSETS AND PATTERNS OF WELLNESS

Stories presented in the helping interview are often negative and full of problems. Carl Rogers, the founder of client-centered counseling, was always able to find something positive in the interview. He considered positive regard and respect for the client essential for future growth. People grow from strength, not from weakness. This is illustrated clearly in neuroscience research. Negative emotions and feelings are located primarily in the amygdala, deep in the brain. Positive emotions (perhaps later in human development) are located in many areas, but the nucleus accumbens sends out signals to the prefrontal cortex, enabling focus on the positive (Ratey, 2008). Some have said (and we agree) that it takes a least three positive comments to balance one negative. Think about your own life experience.

The positive asset search and wellness review are concrete ways to approach positive regard and respect for the client. As you listen to the client, constantly search for strengths and positives, and share your observations. Of course, you do not want to become overly optimistic and minimize the seriousness of the client's situation. However, it is increasingly clear that if you listen to only the sad and negative parts of the client's story, progress and change will be slow and painful.

Clients are "off-balance" when they tend to talk about their problems and what they can't do. The effective interviewer can help them center and feel better about themselves through a strength inventory, discovering what the client is doing right. Some specific, concrete examples of how to engage in a positive asset search include the following:

"Could you tell me a success story that you have had? What was it that you did right?"

"Tell me about a time in the past when someone supported you and what he or she did. What are your currently available support systems?"

"Can you share a time when you helped someone else?"

"What are some things you have been proud of in the past? Now?"

"What do you do well, or what do others say you do well?"

You can also search for external strengths in culture and family:

"Based on your ethnic/racial/spiritual history, can you identify some positive strengths, visual images, and experiences that you have now or have had in the past?"

"Can you recall a friend or family member of your own gender who represents some type of hero in the way he/she dealt with adversity? What did that person do? Can you develop an image of him/her?"

"We all have family strengths despite frequent family concerns. Family can include our extended family, our stepfamilies, and even those who have been special to us over time. For example, some people talk about a special teacher, a neighbor, or an older person who was helpful. Could you tell me about these people in your life and what they mean to you?"

Review the wellness dimensions outlined in Chapter 2 as an additional approach to the search for strengths and positive assets.

4.2
EXAMPLE INTERVIEW: Conflict at Work

KEY CONCEPT QUESTION

▲ How do open and closed questions appear in the context of an interview?

Virtually all of us have experienced conflict on the job—angry or difficult customers, insensitive supervisors, lazy colleagues, or challenges from those we supervise. In the following set of transcripts, we see an employee assistance counselor, Jamila, meeting with Kelly, a junior manager who has a conflict with Peter. The first session illustrates how closed questions can bring out specific facts but can sometimes end in leading the client, even to the point of putting the counselor's ideas into the client's mind.

CLOSED-QUESTION EXAMPLE

Interviewer and Client Conversation	Process Comments
1. **Jamila:** Hi, Kelly. What's happening with you today?	Jamila has talked with Kelly once in the past about difficulties she has had in her early experiences supervising others for the first time. She begins the session with an open question that could also be seen as a standard social greeting.
2. **Kelly:** Well, I'm having problems with Peter again.	Jamila and Kelly have a good relationship. Not all clients are so ready to discuss their issues. More time for developing rapport and trust will be necessary for many clients, even on return visits.

Interviewer and Client Conversation	Process Comments
3. **Jamila:** Is he arguing with you?	Jamila appears interested, is listening and demonstrating good attending skills. However, she asks a closed question, is already defining the issue without discovering Kelly's thoughts and feelings.
4. **Kelly:** [hesitates] Not really; he's so difficult to work with.	Kelly sits back in her chair and waits for the interviewer to take the lead.
5. **Jamila:** Is he getting his work in on time?	See Jamila try to diagnose the problem with Peter by asking a series of closed questions. This is much too early.
6. **Kelly:** No, that's not the issue. He's even early.	
7. **Jamila:** Is his work decent? Does he do a good job?	Jamila is starting to grill Kelly.
8. **Kelly:** That's one of the problems. His work is excellent and he's always there on time. I can't criticize what he does.	
9. **Jamila:** [hesitates] Is he getting along with others on your team?	Jamila frowns and her body tenses as she thinks of what to ask next. Interviewers who rely on closed questions suddenly find themselves running out of questions to ask. They continue searching for another closed question, usually further off the mark.
10. **Kelly:** Well, he likes to go off with Daniel, and they laugh in the corner. It makes me nervous. He ignores the rest of the staff—it isn't just me.	
11. **Jamila:** So, it's you we need to work on. Is that right?	Jamila has been searching for an individual to blame. Jamila relaxes a little as she thinks she is on to something. Kelly sits back in discouragement.
12. **Kelly:** [hesitates and stammers] Well, I suppose so . . . I . . . I . . . really hope you can help me work it out.	Kelly looks to Jamila as the expert. While she dislikes taking blame for the situation, she is also anxious to please and too readily accepts the interviewer's diagnosis.

Closed questions can overwhelm clients and can be used to force them to agree with the interviewer's ideas. Although this example seems extreme, encounters like this are common in daily life and even occur in interviewing and counseling sessions. There is a power differential between clients and counselors. It is possible that an interviewer who fails to listen can impose inappropriate decisions on a client.

OPEN-QUESTION EXAMPLE

The interview is for the client, not the interviewer. Using open questions, Jamila learns Kelly's story rather than the one she imposed with closed questions in the first example. Again, this interview is in the employee assistance office.

Interviewer and Client Conversation	Process Comments
1. **Jamila:** Hi, Kelly. What's happening with you today?	Jamila uses the same easy beginning as in the closed-question example. She has excellent attending skills and is good at relationship building.
2. **Kelly:** Well, I'm having problems with Peter again.	Kelly responds in the same way as in the first demonstration.
3. **Jamila:** More problems? Could you share more with me about what's been happening lately?	Open questions beginning with "could" provide some control to the client. Potentially a "could" question may be responded to as a closed question and answered with "yes" or "no." But in the United States, Canada, and other English-speaking countries, it usually functions as an open question.
4. **Kelly:** This last week Peter has been going off in the corner with Daniel, and the two of them start laughing. He's ignoring most of our staff, and he's been getting under my skin even more lately. In the middle of all this, his work is fine, on time, and near perfect. But he is so impossible to deal with.	We are hearing Kelly's story. The predicted result from open questions is that Kelly will respond with information. She provides an overview of the situation and shares how it is affecting her.
5. **Jamila:** I hear you. Peter is getting even more difficult and seems to be affecting your team as well. It's really stressing you out and you look upset. Is that pretty much how you are feeling about things?	When clients provide lots of information, we need to ensure that we hear them accurately. Jamila summarizes what has been said and acknowledges Kelly's emotions. The closed question at the end is termed a perception check or checkout. Periodically checking with your client can help you in two important ways: (1) It communicates to clients that you are listening and encourages them to continue. (2) It allows the client to correct any wrong assumptions you may have.
6. **Kelly:** That's right. I really need to calm down.	
7. **Jamila:** Let's change the pace a bit. Could you give me a specific example of an exchange you had with Peter last week that didn't work well?	Jamila asks for a concrete example. Specific illustrations of client issues are often helpful in understanding what is really occurring.

Interviewer and Client Conversation	Process Comments
8. **Kelly:** Last week, I asked him to review a bookkeeping report prepared by Anne. It's pretty important that our team understand what's going on. He looked at me like, "Who are you to tell *me* what to do?" But he sat down and did it that day. Friday, at the staff meeting, I asked him to summarize the report for everyone. In front of the whole group, he said he had to review this report for me and joked about me not understanding numbers. Daniel laughed, but the rest of the staff just sat there. He even put Anne down and presented her report as not very interesting and poorly written. He was obviously trying to get me. I just ignored it. But that's typical of what he does.	Specific and concrete examples can be representative of recurring problems. The concrete specifics from one or two detailed stories provide a better understanding of what is really happening. Now that Jamila has heard the specifics, she is better prepared to be helpful.
9. **Jamila:** Underneath it all, you're furious. Kelly, why do you imagine he is doing that to you?	Will the "why" question lead to the discovery of reasons?
10. **Kelly:** [hesitates] Really, I don't know why. I've tried to be helpful to him.	The intentional prediction did not result in the expected response. This is, of course, not unusual. It is likely too soon for Kelly to know why. This illustrates a common problem with "why" questions.
11. **Jamila:** Gender can be an issue; men do put women down at times. Would you be willing to consider that possibility?	Jamila carefully presents her own hunch. But instead of expressing her own ideas as truth, she offers them tentatively with a "would" question and reframes the situation as "possibility."
12. **Kelly:** Jamila, it makes sense. I've halfway thought of it, but I didn't really want to acknowledge the possibility. But it is clear that Peter has taken Daniel away from the team. Until Peter came aboard, we worked together beautifully. [pause] Yes, it makes sense for me. I think he's out to take care of himself. I see Peter going up to my supervisor all the time. He talks to the female staff members in a demeaning way. Somehow, I'd like to keep his great talent on the team, but how when he is so difficult?	With Jamila's help, Kelly is beginning to gain a broader perspective. She thinks of several situations indicating that Peter's ambition and sexist behavior are issues that need to be addressed.
13. **Jamila:** So, the problem is becoming clearer. You want a working team, and you want Peter to be part of it. We can explore the possibility of assertiveness training as a way to deal with Peter. But, before that, what do you bring to this situation that will help you deal with him?	Jamila provides support for Kelly's new frame of reference and ideas for where the interview can go next. She suggests that time needs to be spent on finding positive assets and wellness strengths. Kelly can best resolve these issues if she works from a base of resources and capabilities.

Interviewer and Client Conversation	Process Comments
14. **Kelly:** First, I need to remind myself that I really do know more about our work than Peter. He is new to it. I worked through a similar issue with Jonathan two years ago. He kept hassling me until I had it out with him. He was fine after that. I know my team respects me; they come to me for advice all the time.	Kelly smiles for the first time. She has sufficient support from Jamila to readily come up with her strengths. However, don't expect it always to be that easy. Clients may return to their weaknesses and ignore their assets.
15. **Jamila:** Could you tell me specifically what happened when you sat down and faced Jon's challenge directly?	This "could" question searches for concrete details about when Kelly handled a difficult situation effectively. Jamila can identify specific skills that Kelly can later apply to Peter. At this point the interview can move from problem definition to problem solution.

In this example, we see that Kelly has been given more talk-time and room to explore what is happening. The questions that are focused on specific examples clarify what is happening. We also see that question stems such as *why*, *how*, and *could* have some predictability in expected client responses. The positive asset search is a particularly important part of successful questioning. Issues are best resolved by emphasizing strengths.

You are very likely to work with clients who have similar interpersonal issues wherever you may practice. The previous case examples focus on the single skill of questioning as a way to bring out client stories. Questioning is an extremely helpful skill, but do not forget the dangers of using too many questions.

Complete Interactive Exercise 3: Identifying Questions
Complete Interactive Exercise 4: Intentional Prediction

4.3
INSTRUCTIONAL READING:
Making Questions Work for You

KEY CONCEPT QUESTION

▲ What are some specific concepts and practices that make questions more useful in the interview?

Questions can be facilitative, or they can be so intrusive that clients want to say nothing. Use the ideas presented here to help you define your own questioning techniques and strategies and how questioning fits with your natural interviewing style.

Complete Interactive Exercise 5: Defining Questions, Open vs. Closed Questions

QUESTIONS HELP BEGIN THE INTERVIEW

With verbal clients and a comfortable relationship, the open question facilitates free discussion and leaves plenty of room to talk. Here are some examples:

"What would you like to talk about today?"

"Could you tell me what prompted you to see me?"

"How have things been since we last talked together?"

"The last time we talked you planned to talk with your partner about your sexual difficulties. How did it go this week?"

The first three open questions provide room for the client to talk about virtually anything. The last question is open but provides some focus for the session, building on material from the preceding week. These types of questions will work well for a highly verbal client. However, such open questions may be more than a nontalkative client can handle. It may be best to start the session with more informal conversation—focusing on the weather, a positive part of last week's session, or a current event of interest to the client. You can turn to the issues for this session as the client becomes more comfortable.

THE FIRST WORD OF OPEN QUESTIONS MAY DETERMINE CLIENT RESPONSE

Question stems often, but not always, result in predictable outcomes. *What* questions most often lead to facts.

"What happened?"

"What are you going to do?"

How questions may lead to an exploration of process or feeling and emotion.

"How could that be explained?"

"How do you feel about that?"

Why questions can lead to a discussion of reasons.

"Why did you allow that to happen?"

"Why do you think that is so?"

Use *why* questions with care. While understanding reasons may have value, a discussion of reasons can also lead to sidetracks. In addition, many clients may not respond well because they associate *why* with a past experience of being grilled.

Could, can, or *would* questions are considered maximally open and also contain some advantages of closed questions. Clients are free to say, "No, I don't want to talk about that." *Could* questions suggest less interviewer control.

"Could you tell me more about your situation?"

"Would you give me a specific example?"

"Can you tell me what you'd like to talk about today?"

Give it a try and you'll be surprised to see how effective these simple guidelines can be.

 Complete Interactive Exercise 6: Observation of Questions in Your Daily Interactions

OPEN QUESTIONS HELP CLIENTS ELABORATE AND ENRICH THEIR STORY

A beginning interviewer often asks one or two questions and then wonders what to do next. Even more experienced interviewers can find themselves hard-pressed to know what to do next. To help the session start again and keep it moving, ask an open question on a topic the client presented earlier in the interview.

"Could you tell me more about that?"

"How did you feel when that happened?"

"Given what you've said, what would be your ideal solution to the problem?"

"What might we have missed so far?"

"What else comes to your mind?"

QUESTIONS CAN REVEAL CONCRETE SPECIFICS FROM THE CLIENT'S WORLD

The model question "Could you give me a specific example?" is one of the most useful open questions available to any interviewer. Many clients tend to talk in vague generalities; specific, concrete examples enrich the interview and provide data for understanding action. Suppose, for example, that a client says, "Ricardo makes me so mad!" Some open questions that aim for concreteness and specifics might be

"Could you give me a specific example of what Ricardo does?"

"What does Ricardo do, specifically, that brings out your anger?"

"What do you mean by 'makes me mad'?"

"Could you specify what you do before and after Ricardo makes you mad?"

Closed questions can bring out specifics as well, but even well-directed closed questions may take the initiative away from the client. However, at the discretion of the interviewer, closed questions may prove invaluable:

"Did Ricardo show his anger by striking you?"

"Does Ricardo tease you often?"

"Is Ricardo on drugs?"

Questions like these may encourage clients to say out loud what they have only hinted at before.

QUESTIONS HAVE POTENTIAL PROBLEMS

Questions can have immense value in the interview, but we must not forget their potential problems.

Bombardment/grilling. Too many questions may give too much control to the interviewer and tend to put many clients on the defensive.

Multiple questions. Another form of bombardment, throwing out too many questions at once may confuse clients. However, it may enable clients to select which question they prefer to answer.

Questions as statements. Some interviewers may use questions to sell their own points of view. "Don't you think it would be helpful if you studied more?" This question clearly puts the client on the spot. On the other hand, "What do you think of trying relaxation exercises when you are tense?" might be helpful to get some clients thinking in new ways. Consider alternative and more direct routes of reaching the client. A useful standard is this: If you are going to make a statement, do not frame it as a question.

Why **questions.** *Why* questions can put interviewees on the defensive and cause discomfort. As children, most of us experienced some form of "Why did you do that?" Any question that evokes a sense of being attacked can create client discomfort and defensiveness.

IN CROSS-CULTURAL SITUATIONS, QUESTIONS CAN PROMOTE DISTRUST

If your life background and experience are similar to your client's, you may be able to use questions immediately and freely. If you come from a significantly different cultural background, your questions may be met by distrust and given only grudging answers. Questions place power with the interviewer. A poor client who is clearly in financial jeopardy may not come back for another interview after receiving a barrage of questions from a clearly middle-class interviewer. If you are African American or European American and working with an Asian American or a Latino/a, an extreme questioning style can produce mistrust. If the ethnicities are reversed, the same problem can occur.

Allen was conducting research and teaching in South Australia with Aboriginal social workers. He was seeking to understand their culture and their special needs for training. Allen is naturally inquisitive and sometimes asks many questions. Nonetheless, the relationship between him and the group seemed to be going well. But one day, Matt Rigney, whom Allen felt particularly close to, took him aside and gave some very useful corrective feedback:

> You White fellas! . . . Always asking questions! Let me tell you what goes on in my mind when a White person asks me a question. First, my culture considers many questions rude. But I know you, and that's what you do. But this is what goes on in my mind when you ask me a question. First, I wonder if I can trust you enough to give you an honest answer. Then, I realize that the question you asked is too complex to be answered in a few words. But I know you want an answer. So I chew on the question in my mind. Then, you know what? Before I can answer the first question, you've moved on to the next question!

Allen was lucky he had developed enough trust and rapport that Matt was willing to share his perceptions. Many people of color have said that this kind of feedback represents how they feel about interactions with White people. People with disabilities, gays/lesbians/bisexuals/transgenders/questioning, spiritually conservative

BOX 4.1   National and International Perspectives on Counseling Skills

Using Questions With At-Risk Youth

COURTLAND LEE, PAST PRESIDENT, AMERICAN COUNSELING ASSOCIATION, UNIVERSITY OF MARYLAND

Malik is a 13-year-old African American male who is in the seventh grade at an urban junior high school. He lives in an apartment complex in a lower middle (working) class neighborhood with his mother and 7-year-old sister. Malik's parents have been divorced since he was 6, and he sees his father very infrequently. His mother works two jobs to hold the family together, and she is not able to be there when they come home from school.

Throughout his elementary school years, Malik was an honor roll student. However, since he started junior high school, his grades have dropped dramatically and he expresses no interest in doing well academically. He spends his days at school in the company of a group of seventh- and eighth-grade boys who are frequently in trouble with school officials.

This case is one that is repeated among many African American early teens. But this problem occurs among other racial/ethnic groups as well, particularly those who are struggling economically. And the same pattern occurs frequently even in well-off homes. Many teens are at risk for getting in trouble or using drugs.

While still a boy, Malik has been asked to shoulder a man's responsibilities, picking up things his mother can't do. Simultaneously, his peer group discounts the importance of academic success and wants to challenge traditional authority. And Malik is making the difficult transition from childhood to manhood without a positive male model.

I've developed a counseling program designed to empower adolescent Black males that focuses on personal and cultural pride. The full program focuses on the central question, *"What is a strong Black man?"* While this question is designed for group discussion, it is an important one for adolescent males in general, who might be engaging in individual work. The idea is to use this question to help the youth redefine, in a more positive sense, what it means to be strong and powerful. Some of the related questions that I find helpful include these:

What makes a man strong?
Who are some strong Black men that you know personally? What makes these men strong?
Do you think that you are strong? Why?
What makes a strong body?
Is abuse of your body a sign of strength?
Who are some African heroes or elders that are important to you? What did they do that made them strong?
How is education strength?
What is a strong Black man?
What does a strong Black man do that makes a difference for his people? What can you do to make a difference?

Needless to say, you can't ask an African American adolescent or a youth of any color these questions unless you and he are in a positive and open relationship. Developing sufficient trust to ask these challenging questions may take time. You may have to get out of your office and into the school and community to become a person of trust.

My hope for you as a professional counselor is that you will have a positive attitude when you encounter challenging adolescents. They are seeking models for a successful life, and you may become one of those models yourself. I hope you think about establishing group programs to facilitate development and that you'll use some of these ideas with adolescents to help move them toward a more positive track.

persons, and many others—anyone, in fact—may be distrustful of the interviewer who uses too many questions.

On the other hand, questions can be useful in group discussions to help at-risk youth redefine themselves in a more positive way, as suggested in Box 4.1.

USING OPEN AND CLOSED QUESTIONS WITH LESS VERBAL CLIENTS

Generally in the interview, open questions are preferred over closed questions. Yet it must be recognized that open questions require a verbal client, one who is willing to share with you. Here are some suggestions to encourage clients to talk with you more freely.

Build Trust at the Client's Pace.   A central issue with hesitant clients is trust. Extensive questioning too early can make trust building a slow process with some clients. If the client is required to meet with you or is culturally different from you, he or she may be less willing to talk. Trust building and rapport need to come first, and your own natural openness and social skills are particularly important. With some clients, trust building may take a full session or more.

Search for Concrete Specifics.   Some interviewers and many clients talk in vague generalities. We call this "talking high on the abstraction ladder." This may be contrasted with concrete and specific language, where what is said immediately makes sense. If your client is talking in very general terms and is hard to understand, it often helps to ask, "Could you give me a *specific, concrete* example?"

As the examples become clearer, ask even more specific questions: "You said that you are not getting along with your teacher. What specifically did your teacher say (or do)?" Your chances for helping the client talk will be greatly enhanced when you focus on concrete events in a nonjudgmental fashion, avoiding evaluation and opinion. Following are some examples of concrete questions focusing on specifics.

> Draw out the linear sequence of the story: "What happened first? What happened next? What was the result?"
>
> Focus on observable concrete actions: "What did the other person say? What did he or she do? What did you say or do?"
>
> Help clients see the result of an event: "What happened afterward? What did you do afterward? What did he or she do afterward?" Sometimes clients are so focused on the event that they don't yet realize it is over.
>
> Focus on emotions: "What did you feel or think just before it happened? During? After? What do you think the other person felt?"

Note that each of these questions requires a relatively short answer. These types of open questions are more focused and can be balanced with some closed questions. Do not expect your less verbal client to give you full answers to these questions. You may need to ask closed questions to fill in the details and obtain specific information. "Did he say anything?" "Where was she?" "Is your family angry?" "Did they say 'yes' or 'no'?"

A *leading* closed question is dangerous, particularly with children. In earlier examples, you have seen that a long series of closed questions can bring out the story, but may provide only the client's limited responses to *your* questions rather than what the client really thought or felt. Worse, the client may end up adopting your way of thinking or may simply stop coming to see you.

Complete Case Study: Questions Drawing Out Strengths With Difficult Clients
Complete Group Practice Exercise: Group Practice With Open and Closed Questions

OUR THOUGHTS ABOUT BENJAMIN (From pp. 76–77)

We would likely begin by asking Benjamin what he is thinking about his future after he completes school. We would start with informal conversation about current school events, or something personal we know about Ben. The first question might be, "You'll soon be starting your senior year; will this be different?" Likely he will talk about his feelings and thoughts about the coming year, but soon we predict that discussion about future plans would come up. We would encourage him to elaborate on his responses and explore possibilities.

If Benjamin does not bring up future plans on his own, we'd likely use some of the following questions:

"Beyond meeting class requirements, what are some of your plans for after graduation?"

"How can I help you achieve your goals?"

"What have you done so far about reaching those goals?"

If he focuses on indecision about volunteering for the army, enrolling in a local community college, or attending the state university, we might ask him some of the following questions:

"What about each of these appeals to you?"

"Could you tell me about some of your strengths?"

"If you went to college, what might you like to study?"

"How do finances play a role in these decisions?"

"Are there any negatives about any of these possibilities?"

"How do you imagine your ideal life 10 years from now?"

On the other hand, Benjamin just might look to you for guidance and say, "I don't know, but I guess I better start thinking about it." We might ask him to review his past likes and dislikes for possible clues to the future. Out of these questions, we might see patterns of ability and interest that suggest actions for the future.

"What courses have you liked best in high school?"

"What have been some of your activities?"

"Could you tell me about the jobs you've had in the past?"

"Tell me about your hobbies and what you do in your spare time."

"What gets you most excited and involved?"

"What did you do that made you feel most happy in the past year?"

If Benjamin is uncomfortable in the counseling office, all of these questions might put him off. He might feel that we are grilling him and perhaps even see us as intruding in his world. Usually, getting this type of information and organizing it requires the use of questioning. But questions are only effective if you and the client have a good relationship and are working together.

CHAPTER SUMMARY

Key Points of "Questions"

4.1 Defining Questions

▲ Closed questions focus the interview, provide specific information, and are answered in few words; open questions allow more client talk-time and exploration of client concerns.

▲ Many people have negative experiences with questions. We may have been "put on the spot" or grilled. It becomes key to determine when and how to use questions effectively.

▲ The "What else?" question brings out missing data. It is maximally open and allows the client considerable control.

▲ Questions used to identify strengths can help clients face their issues with more confidence and ability.

4.2 Example Interview: Conflict at Work

▲ Closed questions can bring out specific data, but if they are overused, the interviewer will run out of things to say and so will the client.

▲ Open questions give more control to the client and encourage more client talk-time.

4.3 Instructional Reading: Making Questions Work for You

▲ Questions help begin the interview, elaborate the client's story, and bring out concrete details of the client's world.

▲ Key sentence stems of certain open questions may predict client response. *What* tends to lead to facts, *how* to process and feelings, and *why* to reasons; *could/can/would* were described as maximally open.

▲ Potential difficulties with questions include client grilling and bombardment, the use of questions to make statements, and defensiveness in response to *why* questions.

▲ Questions may be seen as rude and intrusive and may be inappropriate. Sufficient trust needs to be present for questions to work effectively, particularly when multicultural differences are present.

▲ With less verbal clients, be particularly careful not to lead the client into your own frame of reference.

CourseMate and DVD Activities to Build Interviewing Competence

1. Interactive Exercise 1: Open vs. Closed Questions. This exercise will help you recognize open and closed questions.
2. Interactive Exercise 2: Basic Competence. This exercise will help you see how different types of question stems lead to different predictions of what the client might say next.

1. Interactive Exercise 3: Intentional Prediction. This exercise helps you examine intentional prediction. Your task is to read Jamila and Kelly's interview and anticipate or predict what type of statement the client will make following the interviewer's question.

1. Interactive Exercise 4: Defining Questions, Open vs. Closed Questions. Write open and closed questions to elicit further information from five different clients. Can you ask closed questions designed to bring out specifics of the situation? Can you use open questions to facilitate further elaboration of the topic, including the facts, feelings, and possible reasons? What special considerations might be important with each person as you consider age-related multicultural issues?
2. Interactive Exercise 5: Observation of Questions in Your Daily Interactions. Apply the basic question stems *what, how, why,* and *could* in your daily activities, and observe how clients respond differently to each.
3. Case Study: Questions Drawing Out Strengths With Difficult Clients. This exercise explores ways to relate to angry, acting-out youth or adults.
4. Group Practice Exercise: Group Practice With Open and Closed Questions.

Assess your awareness, knowledge, and skills as you conclude the chapter:

1. Flashcards: Use the flashcards to check your understanding of key concepts and facilitate memorization of key information.
2. Self-Assessment Quiz: The quiz will help you assess your current knowledge and prepare for course examinations.
3. Portfolio of Competencies: Evaluate your present level of competence in attending and listening skills using the downloadable Self-Evaluation Checklist. Self-assessment of your attending skills competence demonstrates what you can do in the real world.

ENCOURAGING, PARAPHRASING, AND SUMMARIZING
Skills of Active Listening

CHAPTER 5

ENCOURAGING, PARAPHRASING, AND SUMMARIZING

OPEN AND CLOSED QUESTIONS

ATTENDING BEHAVIOR

ETHICS, MULTICULTURAL COMPETENCE, AND WELLNESS

The possibility is only one sentence away. . . . Our goal is to make the eyes shine!

—*Andrew Zander*

How can these active listening skills help you and your clients?

CHAPTER
GOALS

Clients need to know that the interviewer hears what they say, sees their point of view, and feels their world as they retell their experience. Encouragement, paraphrases, and summarizations are active listening skills that are the heart of the basic listening sequence and help to build empathy. When clients sense that their story is heard, they open up and become more ready for change.

Awareness, knowledge, and skills developed through the concepts of this chapter will enable you to

▲ Help clients talk in more detail about their issues of concern and help prevent the overly talkative client from repeating the same facts. Clarify for the client and you, the interviewer, what is really being said during the session.

▲ Check on the accuracy of what you hear by saying back to clients the essence of their comments and providing periodic summarizations.

Assess your awareness, knowledge, and skills as you begin the chapter:

1. Self-Assessment Quiz: The chapter quiz will help you determine your current level of knowledge. You can take it before and after reading the chapter.
2. Portfolio of Competencies: Before you read the chapter, fill out the downloadable Self-Evaluation Checklist to assess your existing knowledge and competence in attending skills. Then, at the end of the chapter, complete the checklist again to summarize your competencies after study and practice.

5.1
DEFINING ACTIVE LISTENING

KEY CONCEPT QUESTION

▲ **What is active listening?**

Active listening is a communication process that requires intentional participation, decision making, and responding. What we listen to (selective attention) and respond to has a profound influence on how clients talk to us about their concerns. When a client shares with us a lot of information all at once and talks rapidly, we can find ourselves confused and even overwhelmed by the complexity of the story. We need to hear this type of client accurately and sometimes even slow the story down a bit.

▲ EXERCISE 5.1 **Case Scenario**

A client, Jennifer, enters the room and starts talking immediately:

> I really need to talk to you. I don't know where to start. I just got my last exam back and it was a disaster, maybe because I haven't studied much lately. I was up late drinking at a party last night and I almost passed out. I've been sort of going out with a guy for the last month, but that's over as of last night. . . . [pause] But what

really bothers me is that my mom and dad called last Monday and they are going to separate. I know that they have fought a lot, but I never thought it would come to this. I'm thinking of going home, but I'm afraid to. . . .

Jennifer continues for another three minutes in much the same manner, repeating herself, and she seems close to tears. Information is coming so fast that it is hard to follow her. Finally, she stops and looks at you expectantly.

Imagine you are listening to Jennifer's detailed and emotional story. What are you thinking about her at this moment? Write down what you could say and do to help her feel that you empathize with her and understand her concerns, but perhaps also to help her focus. Compare your ideas with the discussion that follows.

When working with Jennifer, a useful first step would be to summarize the essence of Jennifer's several points and say them back to her. As part of this initial response, use a checkout (e.g., "Have I heard you correctly?") to see how accurate your listening was. The **checkout** (sometimes called the "perception check") offers the client a chance to think about what they said and the accuracy of your summary. You could follow this by asking her, "You've talked about many things. Where would you like to start today?"

Choosing to focus first on the precipitating crisis is another possible strategy. We could start with Jennifer's parents' separation, as that seems to be the immediate precipitating crisis, and restate and paraphrase some of her key ideas. Doing this is likely to help her focus on one key issue before turning to the others. The other concerns clearly relate to the parental separation, and will have to be dealt with as well.

▲ EXERCISE 5.2   **What Would You Do?**

How do our thoughts in the preceding paragraphs compare to what you would do?

Active listening demands that you participate fully in the interview by **encouraging, paraphrasing, summarizing,** and helping clients enlarge and enrich their stories. Active listening demands serious attention to empathy. Your goal is to "walk in the client's shoes" and hear small changes in thoughts, feelings, and behaviors. All this will enable you to enter the client's world and worldview more completely. Box 5.1 illustrates that using listening and observing skills with children is just as important as using them with adults.

Encouraging, paraphrasing, and summarizing are basic to empathic understanding and enable you to communicate to clients that they have been heard. When using empathic listening skills, be sure not to mix in your own ideas with what the client has been saying. You say back to clients what you have heard, using their key words. You help clients by distilling, shortening, and clarifying what has been said. Accurate empathic listening is not as common, nor as easy, as it may sound, but its impact is often profound.

Following are the responses you can expect from your client when you use the active listening skills of encouraging, paraphrasing, and summarizing. Remember to use the checkout frequently to obtain feedback on the accuracy of your listening skills.

BOX 5.1   Listening Skills and Children

Using the listening and observing skills with children is just as important as using them with adults. Children too often go through life being told what to do. If we listen to them and their singular constructions of the world, we can reinforce their unique qualities and help them develop a belief in themselves and their own value. Here are a few key comments on the listening skills and children.

Attending	Talk to children at their eye level whenever possible; avoid looking down at them. This may mean sitting on the floor or in small chairs. Be prepared for more topic jumps with children; use attending skills to bring them back to critical issues. They may need to expend excess energy by doing something with their hands; allow them to draw or play with clay as they talk to you, or engage them in a game like Chutes and Ladders or checkers.
Questions and Concreteness	Seek to get the child's perspective, not yours. Children may have difficulty with a general open question, such as "Could you tell me what happened?" Use short sentences, simple words, and a concrete language style. Break down abstract questions into concrete and situational language, using a mix of closed and open questions: "Where were you when the fight occurred?" "What was going on just before the fight?" "Then what happened?" "How did he feel?" "Was she angry?" "What happened next?" "What happened afterward?" In questioning children on touchy issues, be especially careful of closed, leading questions.
Encouraging, Paraphrasing, and Summarizing	These three skills, coupled with good attending and questioning, help children tell their stories. Effective elementary teachers consistently use these skills, especially paraphrasing and encouraging. Observe a competent teacher and identify these distinct microskills.
Other Issues	Provide an atmosphere that is suitable for children by providing small chairs, interesting objects, books, and games. If necessary, head to the playground or nearest gym and play with them there. This is especially useful with kids who can't sit still. Warmth, humor, a smile, an active style, and an actual liking for children are essential. Under stress, children (and many adults) become confused; use names rather than pronouns.

Encouraging	*Predicted Result*
Encourage with short responses that help the client keep talking. These responses may be verbal (repeating key words and short statements) or nonverbal (head nods and smiling).	Clients elaborate on the topic, particularly when encouragers and restatements are used in a questioning tone of voice.
Paraphrasing	*Predicted Result*
Shorten or clarify the essence of what has just been said, but be sure to use the client's main words when you paraphrase. Paraphrases are often fed back to the client in a questioning tone of voice.	Clients will feel heard. They tend to give more detail without repeating the exact same story. If a paraphrase is inaccurate, the client has an opportunity to correct the interviewer.

Summarizing	*Predicted Result*
Summarize client comments and integrate thoughts, emotions, and behaviors. Summarizing is similar to paraphrasing but used over a longer time span.	Clients will feel heard and often learn how the many parts of important stories are integrated. The summary tends to facilitate a more centered and focused discussion. The summary also provides a more coherent transition from one topic to the next or a way to begin and end a full session.

Complete Interactive Exercise 1: Identifying Skills
Complete Interactive Exercise 2: Identification and Classification
Complete Interactive Exercise 3: Basic Competence

▲ EXERCISE 5.3   **Accurate Listening**

Ask a friend or family member to tell you a story (such as a conflict, a positive experience, or a current challenge). Simply sit and listen to what is said, perhaps asking a few questions to enrich and enlarge the story. Say back, as accurately as possible, what you have heard. Ask the person how accurate your summary was and how it felt to be listened to. Here is where you can use the checkout: "Have I heard you correctly?" "Did I miss something you said?" "Is that what you meant?" Write your observations and the other person's reactions.

5.2

EXAMPLE INTERVIEW: They Are Teasing Me About My Shoes

KEY CONCEPT QUESTION

▲ How can the active listening skills be used with children?

All clients have an equal need to know they have been heard, and children are no exception. Many effective elementary teachers constantly say back to students what they have just said. These skills reinforce the conversation and help the children to keep talking from their own frame of reference. Telling your story to someone who hears you accurately is clarifying, comforting, and reassuring. When counseling children, we place even more emphasis on encouraging, paraphrasing, and summarization.

The following example interview is an edited version of a videotaped interview conducted by Mary Bradford Ivey with Damaris, a child actor, role-playing a problem based on a composite of real cases. Damaris is an 11-year-old sixth grader. The session presents a child's problem, but all of us, regardless of age, have experienced nasty teasing and put-downs, often in our closest relationships. Mary first draws out the child's story about teasing and then her thoughts and feelings about the teasing. Mary follows with a focus on the child's strengths, an example of the wellness approach.

Children demand constant involvement, and showing your interest and good humor is even more essential with them. Active listening is especially important with children, as they tend to respond more briefly than adults. On the video of this session, you would see a very involved style with constant encouragement and smiles. This is what we call "warmth," and it is basic to virtually all counseling relationships. Think about it—can you communicate warmth and caring to all your clients?

Interviewer and Client Conversation	Process Comments
1. **Mary:** Damaris, how're you doing?	The relationship between Mary and Damaris is already established; they know each other through school activities.
2. **Damaris:** Good.	She smiles and sits down.
3. **Mary:** I'm glad you could come down. You can use these markers if you want to doodle or draw something while we're talking. I know—you sort of indicated that you wanted to talk to me a little bit.	Mary welcomes the child and offers her something to do with her hands. Many children get restless just talking. Damaris starts to draw almost immediately. You may do better with an active male teen by taking him to the basketball court while you discuss issues. It can also help to have things available for adults to do with their hands.
4. **Damaris:** In school, in my class, there's this group of girls that keep making fun of my shoes, just 'cause I don't have Nikes.	Damaris looks down and appears a bit sad. She stops drawing. Children, particularly the "have-nots," are well aware of their economic circumstances. Some children have used sneakers; Damaris, at least, has newer sneakers.
5. **Mary:** They "keep making fun of your shoes"?	Encourage in the form of a restatement using Damaris's exact key words.
6. **Damaris:** Well, they're not the best; I mean—they're not Nikes, like everyone else has.	Damaris has a slight angry tone mixed with her sadness. She starts to draw again.
7. **Mary:** Yeah, they're nice shoes, though. You know?	It is sometimes tempting to comfort clients rather than just listen. Mary offers reassurance; a simple "uh-huh" would have been more effective. However, reassurance later in the interview may be a very important intervention.
8. **Damaris:** Yeah. But my family's not that rich, you know. Those girls are rich.	Clients, especially children, hesitate to contradict the counselor. Notice that Damaris says, "But. . . ." When clients say, "Yes, but . . . ," interviewers are off track and need to change their style.

Interviewer and Client Conversation	Process Comments
9. **Mary:** I see. And the others can afford Nike shoes, and you have nice shoes, but your shoes are just not like the shoes the others have, and they tease you about it?	Mary backs off her reassurances and paraphrases the essence of what Damaris has been saying using her key words.
10. **Damaris:** Yeah. . . . Well, sometimes they make fun of me and call me names, and I feel sad. I try to ignore them, but still, the feeling inside me just hurts.	If you paraphrase or summarize accurately, a client will usually respond with *yeah* or *yes* and continue to elaborate the story.
11. **Mary:** It makes you feel hurt inside that they should tease you about shoes.	Mary reflects Damaris's feelings. The reflection of feeling is close to a paraphrase and is elaborated in the following chapter.
12. **Damaris:** Mmm-hmm. [pause] It's not fair.	Damaris thinks about Mary's statement and looks up expectantly as if to see what happens next. She thinks back on the basic unfairness of the whole situation.
13. **Mary:** So far, Damaris, I've heard how the kids tease you about not having Nikes and that it really hurts. It's not fair. You know, I think of you, though, and I think of all the things that you do well. I get . . . you know . . . it makes me sad to hear this part because I think of all the talents you have, and all the things that you like to do and—and the strengths that you have.	Mary's brief summary covers most of the important points and Mary also discloses some of her own feelings. Sparingly used self-disclosure can be helpful. Mary begins the positive asset search by reminding Damaris that she has strengths to draw from.
14. **Damaris:** Right. Yeah.	Damaris smiles slightly and relaxes a bit.
15. **Mary:** What comes to mind when you think about all the positive things you are and have to offer?	An open question encourages Damaris to think about her strengths and positives.
16. **Damaris:** Well, in school, the teacher says I'm a good writer, and I want to be a journalist when I grow up. The teacher wants me to put the last story I wrote in the school paper.	Damaris talks a bit more rapidly and smiles.
17. **Mary:** You want to be a journalist, 'cause you can write well? Wow!	Mary enthusiastically paraphrases positive comments using Damaris's own key words.
18. **Damaris:** Mmm-hmm. And I play soccer on our team. I'm one of the people that plays a lot, so I'm like the leader, almost, but . . . [Damaris stops in mid-sentence]	Damaris has many things to feel good about; she is smiling for the first time in the session.

Interviewer and Client Conversation	Process Comments
19. **Mary:** So, you are a scholar, a leader, and an athlete. Other people look up to you. Is that right? So how does it feel when you're a leader in soccer?	Mary has added *scholar* and *athlete* for clarification and elaboration of the positive asset search. She knows from observation on the playground that other children do look up to Damaris. Counselors may add related words to expand the meaning. Mary wisely avoids leading Damaris and uses the checkout, "Is that right?" Mary also asks an open question about feelings. And we note that Damaris used that important word "but." Do you think that Mary should have followed up on that, or should she continue with her search for strengths?
20. **Damaris:** [small giggle, looking down briefly] Yeah. It feels good.	Looking down is not always sadness! The spontaneous movement of looking down briefly is termed the "recognition response." It most often happens when clients learn something new and true about themselves. Damaris has internalized the good feelings.
21. **Mary:** So you're a good student, and you are good at soccer and a leader, and it makes you feel good inside.	Mary summarizes the positive asset search using both facts and feelings. The summary of feeling *good inside* contrasts with the earlier feelings of *hurt inside*.
22. **Damaris:** Yeah, it makes me feel good inside. I do my homework and everything [pause and the sad look returns], but then when I come to school, they just have to spoil it for me.	Again, Damaris agrees with the paraphrase. She feels support from Mary and is now prepared to deal from a stronger position with the teasing. Here we see what lies behind the "but" in 18 above. We believe Mary did the right thing in ignoring the "but" the first time. Now it is obvious that the negative feelings need to be addressed. When Damaris's wellness strengths are clear, Mary can better address those negative feelings.
23. **Mary:** They just spoil it. So you've got these good feelings inside, good that you're strong in academics, good that you're, you know, good at soccer and a leader. Now, I'm just wondering how we can use those good feelings that you feel as a student who's going to be a journalist someday and a soccer player who's a leader. Now the big question is how you can take the good, strong feelings and deal with the kids who are teasing. Let's look at ways to solve your problem now.	Mary restates Damaris's last words and again summarizes the many good things that Damaris does well. Mary changes pace and is ready to move to the problem-solving portion of the interview.

BOX 5.2 Cumulative Stress and Microaggressions: When Do "Small" Events Become Traumatic?

At one level, being teased about the shoes one wears doesn't sound all that serious—children will be children! However, some poor children are teased and laughed at throughout their lives for the clothes they wear. At a high school reunion, Allen talked with a classmate who recalled painful memories, still immediate, of teasing and bullying during school days. Child and adolescent trauma can affect one's whole life experience. These microaggressions she experienced in high school became part of her persona and left her with fewer life possibilities. Small slights become big hurts if repeated again and again. Athletes and "popular" students may talk arrogantly and dismissively about the "nerds," "townies," "hicks," or other out-group. Teachers, coaches, and even counselors sometimes join in the laughter. Over time, these slights mount inside the child or adolescent. Some people internalize their issues in psychological distress; others may act them out in a dramatic fashion—witness the shootings at Columbine High School in Colorado and at other schools throughout the country.

Microaggressions is a term for these small hurts that accumulate and magnify over time (Sue, 2010). Discrimination and prejudice are other examples of cumulative stress and trauma. One of Mary's interns, a young African American woman, spoke of a recent racial insult. At a restaurant she overheard two White people talking loudly about the "good old days" of segregation. Perhaps the remark was not directed at her, but still, it hurt. She related how common racial insults and microaggressions were in her life, directly or indirectly. She could tell how bad things were racially by the size of her phone bill. When an incident

occurred that troubled her, she needed to talk to her sister or parents and seek support. Out of continuing indignities can come feelings of underlying insecurity about one's place in the world (internalized oppression and self-blame) and/or tension and rage about unfairness (externalized awareness of oppression). Either way, the person who is ignored or insulted feels tension in the body, the pulse and heart rate increase, and—over time—hypertension and high blood pressure may result. The psychological becomes physical, and cumulative stress becomes traumatic.

Soldiers at war, veterans, police officers, fire fighters, women who suffer sexual harassment, those who are short or overweight, the physically disfigured through birth or accident, gays and lesbians, and many others are at risk for having cumulative stress build to real trauma or posttraumatic stress.

Be alert for signs of cumulative stress in your clients. Are they internalizing the stressors by blaming themselves? Or are they externalizing and building a pattern of explosive rage and anger? All these people have important stories to tell, and at first these stories may sound routine. The occurrence of posttraumatic stress responses in later life may be alleviated or prevented by your careful listening and support.

Finally, think of yourself as a potential social action agent in your community. What can you do to help groups of clients (such as those described above) deal with microaggressions and stressors more effectively? Understanding broad social stressors is part of being an effective helper—and taking action or organizing groups toward a healthier lifestyle and working with them to take action for betterment represents a challenge for the future.

Mary had a positive and warm relationship and was able to draw out Damaris's story fairly quickly. She moved to positive assets and wellness strengths that make it easier to address client problems and challenges. Clients can solve problems best from their strengths. Too many interviewers draw out discouraging client details and may even fail to notice and use enhancing strengths and supports. At the same time, be sure to give adequate attention to clients' issues and concerns—be positive, but don't minimize what they came to see you about.

As explained in Box 5.2, not all children do as well when they face bullying; in some cases the cumulative effect of teasing, bullying, and cyber-bullying may become traumatic and require extensive interventions, including working with parents and teachers. A firm and clear school policy against all types of bullying can make an immense difference. Bullying no longer should be considered "normal" child behavior. The scars of bullying over time can lead to trauma. At the extreme, when coupled with other life challenges, bullying can lead to depression and even suicide, and we need always to be aware of the multiple interventions that may be required to help the victim. Bullying does not need to happen in any school. Identifying strengths and supports is all the more important with the victimized child, adolescent, or adult. This often means individual counseling, work with the family, and helping the teacher understand what is going on, with special attention to what goes on outside the school in the community.

Complete the following:
- ▲ Video Activity: Toward Multicultural Competence
- ▲ Interactive Exercise 4: Writing Listening Responses
- ▲ Interactive Exercise 5: Encouraging, Paraphrasing, and Summarizing: Marshall and Jesse Interview
- ▲ Case Study: Meridith
- ▲ Weblink Critique

5.3
INSTRUCTIONAL READING: The Active Listening Skills of Encouraging, Paraphrasing, and Summarizing

KEY CONCEPT QUESTIONS

▲ **How do we define the skills of encouraging, paraphrasing, and summarizing, and how do they relate to active listening?**

▲ **What is the relationship between active listening and diversity issues in the interview?**

The active listening skills of encouraging, paraphrasing, and summarizing are different points on a continuum. We move from single words and short phrases to summaries of a section of an interview, or even a series of sessions. In every case, the goal is hearing the client and feeding back what has been said.

Active listening skills help the client clarify issues and move into deeper exploration of concerns. These skills help you make sure that you accurately hear what the client is saying. A proficient demonstration of active listening skills requires you to listen intently, with a nonjudgmental attitude, and clarify what clients say.

A real challenge for most interviewers is to remain nonjudgmental and accepting as we listen to clients, particularly when our internal beliefs and values disagree with theirs. It is all too easy for interviewer behavior to convey judgment and negative attitudes. But a nonjudgmental attitude is key to all interviewing, counseling, and psychotherapy. Listen to clients without evaluating them, either as "good" or "bad." Simply try to hear and accept what they say. You convey your ability to be both neutral and supportive with eye contact (visual), vocal quality, verbal tracking, and body language (3 Vs + B).

Social justice ethics (Chapter 2) and the multicultural movement suggest that interviewers and counselors need to speak up and make judgments in some instances. For example, if a woman denies obvious abuse, the interviewer may have to make a report. If a person has been racially or sexually harassed and blames herself or himself for the problem, again the counselor must do more than pure listening. As a mandated reporter, counselors—like other helping professionals, including doctors, teachers, and ministers—are obligated by law to report neglect or abuse to appropriate authorities. It is critical to inform the client during intake of your legal obligation to report neglect or abuse.

▲ EXERCISE 5.4   **Overtly Racist/Oppresive Behavior**

When a client shows clear racism, sexism, anti-Semitism, or other oppressive thinking in the interview, the interviewer is faced with a challenge. There is an increasing move in the field toward interviewer responsibility to instruct clients who may need to learn more tolerance and respect.

What are your thoughts? _____

What does the current code of ethics suggest? _____

What are you going to do when faced with these challenges? _____

ENCOURAGING

Encouragers are verbal and nonverbal expressions the counselor or therapist can use to prompt clients to continue talking. Encouragers include minimal verbal utterances ("ummm" and "uh-huh"), head nods, open-handed gestures, and positive facial expressions that encourage the client to keep talking. Silence, accompanied by appropriate nonverbal communication, can be another type of encourager. These encouragers are not meant to direct client talk; rather, they simply encourage clients to keep talking.

Repetition of key words can encourage a client and has more influence on the direction of client talk. Consider the following client statement:

> "And then it happened again. The grocery store clerk gave me a dirty look and I got angry. It reminded me of my last job, where I had so much trouble getting along. Why are they always after me?"

The counselor could use a variety of short encouragers in a questioning tone of voice ("Angry?" "Last job?" "Trouble getting along?" "After you?"), and in each case the client would likely talk about a different topic. It is important that you note your selection of single-word encouraging responses as they may direct clients more than you think.

A **restatement** is a type of extended encourager in which the counselor or interviewer repeats short statements, two or more words exactly as used by the client. "The clerk gave you a dirty look." "You got angry." "You had trouble getting along in your last job." "You wonder why they are always after you." Restatements can be used with a questioning tone of voice; they then function much like the single-word encourager. Like short encouragers, different types of restatements lead the client in different directions.

Well-timed encouragers maintain flow and continually communicate to the client that you are listening. All types of encouragers facilitate client talk unless they are overused or used badly. Picking out a key word or short phrase to use as an encourager often leads clients to provide you with important underlying meanings for that word or phrase. Just one well-observed word can open important new avenues in the session. Encouragers and restatements can be the leading edge of vital issues. On the other hand, the use of too many encouragers can seem wooden and unexpressive, whereas too few encouragers may suggest to clients that you are not interested.

Always remember smiling as a key encourager. The warmth and caring you demonstrate may be the most important part of the relationship, even more important than what you say. Facial expression and vocal tone are the nonverbal components of encouraging.

▲ EXERCISE 5.5   **Practice Encouragers and Restatements**

Reread the preceding paragraphs and say the suggested encouragers and restatements aloud. Use different vocal tones and note how your verbal style can facilitate others' talking or stop them cold.

PARAPHRASING

At first glance, paraphrasing appears to be a simple skill, only slightly more complex than encouraging. In encouraging and restating, exact words and phrases are fed back to the client. Paraphrasing covers more of what the client has just said, usually several sentences. Paraphrasing continues to feed back key words and phrases, but catches and distills the essence of what the client has said. Paraphrasing clarifies a confusing client story.

When you paraphrase, the tone of your voice and your body language indicate to the client whether you are interested in listening in more depth or would prefer that the client move on to another topic.

If your paraphrase is accurate, the client is likely to reward you with a "That's right" or "Yes . . . " and then go on to explore the issue in more depth. Once clients know they have been heard, they are often able to move on to new topics. The goal of paraphrasing is to facilitate client exploration and clarify issues.

Accurate paraphrasing will help the client stop repeating a story unnecessarily. Some clients have complex problems that no one has ever bothered to hear accurately, and they literally need to tell their story over and over until someone indicates they have been heard clearly.

How do you paraphrase? Observe the client, hear the client's important words, and use them in your paraphrase much as the client does. You may use your own words, but the main ideas and concepts must reflect the client's view of the world, not yours! An accurate paraphrase usually consists of four dimensions:

1. A *sentence stem* sometimes using the client's name. Names help personalize the session. Examples are: "Damaris, I hear you saying . . . ," "Luciano, sounds like . . . ," "Looks like the situation is. . . ."

2. The *key words* used by the client to describe the situation or person. Include main ideas and exact words that come from clients. This aspect of the paraphrase is sometimes confused with the encouraging restatement. A restatement, however, covers a very limited amount of client talk and is almost entirely in the client's own words.
3. The *essence of what the client has said* in briefer and clearer form. Identify, clarify, and feed back the client's sometimes confused or lengthy talk into succinct and meaningful statements. The counselor has the difficult task of staying true to the client's ideas but not repeating them exactly.
4. A *checkout* for accuracy. The checkout is a brief question at the end of the paraphrase, summary, reflection of feelings, interpretation/reframe, or other microskill. Here you ask the client for feedback on whether the paraphrase (or other skill) was correct and useful. Some sample checkouts are "Am I hearing you correctly?" "Is that close?" "Have I got it right?" It is also possible to paraphrase with an implied checkout by raising your voice at the end of the sentence as if the paraphrase were a question.

In the following example, a brief client statement is followed by key word encouragers, restatements, and a paraphrase that an interviewer might use to encourage client talk.

> "I'm really concerned about my wife. She has this feeling that she has to get out of the house, see the world, and get a job. I'm the breadwinner and I think I have a good income. The children view Yolanda as a perfect mother, and I do too. But last night, we really saw the problem differently and had a terrible argument."

> ▲ *Key word encouragers*: "Breadwinner?" "Terrible argument?" "Perfect mother?"
> ▲ *Restatement encouragers*: "You're really concerned about your wife." "You see yourself as the breadwinner." "You had a terrible argument."
> ▲ *Paraphrase*: "You're concerned about your picture-perfect wife who wants to work even though you have a good income, and you've had a terrible argument. Is that how you see it?"

As always, personalize and make your active listening real. A stem is not always necessary and, if overused, can make your comments seem like parroting. Clients have been known to say in frustration, "That's what I just said; why do you ask?" Again, smiling and warmth make a difference.

SUMMARIZING

Summarizing falls along the same continuum as the key word encourager, restatement, and paraphrase. Summarizing, however, encompasses an even longer period of conversation than paraphrasing; at times it may cover an entire interview or even issues discussed by the client over several interviews.

In summarizing, the interviewer attends to verbal and nonverbal comments from the client over a period of time and selectively attends to key concepts and dimensions, restating them for the client as accurately as possible. A checkout at the end for accuracy is an important part of the summarizing. Following are some examples.

> *To begin a session*: "Let's see, last time we talked about your feelings toward your mother-in-law, and we discussed the argument you had with her when the new baby arrived. You saw yourself as guilty and anxious. Since then you

haven't gotten along too well. We also discussed a plan of action for the week. How did that go?"

Midway in the interview: "So far, I've seen that you felt guilty again when you saw the action plan as manipulative. Yet one idea did work. You were able to talk with your mother-in-law about her garden, and it was the first time you had been able to talk about anything without an argument. You visualize the possibility of following up on the plan next week. Is that about it?"

At the end of the session: "In this interview we've reviewed more detail about your feelings toward your mother-in-law. Some of the following things seem to stand out: First, our plan didn't work completely, but you were able to talk about one thing without yelling. As we talked, we identified some behaviors on your part that could be changed. They include better eye contact, relaxing more, and changing the topic when you start to see yourself getting angry. Does that sum it up?"

Complete Interactive Exercise 6: Toward Intentional Competence
a. Generating Written Encouragers, Restatements, Paraphrases, and Summarizations
b. Practice of Skills in Other Settings
Complete Group Practice Exercise: Group Practice With Active Listening Skills

DIVERSITY AND ACTIVE LISTENING

Periodic encouraging, paraphrasing, and summarizing are basic skills that seem to have wide cross-cultural acceptance. Virtually all your clients like to be listened to accurately. But it may take more time to establish a relationship with clients who are culturally different from you, and knowledge of their languages may help (see Box 5.3).

BOX 5.3   National and International Perspectives on Counseling Skills

Developing Skills to Help the Bilingual Client
AZARA SANTIAGO-RIVERA

It wasn't that long ago that counselors considered bilingualism a "disadvantage." We now know that a new perspective is needed. Let's start with two important assumptions: *The person who speaks two languages is able to work and communicate in two cultures and, actually, is advantaged. The monolingual person is the one at a disadvantage!* Research actually shows that bilingual children have more fully developed capacities and a broader intelligence (Power & Lopez, 1985).

If your client was raised in a Spanish-speaking home, for example, he or she is likely to think in Spanish at times, even though having considerable English skills. We tend to experience the world nonverbally before we add words to describe what

we see, feel, or hear. For example, Salvadorans who experienced war or other forms of oppression *felt* that situation in their own language.

You are very likely to work with clients in your community who come from one or more language backgrounds. Your first task is to understand some of the history and experience of these immigrant groups. Then we suggest that you learn some key words and phrases in their original language. Why? Experiences that occur in a particular language are typically encoded in memory in that language. So, certain memories containing powerful emotions may not be accessible in a person's second language (English) because they were originally encoded in the

BOX 5.3   (Continued)

first language (for example, Spanish). And if the client is talking about something that was experienced in Spanish, Khmer, or Russian, the *key words* are not in English; they are in the original language.

Here is an example of how you might use these ideas in the session:

Social worker: Could you tell me what happened for you when you lost your job?

Maria (Spanish-speaking client): It was hard; I really don't know what to say.

Social worker: It might help us if you would say what happened in Spanish and then you could translate it for me.

Maria: ¡Es tan injusto! Yo pensé que perdí el trabajo porque no hablo el inglés muy bien. Me da mucho coraje cuando me hacen esto. Me siento herido.

Social worker: Thanks; I can see that it really affected you. Could you tell me what you said now in English?

Maria (more emotionally): I said, "It all seemed so unfair. I thought I lost my job because I couldn't speak English well enough for them. It makes me really angry when they do that to me. It hurts."

Social worker: I understand better now. Thanks for sharing that in your own language. I hear you saying that *injusto* hurts and you are very angry. Let's continue to work on this and, from time to time, let's have you talk about the really important things in Spanish, OK?

This brief example provides a start. The next step is to develop a vocabulary of key words in the language of your client. This cannot happen all at once, but you can gradually increase your skills. Here are some Spanish key words that might be useful with many clients:

Respeto: Was the client treated with respect? For example, the social worker might say, "Your employer failed to give you *respeto*."

Familismo: Family is very important to many Spanish-speaking people. You might say, "How are things with your *familia*?"

Emotions (see next chapter) are often experienced in the original language. When reflecting feeling, you could learn and use the following key words.

aguantar: endure	*miedo*: fear
amor: love	*orgullo*: proud
cariño: affection	*sentir*: feel
coraje: anger	

We also recommend learning key sayings, metaphors, and proverbs in the language(s) of your clients. *Dichos* are Spanish proverbs, as in the following examples:

Al que mucho se le da, mucho se le demanda.	The more people give you, the greater the expectations of you.
Más vale tarde que nunca.	Better late than never.
No hay peor sordo que el no quiere oir.	There is no worse deaf person than someone who doesn't want to listen.
La unión hace la fuerza.	Union is strength.

Consider developing a list like this, learn to pronounce them correctly, and you will find them useful in counseling Spanish-speaking clients. Indeed, you are giving them *respeto*. You may wish to learn key words in several languages.

Carlos Zalaquett comments: The Spanish version of *Basic Attending Skills, Las Habilidades Atencionales Básicas: Pilares Fundamentales de la Comunicación Efectiva* (Zalaquett, Ivey, Gluckstern-Packard, & Ivey, 2008), can help both monolingual and bilingual helpers. The attending skills are illustrated with examples provided by Latina/o professionals from different Latin American countries. Using the information and exercises included in the book, you can sharpen your interviewing, counseling, and psychotherapeutic tools to provide effective services to clients who speak Spanish.

North American and European counseling theory and style generally expect the client to get at the problem immediately and may not allow enough emphasis on relationship building. Some traditional Native American Indians, Dene, Pacific Islanders, Aboriginal Australians, and New Zealand Maori may want to spend a full interview getting to know and trust you before you begin. You may find yourself conducting interviews in homes or in other village settings. *Do not expect this to be true of every client who comes from an indigenous background.* If these clients have experienced the dominant culture, they are likely to be more comfortable with your usual style. Nonetheless, expect trust building and rapport to take more time.

Building trust requires learning about the other person's world. In general, if you are actively working as a counselor, it's important to involve yourself in positive community activities that will help you understand your clients better. If you are seen as a person who enjoys yourself in a natural way in the village, the community, and at pow-wows or other cultural celebrations, this will help build general community trust. The same holds true if you are a person of color. It will be helpful for you to visit a synagogue or an all-White church, understand the political/power structure of a community, and view White people as a distinct cultural group with many variations. Each of these activities may help you to avoid stereotyping those who are culturally different from you. Keep in mind the multiple dimensions of the RESPECTFUL model—virtually all interviewing and counseling are cross-cultural in some fashion.

When you are culturally different from your client, self-disclosure and an explanation of your methods may be helpful. For example, if you use only questioning and listening skills, the client may view you as suspicious and untrustworthy. The client may want directions and suggestions for action.

A general recommendation for working cross-culturally is to discuss differences early in the interview. For example, "I'm a White European American and we may need to discuss whether this is an issue for you. And if I miss something, please let me know." "I know that some gay people may distrust heterosexuals. Please let me know if anything bothers you." "Some White people may have issues talking with an African American counselor. If that's a concern, let's talk about it up front." "You are 57 and I [the counselor] am 26. How comfortable are you working with someone my age?" There are no absolute rules here for what is right. A highly acculturated Jamaican, Native American Indian, or Asian American might be offended by the same statements. Also remember that you can, if necessary, refer the client to another colleague if you get the sense that your client is truly uncomfortable and does not trust you.

Some Asian (Cambodian, Chinese, Japanese, Indian) clients from traditional backgrounds may be seeking direction and advice. They are likely to be willing to share their stories, but you may need to tell them why you want to wait a bit before coming up with answers. To establish credibility, there may be times when you have to commit yourself and provide advice earlier than you wish. If this becomes necessary, be assured and confident; just let them know you want to learn more, and that the advice may change as you get to know them better.

It is important to consider differences in gender. Even though there are many exceptions to this "rule," women tend to use more paraphrasing and related listening skills; men tend to use questions more frequently. You may notice in your own classes and workshops that men tend to raise their hands faster at the first question and interrupt more often.

CHAPTER SUMMARY

Key Points of "Encouraging, Paraphrasing, and Summarizing"

CourseMate and DVD Activities to Build Interviewing Competence

5.1 Defining Active Listening

▲ Listening is active. How you selectively attend to clients affects how they tell their stories or discuss their concerns.

▲ Empathy means to experience the client's world and to see things as the client does without mixing in your own thoughts and feelings. "Walk in the other person's shoes."

▲ Encouraging, paraphrasing, and summarizing are key active listening skills that keep the client talking.

1. Interactive Exercise 1: Identifying Skills. Work with a client's statements to determine what skills are used by the interviewer.
2. Interactive Exercise 2: Identification and Classification. Work with a condensed portion of the Mary Bradford Ivey and Damaris interview to categorize each of Mary's leads to practice your ability to classify microskills.
3. Interactive Exercise 3: Basic Competence. How would you respond to the five client comments?

5.2 Example Interview: They Are Teasing Me About My Shoes

▲ Children, adolescents, and some adults will be more comfortable if you provide something for them to do with their hands. Avoid towering over small children; sit at their level. Avoid abstractions, use short sentences and simple words, and focus on concrete, observable issues and behaviors.

▲ Note that the case of Damaris demonstrates effective verbal attending through encouraging, paraphrasing, and summarizing, which help the client explore the issues more effectively. Effective questions are used to bring in new data, organize the discussion, and point out positive strengths.

1. Video Activity: Toward Multicultural Competence. The video clip illustrates the importance of knowledge, awareness, and skills in multicultural areas. The goal in this exercise is to help you think a bit about diversity issues, your response to these issues, and what you might actually say to the client.
2. Interactive Exercise 4: Writing Listening Responses. Your turn to practice what you learned!
3. Interactive Exercise 5: Encouraging, Paraphrasing, and Summarizing: Marshall and Jesse Interview. Here you will find five client comments for you to explore possible interviewer responses and determine which one may be the most helpful.
4. Case Study. Meridith. This interview discusses the loss of power that women sometimes feel in the process of seeking medical assistance for life-threatening illness. Review the case and answer the questions.
5. Weblink Critique: Review different weblinks focusing on bilingualism, nonverbal communication, and attending.

5.3 Instructional Reading: The Active Listening Skills of Encouraging, Paraphrasing, and Summarizing

▲ Encouragers help clients elaborate their stories and include exact words they spoke, uh-huh's, smiles, a warm style, and short comments.

▲ The restatement is a form of encourager, but repeats key words or short phrases back to the client.

1. Interactive Exercise 6: Toward Intentional Competence. This interactive exercise uses two activities to help you master the listening skills.
   a. Generating Written Encouragers, Restatements, Paraphrases, and Summarizations. This activity

Key Points of "Encouraging, Paraphrasing, and Summarizing"

▲ The paraphrase includes (1) a sentence stem, often with the client's name; (2) key words used by the client; (3) the essence of what a client has said in briefer and clarified form; and (4) a checkout for accuracy. The accurate paraphrase will catch the essence of and clarify what the client says.

▲ The summary covers a longer period of the interview. It is also often used to begin or end a session or repeat back to the client what was said in the previous session(s).

▲ Trust is critical when working with clients, particularly those who may be culturally different from you. The same active listening skills are required, but more participation and self-disclosure on your part may be necessary. Trust building occurs when you visit the client's community and learn about cultures different from your own. Best of all is having a varied multicultural group of friends.

CourseMate and DVD Activities to Build Interviewing Competence

allows you to practice the production of these listening skills.

b. Practice of Skills in Other Settings. This activity is designed to help you practice the skills beyond the classroom.

2. Group Practice Exercise: Group Practice With Active Listening Skills.

Assess your awareness, knowledge, and skills as you conclude the chapter:

1. Flashcards: Use the flashcards to check your understanding of key concepts and facilitate memorization of key information.
2. Self-Assessment Quiz: The quiz will help you assess your current knowledge and prepare for course examinations.
3. Portfolio of Competencies: Evaluate your present level of competence in attending and listening skills using the downloadable Self-Evaluation Checklist. Self-assessment of your attending skills competence demonstrates what you can do in the real world.

OBSERVING AND REFLECTING FEELINGS

A Foundation of Client Experiencing

REFLECTION OF FEELING

ENCOURAGING, PARAPHRASING, AND SUMMARIZING

OPEN AND CLOSED QUESTIONS

ATTENDING BEHAVIOR

ETHICS, MULTICULTURAL COMPETENCE, AND WELLNESS

The artistic counselor catches the feelings of the client. Our emotional side often guides our thoughts and action, even without our conscious awareness.

—Allen Ivey

How can observing and reflection of feelings help you and your clients?

CHAPTER
GOALS

Feelings and emotions lie beneath the words, thoughts, and behaviors of all people. The purpose of reflecting feelings is to bring clients' attention to these emotions and make them clear and explicit. The basic listening sequence could not exist without **reflection of feelings**—thoughts and behavior have an emotional base.

Awareness, knowledge, and skills developed through the concepts of this chapter will enable you to

▲ Bring out the richness of the client's emotional world.

▲ Observe that most clients have mixed feelings, thoughts, and behaviors toward themselves, significant others, or events.

▲ Help clients sort through their complex feelings and thoughts.

▲ Center the counselor and client in fundamental emotional experience basic to resolving issues and achieving goals.

Assess your awareness, knowledge, and skill as you begin the chapter:

1. Self-Assessment Quiz: The chapter quiz will help you determine your current level of knowledge now and again after you've finished reading the chapter. You can take it before and after reading the chapter.

2. Portfolio of Competencies: Before you read the chapter, fill out the downloadable Self-Evaluation Checklist to assess your existing knowledge and competence in attending skills. Then, at the end of the chapter, complete the checklist again to summarize your competencies after study and practice.

6.1
DEFINING REFLECTION OF FEELINGS

KEY CONCEPT QUESTIONS

▲ **How is reflection of feelings defined?**

▲ **How are reflection of feelings and paraphrasing skills similar, and how are they different?**

The following description of reflection of feelings includes the responses you can expect from your client when you use this skill. On the other hand, not all clients will be comfortable with exploration of emotions, so be ready for another response to keep the interviewing progressing—be intentional!

Reflection of Feelings	Predicted Result
Identify the key emotions of a client and feed them back to clarify affective experience. With some clients, the brief acknowledgment of feelings may be more appropriate. Often combined with paraphrasing and summarizing.	Clients will experience and understand their emotional state more fully and talk in more depth about feelings. They may correct the interviewer's reflection with a more accurate descriptor.

Complete Video Activity: Reflection of Feelings

Reflection of feelings involves observing emotions, naming them, and repeating them back to the client. Paraphrasing and reflection of feelings are closely related and will often be found together in the same statement, but paraphrasing feeds back key thoughts that the client has just expressed. The important distinction is between the emphasis on content (paraphrase) and emotion (reflection of feelings). Note the content and the feelings expressed by a client named Thomas:

THOMAS: My Dad drank a lot when I was growing up, but it didn't bother me so much until now. [pause] But I was just home and it really hurts to see what Dad's starting to do to my Mum—she's awful quiet, you know. [Looks down with brows furrowed and tense] Why she takes so much, I don't figure out. [Looks at you with a puzzled expression] But, like I was saying, Mum and I were sitting there one night drinking coffee, and he came in, stumbled over the doorstep, and then he got angry. He started to hit my mother and I stopped him. I almost hit him myself, I was so angry. [Anger flashes in his eyes.] I worry about Mum. [A slight tinge of fear seems to mix with the anger in the eyes, and you notice that his body is tensing.]

▲ **EXERCISE 6.1**   **Practice Paraphrasing and Reflection of Feelings**

Paraphrase Thomas's main ideas, the content of his conversation; then focus on emotion and reflect his feelings. Use your intuition and note the main feeling words. Here are two possible sentence stems for your consideration:

▲ Paraphrase: "Thomas, I hear you saying . . ."
▲ Reflection of feelings: "Thomas, sounds like you feel . . ."

PARAPHRASING

Thomas's content includes Dad's history of drinking, Mum's being quiet and taking it, and the difficult situation when Thomas was last home. The paraphrase focuses on the content, clarifies the essence of what the client said, indicates that you heard what was said, and encourages him to move the discussion further.

> "Thomas, your father has been drinking a long time, and your Mum takes a lot. But now he's started to be violent, and you've been tempted to hit him yourself. Have I heard you right?"

In this example, we are focusing on what is happening and seeking to understand the total situation. The key content issue is escalation of violence and the need to protect Thomas's Mum. It will not help the situation if Thomas becomes part of the violence. At this point, the issue is to listen and learn more about the situation; with a joint understanding, we can plan Thomas's actions for the future.

REFLECTION OF FEELINGS

Also during the session, we can focus on emotion in depth and help Thomas work through issues at another level. The first step in eliciting and reflecting feelings is to recognize the key emotional words expressed by the client. When you reviewed Thomas's case, you may have recorded *really hurts*, *angry*, and *worry*. You know with some certainty that the client has these feelings; they have been made explicit. The most basic reflections of feelings would be "It really hurt," "You felt angry," and "You are worried." These reflections of feelings use the client's exact main words.

Next the interviewer must observe the many unspoken feelings or nonverbal cues expressed by the client, even if the client is not fully aware of them. For example, Thomas looked down with brows furrowed and body tense (a likely indication of confusion); anger and fear flashed in his eyes as he was talking about hitting his Dad.

Feelings are also layered, like an onion. Clients may be direct and forthright with single clear emotions, like Thomas, or they may talk about emotional tones such as confused, lost, or frustrated. Intentional listening and reflection of feelings often reveals underlying complex and sometimes conflicting emotions. For example, a client may be frustrated with her or his partner. When the interviewer reflects that frustration, the client may talk about anger at lack of attention, fear of being alone if the relationship breaks up, and lingering deep caring for the partner. As counseling continues, you may discover that the partner is trying to control your client. When the client learns of this control, he or she will most likely experience deep anger at the partner's controlling behavior. Should the relationship end, the client may experience feelings of relief and anticipation of better times. And during all this, the hurt will likely remain constant.

COMBINING PARAPHRASING AND REFLECTION OF FEELINGS

To acknowledge the client's emotions, combine the paraphrase with reflection of feelings by repeating the client's exact stated key feeling words. For example, "Thomas, you're really hurting right now," "You're angry because your Dad hit your Mum," "You're worried that your Dad's drinking is getting worse." Feeding back the feeling words may encourage more detail in the telling of the story.

 Complete Interactive Exercise 1: Paraphrase vs. Reflect Feeling

ADDITIONAL WAYS TO REFLECT FEELINGS AND THE USE OF CHECKOUTS

There are many possibilities for reflecting observed implicit emotions. In each of the following we suggest a checkout so that the accuracy of your observations can be tested with the client. It is also important to be mindful of cultural differences, such as those illustrated in Box 6.1.

Here are three reflections of feelings for your consideration.

1. I hear your anger, but also now, you're hurting and almost fearful about the situation. Am I close to what you are feeling right now? (Explicit anger and accompanying implicit emotions are reflected followed by a checkout for accuracy, thus allowing Thomas to follow up with his own reactions.)
2. Your body language—how you're sitting—gives me a sense that you're tense right now. Stopping your Dad from hitting your Mum brought out a lot of emotion—I see some anger, perhaps even a little fear. Am I right? (The focus here is on nonverbal expression, unspoken feelings. The checkout is particularly important.)

BOX 6.1   National and International Perspectives on Counseling Skills

Does He Have Any Feelings?
WEIJUN ZHANG

Videotaped session: A student from China comes in for counseling, referred by his American roommate. According to the roommate, the client quite often calls his wife's name out loud while dreaming, which usually wakes the others in the apartment, and he was seen several times doing nothing but gazing at his wife's picture. Throughout the session the client is quite cooperative in letting the counselor know all the facts concerning his marriage and why his wife is not able to join him. But each time the counselor tries to identify or elicit his feelings toward his wife, the client diverts these efforts by talking about something else. He remains perfectly polite and expressionless until the end of the session.

No sooner had the practicing counselor in my practicum class stopped the videotape than I heard comments such as "inscrutable" and "He has no feelings!" escape from the mouths of my European American classmates. I do not blame them, for the Chinese student did behave strangely, judged from their frame of reference.

"How do you feel about this?" "What feelings are you experiencing when you think of this?" How many times have we heard questions such as these? The problem with these questions is that they stem from a European American counseling tradition, which is not always appropriate.

For example, in much of Asia, the cultural rationale is that the social order doesn't need extensive consideration of personal, inner feelings. We make sense of ourselves in terms of our society and the roles we are given within the society. In this light, in China, individual feelings are ordinarily seen as lacking social significance. For thousands of years, our ancestors have stressed how one behaves in public, not how one feels inside. We do not believe that feelings have to be consistent with actions. Against such a cultural background, one might understand why the Chinese student was resistant when the counselor showed interest in his feelings and addressed that issue directly.

I am not suggesting here that Asians are devoid of feelings or strong emotions. We are just not supposed to telegraph them as do people from the West. Indeed, if feelings are seen as an insignificant part of an individual and regarded as irrelevant in terms of social importance, why should one send out emotional messages to casual acquaintances or outsiders (the counselor being one of them)?

What is more, most Asian men still have traditional beliefs that showing affection toward one's wife while others are around, even verbally, is a sign of being a sissy, being unmanly or weak. I can still vividly remember when my child was 4 years old, my wife and I once received some serious lecturing on parental influence and social morality from both our parents and grandparents, simply because our son reported to them that he saw "Dad give Mom a kiss." You can imagine how shocking it must be for most Chinese husbands, who do not dare even touch their wives' hands in public, to see on television that American presidential candidates display such intimacy with their spouses on the stage! But the other side of the coin is that not many Chinese husbands watch television sports programs while their wives are busy with household chores after a full day's work. They show their affection by sharing the housework!

3. Your Dad has been drinking for many years. I hear many different feelings—anger, sadness, confusion—and I also hear that you care a lot about both your Mum and your Dad. Am I close to how you feel about what's been going on for a long time? (This is a broader reflection of feelings that summarizes several explicit and implicit feelings and encourages the client to think more broadly.)

Which of the several possibilities suggested for a reflection of feelings is "right"? Any of them could be suitable, if you demonstrate empathy, good listening skills, and are intentionally attuned to your client. Generally, we recommend that you first focus on explicit feelings and use the client's actual emotional words. Later, you can explore the implicit, unspoken feelings.

Complete Weblink Exercises:
- ▲ Empathy and Listening Skills
- ▲ Reflecting Skills
- ▲ Exercise Four: Active Listening

6.2
THE LANGUAGE OF EMOTION

KEY CONCEPT QUESTION

▲ What is the vocabulary of emotion?

People are constantly expressing emotions verbally and nonverbally. General social conversation usually ignores feelings unless they are especially prominent. Prior to interviewing training, we tend to ignore or pay little attention to someone else's emotional experience; we may fail to observe what is happening before our eyes and ears.

▲ EXERCISE 6.2   **Creating a Vocabulary of Emotions**

As a first step toward naming and understanding emotions, it is helpful to establish your own vocabulary of emotions. Focus on four basic feelings—*sad*, *mad*, *glad*, and *scared*. These four emotions are considered the primary emotions, and their commonality, in terms of facial expression and language, has been validated throughout the world in all cultures (Ekman, 2007). In the blanks below, brainstorm emotional words associated with each primary emotion. Think of related emotional words with different intensities. For example, *mad* might lead you to think of *annoyed*, *angry*, and *furious*. Try this exercise and then continue to brainstorm and build your vocabulary of emotions.

Sad	*Mad*	*Glad*	*Scared*
_____	_____	_____	_____
_____	_____	_____	_____
_____	_____	_____	_____

Complete Interactive Exercise 2: Developing a Basic Feeling Vocabulary
Complete Interactive Exercise 3: Increasing Your Feeling Vocabulary

Clients often express emotions in unclear ways, demonstrating mixed and conflicting emotions. You can help them sort through these more complex feelings. Clients may experience caring, anger, fear, and other feelings all at once. A client going through a difficult separation or divorce may express feelings of love toward the partner at one moment and extreme anger the next. Words such as puzzled, sympathy, embarrassment, guilt, pride, jealousy, gratitude, admiration, indignation, and contempt are social emotions that come from primary emotions. How we deal with emotional experience, of course, depends very much on our learning history. Basic emotions appear to be universal across all cultures, but the social emotions appear to be learned from family members and peers. They are made more complex by the multifaceted and challenging world that surrounds us.

You can be especially helpful to clients as they sort out social emotions. Think of how the word *guilt* combines anger toward oneself, sadness, and perhaps even some fear. The feeling of guilt has been learned through social interaction in the family and culture. You have the opportunity through intentional listening and reflection of feelings to aid the client in sorting through mixed emotions. You can enable clients to bring out positive thoughts and a wellness attitude. This is particularly important when you work with clients who use negative thinking.

Love and caring are complex social emotions. The left frontal cortex is the primary location of positive emotional experience (e.g., glad) and is also where our executive decision-making functions lie. Both love and caring represent thoughts and feelings that are expressed *in relationship* with others, although self-love is real as well. We suggest that you think of love as a positive emotion, but with a cognitive component relating to a loved one, while caring is your expression of connection with a client.

▲ EXERCISE 6.3   **Feeling Words Observed in Your First Interview**

Go back to your audio- or videotaped first interview (see Box 1.3 on page 22), transcribe it if you haven't done so already, and underline or highlight every feeling word.

What do you notice or learn?

What may have you missed or overlooked?

What emotions or feelings did you attend to?

What would you do differently as a consequence of your analysis of your first interview?

 Complete Weblink Exercise: Empathy and Listening Skills and Psychological Hugs

6.3
EXAMPLE INTERVIEW: My Mother Has Cancer

KEY CONCEPT QUESTION

▲ How is reflection of feelings integrated into the interview?

Difficult life situations bring us face to face with many emotions. Whether you are dealing with clients who experience physical illness, interpersonal conflict, alcohol or drug abuse, or challenges in the work or school setting, learn how these clients feel about the situation and themselves. The intentional interviewer or counselor is always alert to clients' expression of emotions and knows how to identify and clarify these emotions for them.

The discovery of cancer, AIDS, or other major physical illness brings with it an immense emotional load. Busy physicians and nurses may fail to deal with emotions in their patients or the family members of those who are ill. Illness can be a frightening experience, and family, friends, and neighbors may also have trouble dealing with it. In fact, the diagnosis of a life-threatening disease increases the risk of depression, general anxiety disorder, and suicide. Lorelei Mucci of Harvard University, best known for her wide-ranging research on suicide and cancer of various types, found, for example, that prostate cancer doubles the risk of suicide in affected patients (Fall et al., 2009).

The following transcript illustrates reflection of feelings in action. This is the second session, and Jennifer has just welcomed the client, Stephanie, into the room. After a brief personal exchange of greetings, it is clear, nonverbally, that the client is ready to start immediately.

Interviewer and Client Conversation	Process Comments
1. **Jennifer:** So, Stephanie, how are things going with your mother?	Jennifer knows what the main issue is likely to be, so she introduces it with her first open question.
2. **Stephanie:** Well, the tests came back and the last set looks pretty good. But I'm upset. With cancer, you never can tell. It's hard . . . [pause]	Stephanie speaks quietly and as she talks, her voice becomes even softer. At the word "cancer," she looks down.
3. **Jennifer:** Right now, you're really worried and upset about your mother.	Jennifer uses the emotional word ("upset") used by the client, but adds the unspoken emotion of worry. With "right now," she brings the feelings to here-and-now immediacy. She did not use a checkout. Was that wise?
4. **Stephanie:** That's right. Since she had her first bout with cancer . . . [pause], I've been really concerned and worried. She just doesn't look as well as she used to, she needs a lot more rest. Colon cancer is so scary.	Often if you help clients name their unspoken feelings, they will verbally affirm or nod their head. Naming and acknowledging emotions helps clarify the total situation.
5. **Jennifer:** Scary?	Repeating the key emotional words used by clients often helps them elaborate in more depth.
6. **Stephanie:** Yes, I'm scared for her and for me. They say it can be genetic. She had Stage 2 cancer and we really have to watch things carefully.	Stephanie confirms the intentional prediction and elaborates on the scary feelings. She has a frightened and physically exhausted look on her face.

Interviewer and Client Conversation	Process Comments
7. **Jennifer:** So, we've got two things here. You've just gone through your mother's operation, and that was scary. You said they got the entire tumor, but your Mom really had trouble with the anesthesia, and that was frightening. You had to do all the caregiving, and you felt pretty lonely and unsupported. And the possibility of inheriting the genes is pretty terrifying. Putting it all together, you feel overwhelmed. Is that the right word to use, overwhelmed?	Jennifer summarizes what has been said. She repeats key feelings. She uses a new word, "overwhelmed," which comes from her observations of the total situation. Bringing in a new word to describe emotions needs to be done with care. In this example, Jennifer asked the client if that word made sense, and we see in the next exchange that Stephanie could use "overwhelmed" to discuss what was going on.
8. **Stephanie:** [immediately] Yes, I'm overwhelmed. I'm so tired, I'm scared, and I'm furious with myself. [pause] But I can't be angry; my mother needs me. It makes me feel guilty that I can't do more. [starts to sob]	Stephanie is now talking about her issues at the here-and-now level. Stephanie has not cried in the interview before, and she likely needs to allow herself to cry and let the emotions out. Caregivers often burn out and need care themselves.
9. **Jennifer:** [sits silently for a moment] Stephanie, you've faced a lot and you've done it alone. Allow yourself to pay attention to you for a moment and experience the hurt. [As Stephanie cries, Jennifer comments.] Let it out . . . that's OK.	Stephanie has held it all in and needs to experience what she is feeling. If you are personally comfortable with emotional experience, this ventilation of feelings can be helpful. There will be a need to return to a discussion of Stephanie's situation from a less emotional frame of reference. Jennifer hands Stephanie a box of tissues. A glass of water is available.
10. **Stephanie:** [continues to cry, but the sobbing lessens]	See Box 6.2 for ideas in helping clients deal with deeper emotional experience.
11. **Jennifer:** Stephanie, I really sense your hurt and aloneness. I admire your ability to feel—it shows that you care. Could you sit up a little straighter now and take a deep breath? [pause] How are you doing?	The client sits up, the crying almost stops, and she looks cautiously at the interviewer. She wipes her nose and takes a deep breath. Jennifer did three things here: (1) She reflected Stephanie's here-and-now emotions; (2) she identified a positive asset and strength; and (3) she suggested that Stephanie take a breath. Conscious breathing often helps clients pull themselves together.
12. **Stephanie:** I'm OK. [pause]	She wipes her eyes and continues to breathe. She seems more relaxed now that she has let out some of her emotions. At this point, she can explore the situation more fully—both emotionally and content-wise—as she moves toward decisions.

Interviewer and Client Conversation	Process Comments
13. **Jennifer:** You really feel deeply about what's happened with your mother, and it is so sad. At the same time, I see a lot of strength in you. We've talked earlier about how you supported her both before and after the operation. You did a lot for her. You care, and you also show her how you care by what you do.	Jennifer brings in the positive asset search. In the middle of a difficult time, Stephanie has shown she has the ability to do important things, even in the midst of worry and anxiety. This response from Jennifer reflects feelings, but also summarizes strengths that Stephanie can use to continue to deal with her concerns.

The major skill Jennifer used in this session was reflection of feelings, with a few questions to draw out emotions. Because human change and development are rooted in emotional experience, reflection of feelings is important in all theories of counseling and therapy. Humanistic counselors consider eliciting and reflecting feelings the central skill and strategy of interviewing.

You are most likely beginning your work and starting to discover the importance of reflecting feelings. It may take you some time before you are fully comfortable using this skill because it is seldom a part of daily communication. We suggest you start by first simply noting emotions and then acknowledging them through short reflections indicating that the emotions have been observed. As you gain confidence and skill, you will eventually decide the extent and place of emotional exploration in your helping repertoire.

Reflect on your own personal history and ability to deal with emotions as you learn how to reflect client feelings. If discussing feelings was not common in your past experience, you may have difficulty helping clients explore their issues in depth. The exercises throughout this text may help you gain greater access to your own experiential and emotional world.

▲ **EXERCISE 6.4**   **Your Own Comfort Level With Emotional Expression**

Before moving further, it is important for you to reflect on yourself and your own personal style. How comfortable are you with emotional expression? Was expressing and discussing feelings a common experience in your past? How do you react when others express their feelings? Are you a person who communicates your own feelings?

 Complete Interactive Exercise 4: Acknowledgment of a Feeling

When you practice reflecting feelings, try to use the skill as often as possible. To facilitate your personal and professional development as an interviewer, become aware of and competent in each of the microskills and gradually integrate them naturally. As you develop mastery, it will become easier to combine reflection of feelings with questioning, encouraging, and paraphrasing. The most effective interviewer or counselor—consciously or subconsciously—develops proficiency in the art of tuning in with feelings.

Box 6.2 suggests additional ways to help clients increase or decrease their emotional expressions.

BOX 6.2   Helping Clients Increase or Decrease Emotional Expressiveness

Observe Nonverbals   Breathing directly reflects emotional content. Rapid or frozen breath indicates contact with intense emotion. Also note facial flushing, eye movement, body tension, and changes in vocal tone. Especially, note hesitations. At times you may also find an apparent absence of emotion when discussing a difficult issue. This might be a clue that the client is avoiding dealing with feelings or that the expression of emotion is culturally inappropriate for this client. You can pace clients and then lead them to more expression and awareness of affect. Many people get right to the edge of a feeling, then back away with a joke, change of subject, or intellectual analysis.

Pace Clients
▲ Say to the client that she looked as though she was close to something important. "Would you like to go back and try again?"
▲ Discuss some positive resource that the client has. This base can free the client to face the negative. You as counselor also represent a positive asset.
▲ Consider asking questions. Used carefully, questions may help some clients explore emotions.
▲ Use here-and-now strategies, especially in the present tense: "What are you feeling right now—at this moment?" "What's occurring in your body as you talk about this?" Use the word *do* if you find yourself uncomfortable with emotion: "What do you feel?" or "What did you feel then?" starts to move the client away from here-and-now experiencing.

When Tears, Rage, Despair, Joy, or Exhilaration Occur   Your comfort level with your own emotional expression will affect how a client faces emotion. If you aren't comfortable with a particular emotion, your client will likely avoid it also and you may handle the issue less effectively. It is important to keep a balance between being very present with your own breathing and showing culturally appropriate and supportive eye contact but still allowing room for the client to sob, yell, or shake. You can also use phrases such as these:

I'm here.
I've been there, too.
Let it out . . . that's OK.
These feelings are just right.
I hear you.
I see you.
Breathe with it.

Sometimes it is helpful to keep emotional expression within a fixed time; two minutes is a long time when you are crying. Afterward, helping the person reorient is important. Tools for reorienting the interview include these:

▲ Help the client use slowed, rhythmic breathing.
▲ Discuss the client's positive strengths.
▲ Discuss direct, empowering, self-protective steps that the client can take in response to the feelings expressed.
▲ Stand and walk or center the pelvis and torso in a seated position.
▲ Reframe the emotional experience in a positive way.
▲ Comment that it helps to tell the story many times.

(Continued)

BOX 6.2   Helping Clients Increase or Decrease Emotional Expressiveness (Continued)

Caution	As you work with emotion, there is the possibility of reawakening issues in a client who has a history of painful trauma. When you sense this possibility, decide with the client in advance and obtain permission for the desired depth of emotional experiencing. The beginning interviewer needs to seek supervision and/or refer the client to a more experienced professional.

This box is adapted from a presentation by Leslie Brain, a graduate student at the University of Massachusetts, Amherst.

6.4

INSTRUCTIONAL READING: Becoming Aware of and Skilled With Emotional Experience

KEY CONCEPT QUESTIONS

▲ **Acknowledgment of feelings: Do we always need to explore emotion in depth?**

▲ **Techniques: What specific verbal skills go with reflection of feelings?**

▲ **Emotional complexity: What are mixed emotions, and where might they come from?**

▲ **Observing emotional experience: How do we observe clients' feelings?**

▲ **Positive emotions: How can positive emotions help clients deal with challenging issues?**

▲ **Caution: What care should we use with reflection of feelings?**

Each encounter you have with people throughout the day involves your emotions. Some are pleasant; some are tense, distressing, and conflictual. The interaction may only be with a telemarketer desperate to make a sale, a hurried clerk in a store, or the police as they stop you for speeding. More complex feelings underlie more intimate relationships with those close to you. Awareness of your own and others' feelings can help you move through the tensions of the day gracefully and may enable you to be helpful to others in many small ways. Obviously, your skill in observing and working with emotions is key to your success in interviewing and counseling.

ACKNOWLEDGING FEELINGS

Sometimes a simple, brief acknowledgment of feelings is just as helpful as a full reflection of feelings. In acknowledging feelings, you state the feeling briefly ("You seem to be sad about that," "It makes you happy") and move on with the rest of the conversation. With a harried and perhaps even rude busy clerk or restaurant server, try saying in a warm and supportive tone, "Being that rushed must make you tense."

Often this is met with a relaxed look and an implicit thanks of appreciation. The same structure is used in an acknowledgment as in a full reflection, but with less emphasis and time given to the feeling.

Acknowledging and naming feelings may be especially helpful with children, particularly when they are unaware of what they are feeling. Children often respond well to the classic reflection of feelings, "You feel . . . [sad, mad, glad, scared] . . . because. . . ."

A basic feeling we have toward our parents, family, and best friends is love and caring. This is a deep-seated emotion in most individuals. At the same time, over years of close contact, negative feelings about the same people may also arise. These negative feelings may be buried and may overwhelm and hide positive feelings. Many people want a simple resolution and want to run away from complex mixed emotions. A common task of counselors is to help clients sort out mixed feelings toward significant people in their lives. Ideally the counselor helps the client discover and sort out both positive and negative feelings.

 Complete Interactive Exercise 5: Recognizing Varying Orientations Toward Emotional Expression

THE TECHNIQUES OF REFLECTING FEELINGS

Somewhat like the paraphrase, reflection of feelings involves a typical set of verbal responses that can be used in a variety of ways. The classic reflection of feelings consists of the following dimensions:

1. *Sentence stem.* Choose a sentence stem such as "I hear you are feeling . . . ," "Sounds like you feel . . . ,""I sense you are feeling. . . ." Unfortunately, these sentence stems have been used so often they can sound like comical stereotypes. As you practice, you will want to vary sentence stems and sometimes omit them completely. Using the client's name and the pronoun *you* help soften and personalize the sentence stem.
2. *Feeling label.* Add an emotional word or feeling label to the stem ("Jonathan, you seem to feel bad about . . . ," "Looks like you're happy," "Sounds like you're discouraged today; you look like you feel really down"). For mixed feelings, more than one emotional word may be used ("Maya, you appear both glad and sad . . .").
3. *Context or brief paraphrase.* You may add a brief paraphrase to broaden the reflection of feelings. The words *about, when,* and *because* are only three of many that add context to a reflection of feelings ("Jonathan, you seem to feel bad about all the things that have happened in the past two weeks," "Maya, you appear both glad and sad because you're leaving home").
4. *Tense and immediacy.* Reflections in the present tense ("Right now, you are angry") tend to be more useful than those in the past ("You felt angry then"). Some clients have difficulty with the present tense and talking in the "here and now." Occasionally, a "there and then" review of past feelings can be helpful and feel safer for the client.
5. *Checkout.* Check to see whether your reflection of feelings is accurate. This is especially helpful if the feeling is unspoken ("You feel angry today—am I hearing you correctly?").

Complete Interactive Exercise 6: Facilitating Clients' Exploration of Emotion at Varying Levels

RESPECT FOR DIVERSITY

You need to respect individual and cultural diversity in the way people express feelings. The student from China discussed in Box 6.1 is an example of cultural emotional control. Emotions are obviously still there, but they are expressed differently. Do not expect all Chinese or Asians to be emotionally reserved, however. Their style of emotional expression will depend on their individual upbringing, their acculturation, and other factors. Many New England Yankees may be fully as reserved in emotional expression as the Chinese student described by Weijun Zhang. But again, it would be unwise to stereotype all New Englanders in this fashion.

OBSERVING CLIENT VERBAL AND NONVERBAL FEELINGS

When a client says "I feel sad"—or "glad" or "frightened"—and supports this statement with appropriate nonverbal behavior, identifying emotions is easy. However, many clients present subtle or discrepant messages because they may not be sure how they feel about a person or situation. The counselor may have to identify and label the implicit feelings.

The most obvious technique for identifying client feelings is simply to ask the client an open question: "How do you feel about that?" "Could you explore any emotions that come to mind about your parents?" "What feelings come to mind when you talk about the loss?" With less verbal clients, a closed question in which the counselor supplies the missing feeling word may be helpful: "Does that feel hurtful to you?" "Could it be that you feel angry at them?" "Are you glad?"

At other times the counselor will want to infer, or even guess at, the client's feelings through observation of verbal and nonverbal cues. Discrepancies offer vital clues; they may include discrepancies between what the client says about a person and a slight body movement contradicting the client's words. As many clients have mixed feelings about the most significant events and people in their lives, inference of unstated feelings becomes an important observational skill of the counselor. A client may be talking about caring for and loving parents while holding his or her fist closed. The mixed emotions may be obvious to the observer though not to the client.

Stephanie, in the interview about her mother's cancer, used the following feeling words during the session: *upset, worried, concerned, scared, tired, guilty,* and *anger*. Jennifer, the counselor, reflected those words but also integrated her own observations. We saw Stephanie's reaction to the word *overwhelmed*. The interviewer also emphasized positive emotions, such as caring, and the strengths that the client demonstrated. Stephanie needs to be encouraged to talk through the worries and problems. As the interview moves on, we can anticipate that Stephanie will express fewer difficult emotions and will move to more awareness of positive feelings and strengths.

The intentional counselor also works to help clients label their own emotions and does not respond to every noted emotion. Reflection of feelings must be timed to meet the needs of the individual client. Sometimes it is best simply to recognize the importance of the emotion and keep it in mind for possible comment later.

Complete Video Activity: Allen and Mary Demonstrate the Four Emotional Styles

THE PLACE OF POSITIVE EMOTIONS IN REFLECTING FEELINGS

Positive emotions, whether joyful or merely contented, are likely to color the ways people respond to others and their environments. Research shows that positive emotions broaden the scope of people's attention, expand their repertoires for action, and increase their capacities to cope in a crisis. Research also suggests that positive emotions produce patterns of thought that are flexible, creative, integrative, and open to new information (Gergen & Gergen, 2005). Recently, it has been found that a positive approach actually encourages neuron growth in the cortex, while simultaneously reducing the power of negative neural emotional networks in the amygdala (Likhtik et al., 2008).

Sad, mad, glad, proud, scared—this is one way to organize the language of emotion. But perhaps we need more attention to glad words, such as *pleased, happy, contented, together, excited, delighted, pleasure,* and the like.

▲ EXERCISE 6.5 | **Positive Emotions**

Take a moment to think of specific situations in which you experienced each of these positive emotions. It is very likely that when you think of these situations, you will smile, which will help reduce your overall body tension. It is even likely that your blood pressure will change in a more positive direction. Tension produces damaging cortisol in the brain, and we all need to learn to control our bodies more effectively. Positive imaging is a useful strategy for both you and your clients.

EMOTION AND YOUR BODY

When you experience emotion, your brain signals bodily changes. When you feel sad or angry, a set of chemicals floods your body, and usually these changes will show nonverbally. Emotions change the way your body functions and are a foundation for all our thinking experience (Damasio, 2003). As you help your clients experience more positive emotions, you are also facilitating wellness and a healthier body. The route toward health, of course, often entails confronting negative emotions.

Searching for wellness strengths and positive assets will likely be helpful to you and your clients. Obviously, you need to explore negative and troubling emotions, but if your clients can start from a positive base of emotion, they may be better able to cope with the negative. Following are four examples of how to help clients focus more on positive emotions.

1. *Wellness assessment.* Be sure that you reflect the positive feelings associated with aspects of wellness. For example, your client may feel safety and strength in spirituality, pride in gender and/or cultural identity, caring and warmth from past and/or present friendships, and the intimacy and caring of a love relationship. It would be possible to anchor these emotions early in the interview and then draw on these positive emotions during more stressful moments. Out of a wellness inventory can come a "backpack" of positive emotions and experiences that are always there and can be drawn on as needed.

2. *What's right in the relationship?* Couples with relationship difficulties can be helped if they focus more on the areas where things are going well. What remains good about the relationship? Many couples focus on the 5% of times when they disagree and fail to note the 95% when they have been successful

or enjoyed each other. Some couples respond well when asked to focus on the reasons they got together in the first place. These positive strengths can help them deal with very difficult issues.

3. *Positive homework.* When providing your clients with homework assignments, have them engage daily in activities associated with positive emotions. For example, it is difficult to be sad and depressed while running or walking at a brisk pace. Meditation and yoga are often useful in generating more positive emotions and calmness. Seeing a good movie when one is down can be useful, as can going out with friends for a meal. In short, help clients remember that they have access to joy, even when things are at their most difficult.

4. *Service to others.* Helping others often makes individuals feel good about themselves. When people are discouraged and feeling that they are inadequate, volunteering for a place of worship, soup kitchen, animal shelter, or senior center can improve self-concept and build positive self-oriented feelings. One of the best ways to feel good about oneself is to help others. Altruism works both for the helper and the helpee.

Important caution: Please, do not use these examples as a way to tell your clients that "everything will be OK." Some interviewers and counselors are so afraid of negative emotions that they never allow their clients to express what they really feel. Do not minimize difficult emotions by focusing too quickly on the positive.

SOME LIMITATIONS OF REFLECTING FEELINGS

Reflection has been described as a basic skill of the counseling process, yet it can be overdone. With friends, family, and fellow employees, a quick acknowledgment of feelings ("If I were you, I'd feel angry about that . . ." or "You must be tired today") followed by continued normal conversational flow may be most helpful in developing better relationships. Similarly, with many clients a brief acknowledgment of feelings may be more useful. However, with complex issues, identifying unspoken feelings may be necessary. Sorting out mixed feelings is key to successful counseling, be it vocational interviewing, personal decision making, or in-depth individual counseling and therapy.

Be aware that not all clients will appreciate your comments on their feelings. Exploring the emotional world can be uncomfortable for those who have avoided looking at feelings in the past. An empathic reflection can have a confrontational quality that causes clients to look at themselves from a different perspective; therefore, it may seem intrusive to some. Timing is particularly important with reflection of feelings. Clients tend to disclose feelings only after rapport and trust have been developed. Less verbal clients may find reflection puzzling or may say, "Of course I'm angry; why did you say that?" Some men may believe that expression of feelings is "unmanly." Brief acknowledgment of feelings may be received with appreciation early on and can lead to deeper exploration in later interviews.

Complete Case Study: Becoming Aware of and Skilled With Emotional Experience
Complete Interactive Exercise 7: Identifying Skills. My Mother Has Cancer
Complete Interactive Exercise 8: Naomi, Abusive Relationship
Complete Group Practice Exercise: Group Practice With Reflection of Feeling
Complete Weblink Exercise: Study of Emotions
Complete Weblink Critique

CHAPTER SUMMARY

| Key Points of "Observing and Reflecting Feelings" | CourseMate and DVD Activities to Build Interviewing Competence |

CourseMate and DVD Activities to Build Interviewing Competence

6.1 Defining Reflection of Feelings

▲ Both paraphrasing and refection of feelings feed back to clients what they have been experiencing.

▲ A paraphrase focuses on the verbal content of what the client says; reflection of feelings centers on both verbal and nonverbal emotional underpinnings.

▲ Unspoken feelings may be seen in the client's nonverbal expression, may be heard in the client's vocal tone, or may be inferred from the client's language.

▲ Feelings are layered like an onion. Words like *frustration*, *mixed up*, and *confused* represent conflicting emotions underneath surface words.

▲ Checkouts can help confirm the accuracy of paraphrases and reflections of feelings.

1. Video Activity: Reflection of Feelings. Mary and Sandra demonstrate reflection of feelings. This exercise gives you the opportunity to analyze and comment on reflection of feelings, suggest other ways to accomplish this, and explore your own way of responding to emotional expressions.

2. Interactive Exercise 1: Paraphrase vs. Reflect Feeling. Distinguishing a reflection of feeling from a paraphrase. Practice how to distinguish these skills.

3. Weblink: Empathy and Listening Skills. From *Touching Another Heart* by Lawrence J. Bookbinder, Ph.D., with hotlinks to relevant topics.

4. Weblink: Reflecting Skills. Angelfire.com, with hotlinks to pertinent topics.

5. Weblink: Exercise Four: Active Listening. Center for Rural Studies.

6.2 The Language of Emotion

▲ People constantly feel and express emotion. The proficient interviewer will hear explicit emotions, observe implicit emotions, and feed these emotions back to the client.

▲ Sad, mad, glad, and scared are primary emotions used as root words for building a vocabulary of emotion. They appear to be universal across cultures.

▲ Social emotions (embarrassment, guilt, pride) are modified and built on primary emotions. They are learned in a family and cultural context.

1. Interactive Exercise 2: Developing a Basic Feeling Vocabulary. This exercise gives you an opportunity to develop a basic vocabulary of emotions. Essential to become effective in reflecting feelings.

2. Interactive Exercise 3: Increasing Your Feeling Vocabulary. A good way to amplify and enlarge the vocabulary of emotions initiated in the previous exercise.

3. Weblink: Empathy and Listening Skills and Psychological Hugs. From *Touching Another Heart* by Lawrence J. Bookbinder, Ph.D., with hotlinks to relevant topics.

6.3 Example Interview: My Mother Has Cancer

▲ Because human change and development are rooted in emotional experience, reflection of feelings is important in all theories of counseling and therapy.

▲ Reflection of feelings clarifies the client's emotional state, leads clients in new directions, and results in new discoveries.

▲ It is important to identify positive qualities and emotions to help clients deal more effectively with negative emotions.

▲ In the example interview, Jennifer reflects the main emotional words actually used by the client. She also points out unspoken feelings and checks out with the client whether the identified feeling is accurate. For example, "Is that close to what you feel?"

1. Interactive Exercise 4: Acknowledgment of Feelings. A good exercise to learn and practice brief acknowledgment of feelings.

Key Points of "Observing and Reflecting Feelings"

CourseMate and DVD Activities to Build Interviewing Competence

6.4 Instructional Reading: Becoming Aware of and Skilled With Emotional Experience

▲ Everyone has complex emotions associated with people and events in their lives. Helping clients sort out these feelings is an important part of counseling.

▲ Proficient interviewers must be able to observe and reflect emotional dimensions accurately.

▲ The components of reflection of feelings are (1) a sentence stem, (2) the feeling word, (3) some context, (4) present tense, and (5) a checkout. Beginning with the client's name is helpful, and present tense, here-and-now reflections are often more powerful than a review of past emotions.

▲ Research and clinical experience reveal that special attention to positive emotions can provide clients with strengths to better address their emotional challenges.

▲ Not all clients are comfortable discussing emotion. Reflection of feelings can be overdone, particularly if the client comes from a family or culture that believes emotional expression is inappropriate. A brief acknowledgment of feelings may be more helpful.

1. Interactive Exercise 5: Recognizing Varying Orientations Toward Emotional Expression. Practice your skills to classify emotions according to the developmental level and style they represent.

2. Interactive Exercise 6: Facilitating Clients' Exploration of Emotion at Varying Levels. An exercise to help you discuss different ways to help clients explore their emotions.

3. Video Activity: Allen and Mary Demonstrate the Four Emotional Styles. Observe the masters in action, demonstrating four emotional styles.

4. Case Study: Becoming Aware of and Skilled With Emotional Experience. An exercise to explore different ways of responding to emotional expressions.

5. Interactive Exercise 7: Identifying Skills. My Mother Has Cancer. Classify skills related to reflection of feelings in a negative situation. Note that there are a number of ways to increase the likelihood of emotional expression.

6. Interactive Exercise 8: Naomi, Abusive Relationship.

7. Group Practice Exercise: Group Practice With Reflection of Feeling.

8. Weblink: Study of Emotions. A good place to explore the characteristics, components, and different type of emotions.

9. Weblink Critique: Review and critique of relevant sites related to the study of emotions.

10. Instructional Reading: Becoming Aware of and Skilled With Emotional Experience. An exercise to explore different ways of responding to emotional expressions.

Assess your awareness, knowledge, and skills as you conclude the chapter:

1. Flashcards: Use the flashcards to check your understanding of key concepts and facilitate memorization of key information.

2. Self-Assessment Quiz: The quiz will help you assess your current knowledge and prepare for course examinations.

3. Portfolio of Competencies: Evaluate your present level of competence in attending and listening skills using the downloadable Self-Evaluation Checklist. Self-assessment of your attending skills competence demonstrates what you can do in the real world.

How to Conduct an Interview Using Only Listening Skills

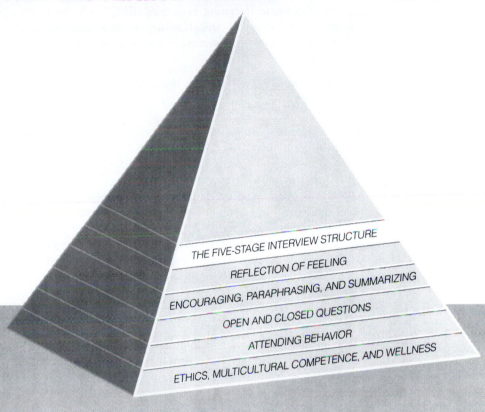

THE FIVE-STAGE INTERVIEW STRUCTURE

REFLECTION OF FEELING

ENCOURAGING, PARAPHRASING, AND SUMMARIZING

OPEN AND CLOSED QUESTIONS

ATTENDING BEHAVIOR

ETHICS, MULTICULTURAL COMPETENCE, AND WELLNESS

The secret of joy in work is contained in one word—excellence. To know how to do something well is to enjoy it.

—*Pearl S. Buck*

How will listening skills and a well-formed interview help you and your clients?

CHAPTER
GOALS

The main goals of this chapter are to examine your intentional competence and to conduct a full interview using the ideas and listening skills presented in previous chapters: the **basic listening sequence (BLS)** and the five stages/dimensions of a well-informed interview (*relationship—story and strengths—goals—restory—action*). Combined, these two skill sets will enable you to complete a full session using only listening skills.

Awareness, knowledge, and skills developed through the concepts of this chapter will enable you to

▲ Define the concept of empathy and rate interviews for empathic understanding.

▲ Understand and use the basic listening sequence (BLS) and discuss its relationship to empathy.

▲ Demonstrate fluid and intentional responsiveness to clients, whether or not your predictions are successful.

▲ Define decisional counseling and how it relates to the five stages/dimensions of the well-formed interview: *relationship—story and strengths—goals—restory—action*.

▲ Conduct a complete interview using only listening skills.

▲ Explore the relationship of Rogerian person-centered counseling to microskills and the five stages of the interview.

Assess your awareness, knowledge, and skills as you begin the chapter:

1. Self-Assessment Quiz: The chapter quiz will help you determine your current level of knowledge. You can take it before and after reading the chapter.

2. Portfolio of Competencies: Before you read the chapter, fill out the download-able Self-Evaluation Checklist to assess your existing knowledge and competence in attending skills. Then, at the end of the chapter, complete the checklist again to summarize your competencies after study and practice.

7.1

KEY CONCEPTS FOR EFFECTIVE INTERVIEWING: Empathic Understanding, the Basic Listening Sequence (BLS), and Predicting Client Interviewing Response

KEY CONCEPT QUESTIONS

▲ **What is empathic understanding, and how does it relate to listening skills?**

▲ **How can you use the basic listening sequence (BLS) to integrate the listening skills with empathic understanding?**

▲ **How do empathy and listening skills relate to prediction of client response?**

▲ **How does diversity affect empathic understanding?**

EMPATHIC UNDERSTANDING: ENTERING THE WORLD OF THE CLIENT

The listening skills of the first section of this book contain the building blocks of empathy and empathic understanding. Carl Rogers (1957, 1961) first brought the importance of empathy to our attention through his person-centered interviewing. He made clear how vital it is to listen carefully, to enter the world of the client, and to communicate that we understand the client's world *as the client sees and experiences it*. **Empathy** is often defined as experiencing the world as if *you* were the client, but with awareness that the client remains separate from you. Many others have elaborated on Rogers's influential definition of empathy (see, for example, Carkhuff, 2000; Egan, 2007; Ivey, Ivey, & Zalaquett, 2010). Current research on brain activity is validating what helping professionals have been saying for a long time, as explained in Box 7.1.

When you respond in an empathic way, you can predict how clients may feel and react.

Empathic Response	*Predicted Result*
Experiencing the client's world as if you were the client; understanding his or her key issues and feeding them back to clarify experience. This requires attending skills and using the important key words of the client, but distilling and shortening the main ideas. In additive empathy, the interviewer may add meaning and feelings beyond those originally expressed by the client. If done ineffectively, it may subtract from the client's experience.	Clients will feel understood and move on to explore their issues in more depth. Empathy is best assessed by clients' reaction to a statement.

Empathic understanding and careful listening are important in all fields where human communication is central. When you go to see your physician, you want the best deductive skills, but you also want the doctor to listen to your story and be able to relate to it. It is vital to use the listening skills, but we also need to use them with a desire to understand the client and the way the client experiences the world.

Part of empathy is *communicating that you understand*. The paraphrase, reflection of feeling, and summarization are particularly important skills to show clients that you are seeking to be a part of their world with them.

▲ EXERCISE 7.1   **Empathic Understanding**

Review audio- or videotaped interviews and assess the level of empathic understanding demonstrated by the interviewer. In addition, examine your own tapes or discs and consider how accurately you are listening. Rate yourself and other interviewers you observe on the following three-level empathy scale.

BOX 7.1 Empathy, Brain, and Developmental Research

Empathy is not just an abstract idea—empathy is identifiable and measurable in the physical brain. Fascinating research on brain activity validates what the helping field has been saying for years. "The basic building blocks [of empathy] are hardwired into the brain and await development through interaction with others. . . . Empathy [is] an intentional capacity" (Decety & Jackson, 2004, pp. 71, 93).

Volunteer couples experienced a painful electric shock and then observed their loved ones experiencing the same shock. When the shock was delivered, brain scans revealed that the painful stimulus appeared in the cognitive pain-processing network of *both* participants, even though only the shocked partner's brain registered actual physical pain (Singer et al., 2004). This finding has been replicated in several other experiments.

What we learn here is that the empathic person's brain responds to another's experience, even though the observer does not physically feel the other's pain. These responding cells are called "mirror neurons" and are the physical basis of empathy. Many studies over the years support this central point. For example, children around their second year indicate concern for others cognitively, emotionally, and behaviorally when they notice another child having a problem and often try to help. Zhan-Waxler and his group were the first to notice this in 1992. Since then the developmental base of empathy has been well established. You may have seen two young children playing together when one falls and starts crying. Even though the second child has not been hurt, he or she also cries. This ability

to observe the feelings of others could be considered the developmental roots of empathic understanding.

Awareness of self, awareness of others, and the ability to differentiate yourself from the client are essential for empathic understanding. Young children may empathize, but in the situations described above, they have not yet separated themselves from what they have seen. Our task as counselors and therapists is to resonate or empathize with the client's story, but we also need to separate ourselves from that client. In addition, we need to work intentionally with an array of skills and strategies that can help the client to grow further. That client growth will be evident not only in behavior but also in measurable aspects of brain functioning.

A particularly fascinating study found that client and therapist perceptions of times of strong empathic communication correlated closely with physiological responses. When real understanding occurred between client and counselor, their bodies were also in tune, as measured by skin conductance in 20 pairs of clients and counselors (Marci, Ham, Moran, & Orr, 2007).

Conversely, brain research reveals that the antisocial, criminal personality has a reduced ability to appreciate the emotions of others, a counseling fact that is well known (Blair, 2001). Their mirror neurons tend not to fire when they observe the experiences of others. In fact, a child diagnosed with clear conduct disorder or a teen recognized as seriously antisocial has areas of the brain that take pleasure in seeing others suffer. In effect, they get a "high" from the pain they inflict.

Level 1. **Subtractive empathy**: The interviewer response gives back to the client less than what the client has said and perhaps even distorts or misunderstands the client. In this case, the listening skills are used inappropriately and take the client off track. Focusing solely on the negative also leads to subtractive empathy.

Level 2. **Basic empathy**: Interviewer responses are roughly interchangeable with those of the client. The interviewer is able to say back accurately what the client has said. Skilled and intentional use of the basic listening sequence is a way to demonstrate basic empathy.

Level 3. **Additive empathy**: Interviewer responses add a link to something the client has said earlier or a new idea or frame of reference that helps the client see a new perspective. Wellness and the positive asset search can be vital parts of additive empathy. Skilled use of listening skills and/or influencing skills (see Appendix I) enable an interviewer to become additive.

Also, you can rate the interviewing skills to establish the quality of empathic understanding. A 5-point scale for examining level of counselor's empathic response is shown below; its related constructs are presented in Ivey, Ivey, and Zalaquett, 2010.

Level	1	2	3	4	5
	Subtractive		Interchangeable		Additive

It is easier to be empathic if you have experienced similar issues in your own life. If you come from a family that experienced alcohol or substance abuse, you may have a special understanding of where this type of client is "coming from." Women in rape support groups find empathy and understanding when they hear others tell a story similar to their own. Cancer survivors may feel more trust and understanding from someone who has had experiences parallel to their own. But someone else's experience, even if similar, is not your experience. Just because you have "been there" does not necessarily mean that you are automatically empathic. For example, each soldier's experience of war will be unique.

Empathy requires two things. The first requirement is to be with the client; intentional listening is your key to entering the client's experience as fully as possible. But it is equally important to be aware that the client's world is not your world. Do not lay "your thing" on the client.

Box 7.2 speaks to one aspect of life experience that is seldom considered—the culture of those who have experienced infertility. The question needs to be raised—"If you haven't *been there*, how empathic can you really be?" And if you have not had similar experiences to those of the client, how can you still be an effective helper?

BOX 7.2 National and International Perspectives on Counseling Skills

 Is Empathy Always Possible?
KATHRYN M. QUIRK, M.A.

As beginning students in counseling, one of the first concepts we run into is empathy—experiencing and understanding the world of the client. Certainly this is core to the helping interview.

But as I read my text (not this one), I felt increasingly uncomfortable. Was this almost magical concept really possible? I'll tell you why. First the happy ending. I am now the happy mother of a lovely child, the darling of my life. But Ryan did not come easily and my husband and I needed the help of three fertility clinics.

The "simpler" strategies of getting pregnant failed for three years. Those years were agonizing, but only a sample of the trauma we were to face (yes, dear reader, going through fertility procedures meets the full definition of trauma). We then moved to complicated in vitro procedures involving Petri dishes and surgery. The first three procedures failed and the fourth resulted in a pregnancy that ended when twins died after three months. I don't like the word "fetus"—and grieving for lost babies was horrible.

(Continued)

BOX 7.2   National and International Perspectives on Counseling Skills (Continued)

We moved to a new clinic and our fifth try was fantastically successful.

How does all this relate to empathy? I recall a pleasant and expert nurse who counseled a group of us experiencing primary infertility. She was helpful and had good suggestions, but when things got emotional and we cried, she simply didn't get it. She would say that she understood and knew what we were going through. But let's face it; *she hadn't been there herself.* How could she truly understand the physical pain or the feelings of failure, shame, and hopelessness? She didn't understand our loneliness and, perhaps worst of all, the crushed hopes. How could she understand what we were really feeling? I resented it when she said she understood when she clearly did not and could not. Fortunately, those in the group who had "been there" supplied the needed empathy and support.

Does this mean that if you haven't experienced the inner world and actual experience of the client, you can't be empathic? At first I thought that understanding my experience was impossible except for those who had experienced what I had gone through. However, I've softened my thoughts somewhat as I learn about and think about good counseling. I still feel that *being there* is what serves as a foundation for the deepest empathy. But the nurse could have provided a deeper empathy than she did if she had admitted openly that she understood our feelings and experience only partially. She did, after all, have more experience listening to people with pregnancy challenges than we did. She did have something to offer.

Failing to discuss and admit that she was different from us suggested to me and to others in the group that she did not understand. What could she have

done? First, I think she should have said early in her work with us that she herself had not experienced the difficulties that we were going through and admitted that her understanding and empathy were therefore only partial. But she could have pointed out that she understood pain and loss and perhaps even shared some of her own difficult experience. Saying this and also outlining her experience and knowledge would have developed more trust and given us all a deeper feeling that she was an empathic person.

As I've gone through my counseling program, I've increasingly become aware that I too will have problems with being truly empathic and communicating the understanding I do have. When I meet clients who are different from me multiculturally (e.g., race, sexual orientation, religious commitment), I now know that I need to discuss these issues upfront. And I have the obligation to learn as much about the cultural background of these clients as I can. To maximize my empathic potential, I need to read, get out in the community where these people live, and participate with them when I can.

This also holds true for me when I work with alcoholics, cancer survivors, and those who have been raped. I haven't been there, but I have a responsibility to learn more about those whose life experience is different from mine.

Empathy is clearly important, but it is not learned just from classes and books. We all need to examine the human experience and become more fully aware of the life of those around us. And a special P.S. needs to be added—we have two more wonderful sons: Charlie, who arrived in July 2006, and Tegan, shown on the dedication page of this book, who arrived in October 2009.

 Complete Interactive Exercise 1: I Don't Seem to Fit In: Predicting Results, the Basic Listening Sequence, and Empathic Understanding

THE BASIC LISTENING SEQUENCE

Interviewing is not a chance process. Empathic understanding can be increased through observation and careful listening. Observations of interviews in counseling and therapy as well as in management, medicine, and other settings have revealed a common thread of skill usage. The basic listening sequence, or BLS, helps you understand the basic structure of the client's story by effectively utilizing all of the listening skills, including using open and closed questions, encouraging, paraphrasing, reflecting feelings, and summarization. These skills have been discussed separately in previous chapters; now their full impact in the interview will be realized when using them together. When you use the sequence of listening skills, you can predict how clients are likely to respond.

Basic Listening Sequence	Predicted Result
Select and practice all elements of the basic listening sequence: using open and closed questions, encouraging, paraphrasing, reflecting feelings, and summarizing. These are supplemented by attending behavior and client observation skills.	Clients will discuss their stories, issues, or concerns, including the key facts, thoughts, feelings, and behaviors. Clients will feel that their stories have been heard.

 Complete Interactive Exercise 2: The BLS in Many Settings

With the basic listening sequence, you have the opportunity to obtain feedback on the accuracy of your listening through the checkout. The client will let you know how well you have listened.

The BLS integrates several key skills. Many successful interviewers begin their sessions with an open question followed by closed questions for diagnosis and clarification. The paraphrase catches the essence of what the client is saying, and the reflection of feeling examines key emotions. These skills are followed by a summary of the concern expressed by the client. Encouragers may be used throughout the interview to help evoke details, while attending and observation remain an underlying part of the entire process.

The BLS, used effectively, is predicted to bring out client stories, including facts, thoughts, feelings, and behaviors. Summarization is particularly useful in bringing order and making sense of client conversation. The skills of the BLS need not be used in any specific sequence, although the basic sequence (questions, encouragers, paraphrases, reflection of feeling, summarization) does appear regularly in effective interviews. Each person needs to adapt these skills to meet the requirements of the client and the situation. The effective interviewer uses client observation skills to note client reactions and intentionally *flexes* to provide the support the client needs.

For the beginning counselor or interviewer, mastery of the BLS is critical, as these skills can be used in many different situations. It is not unusual for a person knowledgeable in the concepts of intentional interviewing to be conducting career counseling at a college in the morning, training parents in communication skills in the afternoon, and working as a management consultant on group meeting skills in the evening (see Table 7.1).

Counseling and the interview can be difficult experiences for some clients. They have come to discuss their problems and resolve conflicts, so the session can rapidly become a depressing litany of failures and fears. It becomes important to use the BLS to help the client identify assets and resources. To ensure a more optimistic and directed interview, use the positive asset search and wellness approach. Rather than just ask about problems, the intentional interviewer seeks to point out the client's positives and strengths. Even in the most difficult situation, it is possible to find good things about the client and resources for later problem resolution. Emphasizing positive assets also gives the client a sense of personal worth as he or she talks with you.

▲ **TABLE 7.1** Three Examples of the Basic Listening Sequence

Skill	Counseling	Management	Medicine
Open question	"Could you tell me what you'd like to talk to me about . . ."		
Closed question	"Did you graduate from high school?" "What specific careers have you looked at?"	"Who was involved with the production line problem?" "Did you check the main belt?"	"Is the headache on the left side or on the right? How long have you had it?"
Encouragers	Repetition of key words and restatement of longer phrases.		
Paraphrases	"So you're considering returning to college."	"Sounds like you've consulted with almost everyone."	"It looks like you feel it's on the left side and may be a result of the car accident."
Reflection of feeling	"You feel confident of your ability but worry about getting in."	"I sense you're upset and troubled by the supervisor's reaction."	"It appears you've been feeling very anxious and tense lately."
Summarization	In each case the effective counselor, manager, or physician summarizes the story from the client's point of view *before* bringing in the interviewer's point of view.		

INTENTIONALITY, PREDICTABILITY, AND FLEXIBILITY: THE IVEY TAXONOMY

Each chapter provides brief definitions of the skills and predicted outcomes. A full presentation of the skills included in the Ivey Taxonomy can be found in Appendix I. These definitions and their predicted responses are given because effective empathic interviewing requires intentional practice. By studying, practicing, and experiencing you will gain the capacity to automatically anticipate what happens as a result of your actions. It is amazing what you can do once you master skills and use them intentionally. Skills that once had to be studied now have become part of you. However, remain aware that *predictions of what the client is going to do next are always tentative.* Be ready to flex intentionally when the unexpected occurs. Flexibility means that if something you try doesn't work, you don't try more of the same—try something different.

Knowing the skills and expected outcomes will help you to use them intentionally and determine their effect, as well as select a different skill if the first one doesn't work. Again, as you gain experience, flexibility with the skills described here will become an automatic and natural part of your work.

 Watch Video: Reflection of Feelings

DIVERSITY, EMPATHIC UNDERSTANDING, AND THE BASIC LISTENING SEQUENCE

Diversity will always characterize the mainstream of interviewing and counseling. For every interview, you will encounter people with varying individual and cultural backgrounds, and you must factor in the many issues of multiculturalism (e.g., ethnicity/race, people with disabilities, sexual orientation, spirituality/religion). In effect, all interviewing is multicultural.

While attending and listening are skills, empathy is more an art form. You will want to use the skills of this book differently as you generate empathic understanding and that all-important working relationship. We have mentioned the importance of playing games with children as part of developing empathy. Having tissues close by in your office is a sign of your empathic caring and makes tears more natural and less embarrassing. Your willingness to flex and behave differently with varying clients is key. Clearly, even within one cultural or economic group, empathy plays itself out differently.

The chapter on attending behavior pointed out differences in verbal tracking, vocal tone, visuals, and body language among different cultures. Yes, traditional Arabs tend to talk more closely face to face, but if you adopted that style and are not Arab, you likely would be seen as artificial. At the same time, it seems only logical that you sit closer and perhaps even pay closer attention than usual. By contrast, Allen recalls speaking with a traditional Aboriginal at a workshop. She had a quiet voice and was hard to hear. Allen kept coming closer to hear her words and she ended up in a corner looking down. Embarrassed at what he had done, he backed off. Allen was not empathic in that situation. He should have said, "I'm a little hard of hearing, do you mind if I step closer?" In this example, the person outside of the culture recognizes difference and may share this with the client, while simultaneously trying to meet the client halfway.

One client likes direct eye contact; another is more comfortable if you look away. One client wants to solve problems instantly and requires you to start immediately. Another may not trust you, and you find yourself meeting with her or him several times before anything significant happens. But when it does, you have communicated patience and understanding. *Listening is love!*

Intentional competence requires flexibility and the ability to move and change with constantly changing client needs. Our clients will constantly vary, and we need to be careful when anticipating specific results from our interventions. What "works" as expected one time may not the next.

You have already covered the basic sections of the taxonomy—ethics, multicultural competence, wellness, attending behavior, and the basic listening sequence. In addition, we suggest that you examine the summary of the five-stage interview discussed in more detail in the following section.

Complete Weblink Exercise 1: Listening Self-Assessment
Complete Weblink Exercise 2: Instruction Guide #15

7.2
INSTRUCTIONAL READING: Decisional Counseling and the Five Stages/Dimensions of the Well-Formed Interview

KEY CONCEPT QUESTIONS

▲ **What is decisional counseling, and how does it relate to the structure of the interview?**

▲ **How does person-centered counseling relate to the structure of the interview?**

▲ **What are the five stages of the well-formed interview?**

▲ **Do we always need to follow the same five-stage order when conducting an interview?**

▲ **What are some multicultural issues in the five-stage interview?**

You are about to be asked to complete a full decisional interview using only the attending, observation, and listening skills stressed so far in this chapter. Here, you will see how the skills are integrated in a full interview.

DECISION MAKING AS A BASIS FOR STRUCTURING THE INTERVIEW

Decisional counseling may be described as a practical model that recognizes decision making and problem solving as undergirding most—perhaps all—systems of counseling. The basic point is simple: Whether clients need to choose among careers, decide to have a child, deal with conflict, or recover from depression, they are always making decisions. Whatever theory one operates from, eventually clients will be making decisions.

Many see Benjamin Franklin as the originator of the systematic decision-making model. He suggested three stages of problem solving: (1) identify the problem clearly, (2) generate alternative answers, and (3) decide what action to take. Another term for decisional counseling is problem-solving counseling. The essential issue is the same regardless of the terms we use: How can we help clients work through issues and come up with new answers?

The goal in decisional counseling is to facilitate decision making and to consider the many *traits and factors* underlying any single decision. Trait-and-factor theory has a long history in the counseling field, dating back to Frank Parsons's development of the Boston Vocational Bureau in 1908. Parsons pointed out that in making a vocational decision the client needs to (1) consider personal traits, abilities, skills, and interests; (2) examine the environmental factors (opportunities, job availability, location, and so on); and (3) develop "true reasoning on the relations of these two groups of facts" (Parsons, 1909/1967, p. 5). Since that time, proponents of trait-and-factor theory have searched for the many dimensions that underlie "true reasoning" and decision making.

Gradually trait-and-factor theory came to be seen as limited, and new decisional and problem-solving models have arisen. Several modern and systematic approaches to decisional counseling have been developed (Brammer & MacDonald, 2002; Chang, D'Zurilla, & Sanna, 2004; D'Zurilla, 1999; Egan, 2007; Ivey & Ivey, 2007; Ivey & Matthews, 1984; Ivey, D'Andrea, & Ivey, 2012; Janis & Mann, 1977).

Decisional counseling will be examined in considerably more depth in Chapter 13, thus giving you a solid framework for integrating the five-stage model in interviewing practice.

THE FIVE-STAGE MODEL FOR STRUCTURING THE INTERVIEW

The five-stage model of the interview, which includes *relationship—story and strengths—goals—restory—action*, develops and expands earlier decision-making frameworks. In Ben Franklin's terms, the *story* focuses on defining the problem carefully, *restory* on generating alternatives, and *action* on deciding what to do. Franklin and trait-and-factor theory gave insufficient attention to the relationship of interviewer and client. It is critical to establish a *relationship* with the client and to maintain this relationship throughout the interview. As part of relationship building, basic issues of confidentiality and ethical concerns need to be discussed. It is also important to structure the session so that everyone knows what is expected; thus the five-stage structure gives attention to *goals*, both preliminary and revised.

Along with defining client issues, problems, and concerns, the five-stage model speaks of drawing out the client's story and strengths. A single-minded problem-based approach may miss potential positive resources available to the client. Goals are important in both the traditional and decisional model. The interview needs direction and purpose. Based on the client's strengths you may then explore novel solutions, generate new ideas, and create new stories. Action—taking new ideas home to the real world—is critical to effect change. Interviewers with a problem orientation focus often fail to emphasize this critical dimension of change. *If the client takes home and actually practices the thoughts, feelings, and behaviors discovered in the session, your interview has clearly made a difference.*

At this point, please review Table 7.2, which summarizes the five stages of the interview in detail. Note that listening skills are central at each stage.

After you have mastered the five stages in a linear fashion, consider them an important checklist to ensure that you have covered all bases in any session. However, it is not always necessary to follow them in a specific order. Many clients will discuss issues in varying ways, often moving from one stage to another and then back again. New information revealed in later interviewing stages may result in the need for more data about the basic story, thus redefining client concerns and goals in a new way.

▲ **TABLE 7.2** The Five Stages/Dimensions of the Microskills Interview

At the core of the five-stage structure are the positive asset search and the wellness approach (see Chapter 2).			
Stage/Dimension	*Function and Purpose*	*Commonly Used Skills*	*Predicted Result*
1. *Relationship*. Initiate the session. Develop rapport and structuring. "Hello, what would you like to talk about today?"	Build a working alliance and enable the client to feel comfortable with the interviewing process. Explain what is likely to happen in the session or series of interviews, including informed consent and ethical issues.	*Attending, observation skills, information giving* to help *structure* the interview. If the client asks you questions, you may use *self-disclosure*.	The client feels at ease with an understanding of the key ethical issues and the purpose of the interview. The client may also know you more completely as a person and professional.
2. *Story and Strengths*. Gather data. Draw out client stories, concerns, problems, or issues. "What's your concern?" "What are your strengths and resources?"	Discover and clarify why the client has come to the interview and listen to the client's stories and issues. Identify strengths and resources as part of a wellness approach.	*Attending* and *observation* skills, especially *the basic listening sequence* and the *positive asset search*.	The client shares thoughts, feelings, and behaviors; tells the story in detail; presents strengths and resources.

(Continued)

TABLE 7.2   The Five Stages/Dimensions of the Microskills Interview (Continued)

At the core of the five-stage structure are the positive asset search and the wellness approach (see Chapter 2).			
Stage/Dimension	*Function and Purpose*	*Commonly Used Skills*	*Predicted Result*
3. *Goals*. Set goals mutually. "What do you want to happen?"	In brief counseling (Chapter 14), goal setting is fundamental, and this stage may be part of the first phase of the interview.	*Attending skills*, especially the *basic listening sequence*; certain *influencing skills*, especially *confrontation* (Chapter 8), may be useful.	The client will discuss directions in which he or she might want to go, new ways of thinking, desired feeling states, and behaviors that might be changed. The client might also seek to learn how to live more effectively with situations or events that cannot be changed at this point (rape, death, an accident, an illness). A more ideal story ending might be defined.
4. *Restory*. Explore alternatives, confront client incongruities and conflict, restory. "What are we going to do about it?" "Can we generate new ways of thinking, feeling, and behaving?"	Generate at least *three* alternatives that may resolve the client's issues. Creativity is useful here. Seek to find at least three alternatives so that the client has a choice. One choice at times may be to do nothing and accept things as they are.	*Summary* of major discrepancies with a supportive *confrontation*. More extensive use of *influencing skills*, depending on theoretical orientation (e.g., *interpretation, reflection of meaning, feedback*). But this is also possible using only *listening skills*. Use *creativity to* solve problems.	The client may reexamine individual goals in new ways, solve problems from at least those alternatives, and start the move toward new stories and actions.
5. *Action*. Plan for generalizing interview learning to "real life." "Will you do it?"	Generalize new learning and facilitate client changes in thoughts, feelings, and behaviors in daily life. Commit the client to homework and action. As appropriate, plan for termination of sessions.	*Influencing skills*, such as *directives* and *information/explanation*, plus *attending and observation skills* and the *basic listening sequence*.	The client demonstrates changes in behavior, thoughts, and feelings in daily life outside of the interview.

The circle of interviewing in Figure 7.1 reminds us that helping is a mutual endeavor between client and counselor. We need to be flexible in our use of skills and strategies. A circle has no beginning or end; it is a symbol of an egalitarian relationship in which interviewer and client work together. The hub of the interviewing circle is wellness and the positive asset search, important in all stages and dimensions.

STAGE 1. RELATIONSHIP—INITIATING THE SESSION: RAPPORT, TRUST BUILDING, STRUCTURING, AND PRELIMINARY GOALS ("HELLO")

Introducing the interview and building rapport are most important in the first interview, and they will remain central in all subsequent sessions. Most interviews begin

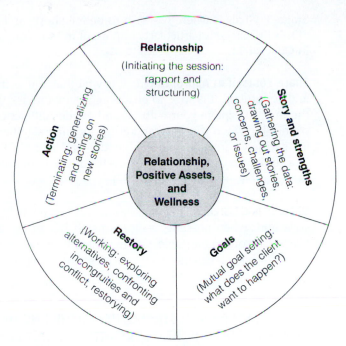

FIGURE 7.1 Circle of interviewing stages

with some variation of "Could you tell me how I might be of help?" or "What would you like to talk about today?" "Hello, Lynette." "It's good to meet you, Marcus." A prime rule for establishing rapport is to use the client's name and repeat it periodically throughout the session. Once you have completed the necessary structuring of the session (informing the client about legal and ethical issues), many clients are immediately ready to launch into a discussion of their issues. These clients represent instant trust, and it is our duty to honor that especially carefully and continue to work on relationship issues throughout the session.

Some interviewing situations require more extensive attention and time to the rapport stage than others. Rapport building can be quite lengthy and blend into treatment. For example, in Reality Therapy with a delinquent youth, playing Ping-Pong or basketball and getting to know the client on a personal basis may be part of the treatment. It may take several sessions before those who are culturally different from you develop real trust.

Structure.   Structuring the session includes informed consent and ethical issues, as outlined in Chapter 2. Clients need to know their rights and the limitations of the session. If this is part of an ongoing series of sessions, you can help maintain continuity by summarizing and integrating past interviews with the current session.

Listen for Preliminary Goals.   Although more definitive goal setting happens at Stage 3, initial goals are helpful during the first session(s). These early goals provide an initial structure for you as you seek to understand and empathize with the client. The early goals are revised and clarified after a fuller story has been brought out.

Stage 3: Goals is where we wish to define short- and long-term goals. Goal setting at this first stage is particularly important in brief counseling and coaching, described in Chapter 14.

Share Yourself as Appropriate.   It is vital that you as interviewer be open, authentic, and congruent. It is wise to encourage clients to ask you questions, and this is also the time to explore your cultural and gender differences. What about cross-cultural counseling when your race and ethnicity differ significantly from clients? Authorities increasingly agree that cultural, gender, and ethnic differences need to be addressed in a straightforward manner relatively early in counseling, often in the first interview (for example, see Sue & Sue, 2007).

Observe and Listen.   The first session tells you a lot about the client. Note the client's style and, when possible, seek to match her or his language. As the relationship becomes more comfortable, you may note that you and the client have a natural mirroring of body language. This indicates that the client is clearly ready to move on to telling the story and finding strengths (Stage 2).

STAGE 2. STORY AND STRENGTHS—GATHERING DATA: DRAWING OUT STORIES, CONCERNS, AND STRENGTHS ("WHAT IS YOUR CONCERN?" "WHAT ARE YOUR STRENGTHS AND RESOURCES?")

Draw Out the Client's Story.   What are the client's thoughts, feelings, and behaviors related to her or his concern? We draw out stories and concerns using the skills of the basic listening sequence. Open and closed questions will help define the issue as the client views it. Encouragers, paraphrases, and checkouts will provide additional clarity and an opportunity for you to verify whether you heard correctly. Reflection of feeling will provide understanding of the emotional underpinnings. Finally, summarization provides a good way to put the client's conversation into an orderly format.

Elaborate the Story.   Next, explore related thoughts, feelings, and interactions with others. Gather information and data about clients and their perceptions. The basic journalistic outline of *who, what, where, when, how,* and *why* provides an often useful framework to make sure you have covered the most important items. In your attempts to define the central client concerns, always ask yourself, what is the client's real-world and current story?

Draw Out Strength and Resource Stories.   Clients grow best when we identity what they can do rather than what they can't do. Don't focus just on the difficulties and problems. The positive asset search should be part of this stage of the interview. This might be the place for a comprehensive wellness search, as described in Chapter 2.

Failure to Treat.   Failure to treat can be the cause of a malpractice suit. This most often occurs when counselors fail to draw out stories and clients are unclear about what they really want, often resulting in no clear goals for the interviews. When the story is clear and strengths are established, it is time to review and clarify goals. Clients who participate in goal setting and understand the reasons for your helping interventions may be more likely to participate in the process and be more open to change.

STAGE 3. GOALS—MUTUAL GOAL SETTING ("WHAT DO YOU WANT TO HAPPEN?")

Mutuality and an Egalitarian Approach.   Your active involvement in client goal setting is important. *If you and the client don't know where the interview is going, you may end up somewhere else!* Too often the client and counselor assume they are working toward the same outcome when actually each of them wants something different. A client may be satisfied with sleeping better at night, but the counselor wants complete personality reconstruction. The client may want brief advice about how to find a new job, whereas the counselor wants to give extensive vocational testing and suggest a new career.

Refining and Making Goals More Precise.   It is wise to search for a general interview goal early in the relationship, as it provides a glue or purpose between you and the client. At this point, it helps to review that early goal, divide it into subgoals if necessary, and make it truly clear and doable. It has been suggested that if you don't have a goal, you're just complaining. Authorities on brief counseling (Chapter 14) favor setting goals in the first part of the interview, along with relationship building. With high school discipline problems, less verbal clients, and members of some cultural groups, setting up a clear joint goal may be the key factor in relationship building. If you adapt your interviewing style to each client, you will have a better chance to succeed.

Summarizing the Differences Between the Present Story and the Preferred Outcome.   Once the goal has been established, a brief summary of the original presenting concern as contrasted with the defined goals can be very helpful. Consider the model sentence below as a basic beginning to working through client issues. This is your opening to Stage 4, restorying.

> "On one hand, your problem/concern/issue is [summarize the situation briefly], but on the other hand, your goal is [summarize the goal]. What occurs to you as possibilities for resolution?"

Define both the problem and the desired outcome in the client's language. The summary confrontation should list several alternatives that the client has considered. The client ideally should generate more than one possibility before moving on to Stage 4. You may want to use hand movements, as if balancing the scales, to present the real and the ideal. Using such physical movements can add clarity to the summary confrontation of key issues.

Define a goal, make the goal explicit, search for assets to help facilitate goal attainment, and only then return to examine the nature of the concern.

STAGE 4. RESTORY—WORKING: EXPLORING ALTERNATIVES, CONFRONTING CLIENT INCONGRUITIES AND CONFLICT, RESTORYING ("WHAT ARE WE GOING TO DO ABOUT IT?")

Starting the Exploration Process.   How does the interviewer help the client work through new solutions? Summarize the client conflict as just described in the model sentence in the preceding section, "On the one hand. . . ." But be sure that the summarization of the issue is complete and that the facts of the situation and the client's thoughts and feelings are part of this summary.

Use the basic listening sequence to facilitate the client's resolution of the issue(s). Imagine a school counselor talking with a teen who has just had a major showdown with the principal. Establish rapport, but expect the teen to challenge you; he or she likely expects you to support the principal. Do not judge, but gather data from the teen's point of view. If you have developed rapport (Stage 1) and listened during data gathering (Stage 2), the teen will search for solutions in a more positive way. Follow by asking what he or she would like to have happen in terms of a positive change. Work with the teen to find a way to "save face" and move on.

Encourage Client Creativity.   Your first goal in restorying is to encourage your clients to discover their own solutions. To explore and create with the teen above, listen well and use summarization—"You see the situation as . . . and your goal is. . . . The principal tells a different story and his goal is likely to be. . . ." If you have developed rapport and listened well, many teens will be able to generate ideas to help resolve the situation.

The basic listening sequence and skilled questioning are useful in facilitating client exploration of answers and solutions. Here are some useful questions to assist client problem solving. The last two focus on a wellness approach and would be common in brief counseling.

- ▲ "Can you brainstorm ideas—just anything that occurs to you?"
- ▲ "What other alternatives can you think of?"
- ▲ "Tell me about a success that you have had."
- ▲ "What has worked for you before?"
- ▲ "What part of the problem is workable if you can't solve it all right now?"
- ▲ "Which of the ideas that we have generated appeals to you most?"
- ▲ "What are the consequences of taking that alternative?"

Interviewing, counseling, and psychotherapy all try to resolve issues in clients' lives in a similar fashion. The counselor needs to establish rapport, define the problem, and help the client identify desired outcomes.

Relate Client Issues and Concerns to Desired Outcomes.   The distinction between the problem and the desired outcome is the major incongruity that may be resolved in three basic ways. First, the counselor uses attending skills to clarify the client's frame of reference and then feeds back a summary of client concerns and the goal. Often clients generate their own synthesis and resolve their challenges. Second, interviewers can use information, bibliotherapy, and psychoeducational interventions to help clients generate new answers. Third, if clients do not generate their own answers, the interviewer can use interpretation, self-disclosure, and other influencing skills to resolve the conflict. Finally, in systematic problem solving and decision making, counselor and client generate and brainstorm alternatives for action and set priorities among the most promising possibilities.

Aim for a Decision and a New Story.   This exploration/brainstorming/testing of theoretical strategies facilitates client decision making and the generation of a new story. Once a decision has been made or a new workable story developed, see that plans are made to put these ideas into action in the real world. You also need to help clients generalize feelings, thoughts, and behaviors and a plan for action beyond the interview itself.

STAGE 5. ACTION—CONCLUDING: GENERALIZING AND ACTING ON NEW STORIES ("WILL YOU DO IT?")

The complexities of life are such that taking a new behavior back to the home setting is difficult. How do we generalize thoughts, feelings, and behavior to daily life? Some counseling theories work on the assumption that behavior and attitude change will come out of new unconscious learning; they "trust" that clients will change spontaneously. This indeed can happen, but there is increasing evidence that planning for change greatly increases the likelihood that it will actually happen.

Consider the situation with the teen in conflict with the principal. Some good ideas may have been generated, but unless the teen follows up on them, nothing is likely to change in the conflict situation. Find something that "works" and leads to changes in the repeating behavioral problems. As you read through the list of generalization suggestions below, consider what you would do to help this teen and other clients restory and change their thoughts, feelings, and behaviors.

Change does not always come easily, and many clients revert to earlier, less intentional behaviors. Work to help the client plan for change to ensure real-world relevance. Following are some techniques that interviewers and counselors have used to facilitate the transfer of learning from the interview.

Homework and Journaling. Assigning homework so that the effect of the interview continues after the session ends has become increasingly standard. Some use "personal experiments" with clients who do not like the idea of doing "homework." Negotiate specific tasks for the client to try during the week after the session ends. Use very specific and concrete behavioral assignments, such as "To help your shyness, you agreed to approach one person after church/synagogue/mosque and introduce yourself." Ask the client to keep a journal of key thoughts and feelings during the week; this can become the basis of the follow-up session. Another possibility is paradoxical intention: "Next week, I want you to deliberately do the same self-defeating behavior that we have talked about. But take special notice of how others react and how you feel." This helps the client become much more aware of what he or she is doing and its results.

Role-playing. The client can practice the new behavior in a role-play with the interviewer. Role-playing emphasizes the specifics of learning and increases the likelihood that the client will recognize the need for the new behavior after the session is over. In the case of the teen, you could become the principal and role-play the forthcoming meeting with the principal. Furthermore, you can role-play what to do if the meeting does not go as expected.

Follow-up and Support. Ask the client to return for further sessions, each with a specific goal. The counselor can provide social and emotional support through difficult periods. Follow-up is a sign that you care. Use the telephone for behavior maintenance checks. Using e-mail is possible, but may result in loss of your privacy. Many counselors and therapists call clients but do not give them their e-mail. If you work in an agency and give a phone number to a client, use an office number that is always attended.

Contract for Change. These suggestions are only a few of the possibilities for developing and maintaining client change. Each individual will respond differently to these techniques, and client observation skills will help determine which technique or set of techniques is most likely to be helpful. For maximal impact and behavior transfer, we suggest a combination of several techniques and strategies over time. Behaviors and attitudes learned in the interview do not necessarily transfer to daily life without careful planning. Ask your client at the close of the interview, "Will you do it?" as a form of contract between the two of you for the future. Often a written contract or plan can be helpful.

Relapse Prevention. Relapse Prevention (RP) (Marlatt & Donovan, 2007) is a cognitive-behavioral approach used to identify and prevent high-risk situations threatening the sustainability of change. Its goal is to guide the client to identify potential components of a personal relapse and then to develop specific strategies for coping with those components. RP raises awareness of triggers for negative behaviors such as interpersonal conflict, social pressure, and negative emotional states; it helps clients make lifestyle changes, reduce the likelihood of encountering high-risk situations, and implement rehearsed coping strategies when encountering such situations. For example, a person who overeats to reduce tension associated with stressful interpersonal relations might be encouraged either to develop more positive relations or to learn stress-management techniques to better cope with stressful relationships. RP is discussed in more detail with specific suggestions in Chapter 14.

 Complete Case Study Activity: Couple Conflict

7.3
TAKING NOTES IN THE INTERVIEW

KEY CONCEPT QUESTION

▲ What about taking notes in the interview?

Beginning helpers typically question whether or not they should take notes during the interview, and it's not hard to find opinions for or against note taking. So, we are going to share our opinions based on our experience, recognizing the individuals vary in their thoughts. Most important, follow the directions of your agency.

Intentionality in interviewing requires accurate information. Therefore, listen intentionally and take notes. You and your client can usually work out an arrangement suitable for both of you. If you personally are relaxed about note taking, it will seldom become an issue. If you are worried about taking notes, it likely will be a problem. When working with a new client, obtain permission early about taking notes. We suggest that any case notes be made available to clients and in practice sessions. Volunteer client feedback on the interview can be most helpful in thinking about your own style of helping. Using your own natural style, you might begin:

> "I'd like to take a few notes while you talk. Would that be OK? I'll write down your exact key words as important reference points for both of us. I'll make a copy of the notes before you leave, if you wish. As you know, all notes in your file are open to you at any time."

In-session note taking is often most helpful in the initial portions of interviewing and counseling and less important as you get to know the client better. Audiotaping and videotaping the session follow the same guidelines. If you are relaxed and provide a rationale to your client, making this type of record of the interview generally goes smoothly. Some clients find it helpful to take audio recordings of the session home and listen to them, thus enhancing learning from the interview. There is nothing wrong with *not* taking notes in the session, but records are very important and we recommend writing session summaries shortly after the interview finishes.

HIPAA (Health Insurance Portability and Accountability Act) legal requirements regarding note taking are detailed and specific, but not always clear. There are rules that give certain aspects of counseling more detailed protection than general medical records, but these rules are sometimes written vaguely, and they mention the possibility of maintaining dual records of psychotherapy. The agency you work with in practicum or internship can guide you in this area. You will also find Zur's (2005) *The HIPAA Compliance Kit* and the weblinks listed in the chapter summary helpful.

7.4
EXAMPLE INTERVIEW: I Can't Get Along With My Boss

KEY CONCEPT QUESTION

▲ How is a five-stage interview conducted?

It requires a verbal, cooperative client to work through a complete interview using only listening skills. This interview has been edited to show portions that demonstrate skill usage and levels of empathy. Robert, the client, is 20 and a part-time student who is in conflict with his boss at work. Machiko, the counselor, finds him relatively verbal and willing to work on the problem with her assistance.

STAGE 1: RELATIONSHIP

Counselor/Client Statement	Process Comments
1. **Machiko:** Robert, do you mind if we tape this interview? It's for a class exercise in interviewing. I'll be making a transcript of the session, which the professor will read. Okay? We can turn the recorder off at any time. I'll show you the transcript if you are interested. I won't use the material if you decide later you don't want me to use it. Could you sign this consent form?	Machiko opens with a closed question followed by structuring information. It is critical to obtain client permission and offer client control over the material before taping. As a student you cannot legally control confidentiality, but it is your responsibility to protect your client.
2. **Robert:** Sounds fine; I do have something to talk about. Okay, I'll sign it. [Pause as he signs]	Robert seems at ease and relaxed. As the taping was presented casually, he is not concerned about the use of the recorder. Rapport was easily established.

Counselor/Client Statement	Process Comments
3. **Machiko:** What would you like to share?	The open question, almost social in nature, is designed to give maximum personal space to the client.
4. **Robert:** My boss. He's pretty awful.	Robert indicates clearly through his non-verbal behavior that he is ready to go. Machiko observes that he is comfortable and decides to move immediately to gather data (Stage 2). With some clients, several interviews may be required to reach this level of rapport.

STAGE 2: STORY AND STRENGTHS

Counselor/Client Statement	Process Comments
5. **Machiko:** Could you tell me about it?	This open question is oriented toward obtaining a general outline of the problem the client brings to the session.
6. **Robert:** Well, he's impossible.	Instead of the expected general outline of the concern, Robert gives a brief answer. The predicted consequence didn't happen.
7. **Machiko:** Impossible? . . . Go on. . .	Encourager. Intentional competence requires you to be ready with another response. Tone of voice is especially important here in communicating to the client.
8. **Robert:** Well, he's impossible. Yeah, really impossible. It seems that no matter what I do he is on me, always looking over my shoulder. I don't think he trusts me.	We are seeing the story develop. Clients often elaborate on the specific meaning of a concern if you use the encourager. In this case the prediction holds true.
9. **Machiko:** Could you give me a more specific example of what he is doing to indicate he doesn't trust you?	Robert is a bit vague in his discussion. Machiko asks an open question eliciting concreteness in the story.
10. **Robert:** Well, maybe it isn't trust. Like last week, I had this customer lip off to me. He had a complaint about a shirt he bought. I don't like customers yelling at me when it isn't my fault, so I started talking back. No one can do *that* to me! And of course the boss didn't like it and chewed me out. It wasn't fair.	As events become more concrete through specific examples, we understand more fully what is going on in the client's life and mind.

Counselor/Client Statement	Process Comments
11. **Machiko:** As I hear it, Robert, it sounds as though this guy gave you a bad time and it made you angry, and then the boss came in.	Machiko's response is relatively similar to what Robert said. Her paraphrase and reflection of feeling represents basic interchangeable empathy.
12. **Robert:** Exactly! It really made me angry. I have never liked anyone telling me what to do. I left my last job because the boss was doing the same thing.	Accurate listening often results in the client's saying "exactly" or something similar.
13. **Machiko:** So your last boss wasn't fair either?	Machiko's vocal tone and body language communicate nonjudgmental warmth and respect. She brings back Robert's key word *fair* by paraphrasing with a questioning tone of voice, which represents an implied checkout. This is an interchangeable empathic response (Level 2).

The interview continues to explore Robert's conflict with customers, his boss, and past supervisors. There appears to be a pattern of conflict with authority figures over the past several years. This is a common pattern among young males in their early careers. After a detailed discussion of the specific conflict situation and several other examples of the pattern, Machiko decides to conduct a positive asset search to discover strengths.

Counselor/Client Statement	Process Comments
14. **Robert:** You got it.	
15. **Machiko:** Robert, we've been talking for a while about difficulties at work. I'd like to know some things that have gone well for you there. Could you tell me about something you feel good about at work?	Paraphrase, structuring, open question, and beginning positive asset search.
16. **Robert:** Yeah; I work hard. They always say I'm a good worker. I feel good about that.	Robert's increasingly tense body language starts to relax with the introduction of the positive asset search. He talks more slowly.
17. **Machiko:** Sounds like it makes you feel good about yourself to work hard.	Reflection of feeling, emphasis on positive regard (Level 2 empathy).
18. **Robert:** Yeah. For example, . . .	

Robert continues to talk about his accomplishments. In this way Machiko learns some of the positives Robert has in his past and not just his problems. She has used the basic listening sequence to help Robert feel better about himself. Machiko also learns that Robert has positive assets, such as determination and willingness to work hard, to help him resolve his own problems.

STAGE 3: GOALS

Counselor/Client Statement	Process Comments
19. **Machiko:** Robert, given all the things you've talked about, could you describe an ideal solution? How would you like things to be?	Open question. The addition of a new possibility for the client represents additive empathy (Level 3). It enables Robert to think of something new.
20. **Robert:** Gee, I guess I'd like things to be smoother, easier, with less conflict. I come home so tired and angry.	
21. **Machiko:** I hear that. It's taking a lot out of you. Tell me more specifically how things might be better.	Paraphrase, open question oriented toward concreteness.
22. **Robert:** I'd just like less hassle. I know what I'm doing, but somehow that isn't helping. I'd just like to be able to resolve these conflicts without always having to give in.	Robert is not as concrete and specific as anticipated. But he brings in a new aspect of the conflict—giving in.
23. **Machiko:** Give in?	Encourager.
24. **Robert:** Yeah.	

Machiko learns another dimension of Robert's conflict with others. Subsequent use of the basic listening sequence brings out this pattern with several customers and employees. As new data emerge in the goal-setting process, you may find it necessary to change the definition of the concern and perhaps even return to Stage 2 for more data gathering.

Counselor/Client Statement	Process Comments
25. **Machiko:** So, Robert, I hear two things in terms of goals. One, that you'd like less hassle, but another, equally important, is that you don't like to give in. Have I heard you correctly?	Machiko uses a summary to help Robert clarify his problem, even though no resolution is yet in sight. She checks out the accuracy of her hearing (Level 3 additive empathy).
26. **Robert:** You're right on, but what am I going to do about it?	

STAGE 4: RESTORY

Counselor/Client Statement	Process Comments
27. **Machiko:** So, Robert, on the one hand I heard you have a long-term pattern of conflict with supervisors and customers who give you a bad time. On the other hand, I also heard just as loud and clear your desire to have less hassle and not give in to others. We also know that you are a good worker and like to do a good job. Given all this, what do you think you can do about it?	Machiko remains nonjudgmental and appears to be very congruent with the client in terms of both words and body language. In this major empathic Level 3 summary, she distills and clarifies what the client said.
28. **Robert:** Well, I'm a good worker, but I've been fighting too much. I let the boss and the customers control me too much. I think the next time a customer complains, I'll keep quiet and fill out the refund certificate. Why should I take on the world?	Robert talks more rapidly. He, too, leans forward. However, his brow is furrowed indicating some tension. He is "working hard."
29. **Machiko:** So one thing you can do is keep quiet. You could maintain control in your own way, and you would not be giving in.	Paraphrase, Level 3 additive empathy. Machiko is using Robert's key words and feelings from earlier in the interview to reinforce his present thinking. But she waits for Robert's response.
30. **Robert:** Yeah, that's what I'll do, keep quiet.	He sits back, his arms folded. This suggests that the "good" response above was in some way actually subtractive. There is more work to do.
31. **Machiko:** Sounds like a good beginning, but I'm sure you can think of other things as well, especially when you simply can't be quiet. Can you brainstorm more ideas?	Machiko gives Robert brief feedback. Her open question is a Level 3 response adding to the interview. She is aware that his closed nonverbals suggest more is needed.

Clients are often too willing to seize the first idea as a way to agree and avoid looking fully at issues. It is helpful to use a variety of questions and listening skills to further draw out the client. Later in the interview, Robert was able to generate two other useful suggestions: (1) to talk frankly with his boss and seek his advice; and (2) to plan an exercise program to blow off steam and energy. In addition, Robert began to realize that his problem with his boss was only one example of a continuing problem with anger. He and Machiko discussed the possibilities of continuing their discussions or for Robert to visit a professional counselor to work with him on anger management. Robert decided he'd like to talk with Machiko a bit more. A contract was made: If the situation did not improve within 2 weeks, Robert would seek professional help.

STAGE 5: ACTION—GENERALIZATION AND TRANSFER OF LEARNING

Counselor/Client Statement	Process Comments
32. **Machiko:** So you've decided that the most useful step is to talk with your boss. But the big question is "Will you do it?"	Paraphrase, open question.
33. **Robert:** Sure, I'll do it. The first time the boss seems relaxed.	
34. **Machiko:** As you've described him, Robert, that may be a long wait. Could you set up a specific plan so we can talk about it the next time we meet?	Paraphrase, open question. To generalize from the interview, it is important to encourage specific and concrete action in your client.
35. **Robert:** I suppose you're right. Okay, occasionally he and I drink coffee in the late afternoon at Rooster's. I'll bring it up with him tomorrow.	A specific plan to take action on the new story is developing.
36. **Machiko:** What, specifically, are you going to say?	Open question, again eliciting concreteness.
37. **Robert:** I could tell him that I like working there, but I'm concerned about how to handle difficult customers. I'll ask his advice and how he does it. In some ways, it worries me a little; I don't want to give in to the boss . . . but maybe he will have a useful idea.	Robert is able to plan something that might work. With other clients, you may role-play, give advice, actually assign home-work. You will also note that Robert is still concerned about "giving in."
38. **Machiko:** Would you like to talk more about this the next time we meet? Maybe through your talk with your boss we can figure out how to deal with this in a way that makes you feel more comfortable. Sounds like a good contract. Robert, you'll talk with your boss and we'll meet later this week or next week.	Open question, structuring. If Robert does talk to his boss and listens to his advice—and actually changes his behavior—then this interview could be rated holistically at Level 3. If not, then a lower rating is obvious.

 Complete Interactive Exercise 3: The Impossible Boss

It would have been wise to specify the follow-up contract even more precisely, but this would most likely entail the use of influencing skills, prescribing homework, and so forth. You will find that concreteness and specificity are very important in assisting clients to make and act on decisions. It was an especially important response when Machiko asked Robert what he was specifically going to do.

You may find it challenging to work through the systematic five-stage interview using only attending, observation, and the basic listening sequence, yet it can be done. It is a useful format to use with individuals who are verbal and anxious to resolve their own issues. You will also find this decisional structure useful with resistant clients who want to make their own decisions. By acting as a mirror and asking questions, you can encourage many of your clients to find their own direction. More information on decisional counseling will be presented in Chapter 13.

7.5

PERSON-CENTERED COUNSELING AND THE WELL-FORMED INTERVIEW

Theoretically and philosophically, the interview style using only listening skills is related to Carl Rogers's person-centered counseling (Rogers, 1957). Rogers developed guidelines for the "necessary and sufficient conditions of therapeutic personality change." The empathic constructs described in this chapter are derived from his thinking. In the Machiko-Robert transcript, you saw a decisional model combined with a modified person-centered approach. Respect for the client's ability to find her or his own unique direction is implicit in conducting an interview without using information, advice, and influencing skills.

Rogers originally was opposed to the use of questions but in later life modified his position so that in some interviews a very few questions might be asked. These would be quite open and as "nondirective" as possible. "What is your goal?" and "What meaning does that have for you?" are two examples of very open questions that tend not to box the client into the interviewer's perspective or theory.

It was Carl Rogers who truly brought the ideas of empathic understanding to the interviewing process. Many would say that he *humanized* counseling and psychotherapy by stressing the importance of relationship, respect, authenticity, and positive regard. This is also called the "working alliance," a useful term as it stresses the way we need to work with, rather than on, the client. A good relationship and working alliance may be in itself sufficient to produce positive change.

You will encounter much more about the person-centered theory in future study. However, at this time you can gain a beginning understanding of some aspects of the approach by focusing on listening skills, using as few questions as possible. In the process, think constantly of "being in the client's shoes" and experiencing the world as he or she does. If you try this exercise, you will gain a better understanding what Carl Rogers is seeking—to discover the client's outer and inner world as the client experiences it.

Review Website: Carl Rogers: Home
Review Website: Carl Rogers: Streaming Video
Watch Windows Media files of videotapes and films of Carl Rogers's discussions, presentations, and therapy sessions
Review Website: Carl Rogers: Links to other PCA sites

CHAPTER SUMMARY

| Key Points of "How to Conduct an Interview Using Only Listening Skills" | CourseMate and DVD Activities to Build Interviewing Competence |

7.1 Key Concepts for Effective Interviewing: Empathic Understanding, the Basic Listening Sequence (BLS), and Predicting Client Interviewing Response

▲ Empathy is defined as experiencing the world as if you were the client, but with awareness that the client remains separate from you. Be with the client, but be aware that the client's world is not your world.

▲ Three levels of empathic understanding are (1) subtractive (where you fail to hear the client's story fully), (2) basic empathy (where your responses are accurate and interchangeable with what the client says), and (3) additive (where you contribute to client understanding at a new and deeper level).

▲ The basic listening sequence (BLS) is built on attending and observing the client; the key skills are using open and closed questions, encouraging, paraphrasing, reflecting feelings, and summarizing.

▲ When we listen to clients using the BLS, we want to discover the overall background of the client's story and learn about the facts, thoughts, feelings, and behaviors that go with that story.

▲ The Ivey Taxonomy (Appendix I) is an expansion of the microskills hierarchy. When we use a specific microskill, we can anticipate what the client may do or say.

▲ Clients will often say or do something unexpected. Intentional competence requires you to flex and generate a new alternative for helping when the first skill or strategy produces an unexpected result.

▲ Diversity includes multiple dimensions. To remain effective when you are culturally different from the client, it is important to demonstrate empathic understanding while simultaneously being aware that you cannot enter the full world of the client's experience.

1. Interactive Exercise 1: I Don't Seem to Fit In: Predicting Results, the Basic Listening Sequence, and Empathic Understanding.
2. Interactive Exercise 2: The BLS in Many Settings. Practice writing responses that represent the BLS.
3. Video Activity: Reflection of Feelings. Mary and Sandra's demonstrate reflection of feelings. This exercise gives you the opportunity to analyze and comment on reflection of feelings, suggest other ways to accomplish this, and explore your own way of responding to emotional expressions.
4. Weblink Exercise 1: Listening Self-Assessment. Low-cost, computer-scored listening test, posted by the Brandt Management Group, 1999.
5. Weblink Exercise 2: Instruction Guide #15, Counseling. United States Air Force Auxiliary, Civil Air Patrol Instructors Guides.

7.2 Instructional Reading: Decisional Counseling and the Five Stages/ Dimensions of the Well-Formed Interview

▲ Decisional counseling is a practical model that recognizes that clients are always making decisions.

▲ Regardless of varying counseling and therapy theories, most interviews involve making some sort

1. Case Study: Couple Conflict.

Key Points of "How to Conduct an Interview Using Only Listening Skills"

CourseMate and DVD Activities to Build Interviewing Competence

of decision, including defining the problem, defining the goal, and selecting from alternatives.

▲ The five stages of the well-formed interview are (1) relationship, (2) story and strengths, (3) goals, (4) restory, and (5) action.

▲ Following the five stages in order is often helpful to you and the client, but be flexible and move with the client's needs and interests.

▲ Consider the five stages a checklist to be covered in each session.

7.3 Taking Notes in the Interview

▲ If you personally are relaxed about note taking, it will seldom become an issue. If you are worried about taking notes, it likely will be a problem.

▲ When you wish to take notes or audiotape or videotape a session, obtain permission from the client early in the session. Make all records available to the clients, even volunteer clients.

1. Weblink 1: Progress Notes and Psychotherapy Notes.
2. Weblink 2: Alaska Bar Association/The Sticky Wicket of Psychotherapy Notes.
3. Weblink 3: Psych Central/Psychotherapy Notes and HIPAA.
4. Weblink 4: Columbia University Medical Center/HIPAA Policies.
5. Weblink 5: Use or Disclosure of Psychotherapy Notes.

7.4 Example Interview: I Can't Get Along With My Boss

▲ Machiko was able to complete a full interview with some success using only the basic listening skills.

▲ She was able to maintain her skills at basic interchangeable empathy (Level 2), but she also had several additive (Level 3) responses.

▲ When her responses did not achieve predicted results, she was able to flex intentionally and try another skill.

▲ Helping the client make specific plans for generalization of new learning is particularly important. Often interviewers and clients are tired at the end of the session and fail to take this final step.

1. Interactive Exercise 3: The Impossible Boss. Review and analyze Machiko and Robert's interview, and score your responses.

7.5 Person-Centered Counseling and the Well-Formed Interview

▲ The interview using only listening skills was developed by Carl Rogers.

▲ Empathic constructs are derived from his person-centered counseling theory.

▲ Respect for the client's ability to find her or his own unique direction is basic to his thinking.

▲ "What is your goal?" and "What meaning does that have for you?" are two examples of very open questions that Rogers may use.

1. Website: Carl Rogers: Home. CarlRogers.info serves as a gateway to the intellectual work of Dr. Carl R. Rogers, the creator of person-centered counseling.
2. Website: Carl Rogers: Streaming Video.
3. Windows Media files of videotapes and films of Carl Rogers's discussions, presentations, and therapy sessions.
4. Website: Carl Rogers: Links to other PCA sites.

Key Points of "How to Conduct an Interview Using Only Listening Skills"

CourseMate and DVD Activities to Build Interviewing Competence

▲ Rogers *humanized* counseling and psychotherapy by stressing the importance of relationship, respect, authenticity, and positive regard.

▲ "Working alliance" is a useful term that stresses we need to work with, rather than on, the client.

▲ A good relationship and working alliance may be in itself sufficient to produce positive change.

▲ You can gain some understanding of the Rogerian approach by focusing on listening skills and using as few questions as possible.

▲ Think constantly of "being in the client's shoes" and experiencing the world as he or she does.

5. Weblinks to other sites connected with Carl Rogers and the person-centered approach.

Assess your awareness, knowledge, and skills as you conclude the chapter:

1. **Flashcards:** Use the flashcards to check your understanding of key concepts and facilitate memorization of key information.

2. **Self-Assessment Quiz:** The quiz will help you assess your current knowledge and prepare for course examinations.

3. **Portfolio of Competencies:** Evaluate your present level of competence in attending and listening skills using the downloadable Self-Evaluation Checklist. Self-assessment of your attending skills competence demonstrates what you can do in the real world.

HELPING CLIENTS GENERATE NEW STORIES THAT LEAD TO ACTION
Influencing Skills and Strategies

SECTION

If you change the way you look at things, the things you look at will change.

—*Bill Alden*

The first two sections of this book focused on listening. Never forget the watchwords *When in doubt, listen!* and *Love is listening!* But now we turn to becoming more actively involved in the change process. The goal of the *relationship—story and strengths—goals—restory—action* model is to help clients find new stories and use them for growth and change. Based on a solid empathic relationship, we draw out and listen to the client's story, find positive strengths, and help generate new stories that lead to change and action. The five-stage interview provides a structure that, for many clients, will be complete in itself—simply telling the story in a positive, supportive atmosphere is often sufficient for positive change to occur. Because clients have changed the way they look at things, the things they look at have changed as well.

Look carefully at the word **influencing.** A web definition is "a power affecting a person, thing, or course of events, especially one that operates without any direct or apparent effort" (*Free Dictionary*, 2011). That particular definition is helpful because it (1) reminds us that interviewers and counselors are in a position of power relative to clients, even when we seek to be egalitarian, and (2) emphasizes that we can exert influence without "any apparent effort." Listening, in that sense, becomes an influencing skill. Listen carefully, with love.

The influencing skills and strategies of Section III present many specifics for helping clients restory their lives, develop new ways of thinking and feeling, and taking action to achieve desired goals. We find that both counseling students and professionals can become so involved with the change process that they forget to listen and start using their power to meet their own needs rather than those of the client.

Throughout this section, think of counseling and therapy as a creative process. Creativity is the root of change. The famed theologian/philosopher Paul Tillich spoke of the "creation of the *New.*" To help clients change and create the *New*, we need to listen to the client's story and summarize it fully, recognizing the emotional base of client concerns. An empathic approach is key. Then, if necessary, consider the influencing skills as additional strategies to encourage client intentionality, for intentionality is another way to think about the creative individual.

Chapter 8, "The Skill of Confrontation," presents what may be the most important agent of change in the interview after the listening skills. When confronting, you observe discrepancies and conflict in the client, summarize and reflect them, then facilitate resolving the client's issues. Included in this chapter is the *Client Change Scale*, which enables you to assess your client's responses and evaluate larger developmental change over a series of interviews.

Chapter 9, "Focusing the Interview," extends the concept of confrontation and illustrates how multiple frameworks for examining stories and personal issues can result in breakthroughs in client understanding. Too many counselors and interviewers focus narrowly on their clients' problems and miss contextual issues such as the impact of family, significant others, and multicultural and social factors.

Chapter 10, "Reflection of Meaning and Interpretation/Reframing," discusses skills that are complex and rich. The use of these closely related skills provides a deeper understanding of each client's issues and history. Both skills are concerned with finding new perspectives on life and its meaning. Reflection of meaning tends to focus more on values and critical issues underlying overt behavior; these discoveries come from the client rather than the interviewer. Interpretation and reframing tend to come from the interviewer's personal or theoretical perspective and provide clients with new ways to understand themselves and their situations.

Chapter 11, "Self-Disclosure and Feedback," explores two specific strategies for facilitating client change. Effective use of appropriate self-disclosure and feedback makes the interview more immediate and personal, and helps clients' creation of new stories.

Chapter 12, "Logical Consequences, Information/Psychoeducation, and Directives," provides you with the basics of three additional influencing skills. Each of these skills has specific strategies that can help clients create new stories and generalize and act on what is discovered in the session. All of these involve active participation by the interviewer.

Do not expect to achieve competence in all the concepts, skills, and strategies of this section immediately. Confrontation, for example, is a complex skill that even the most experienced counselor or therapist can always improve. The learning in interviewing, counseling, and therapy never ends. *Be patient, listen, learn new things constantly, and practice new skills and ideas until you can use them effectively in the interview.*

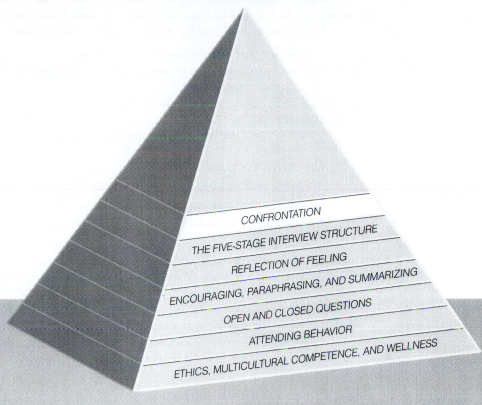

CONFRONTATION

THE FIVE-STAGE INTERVIEW STRUCTURE

REFLECTION OF FEELING

ENCOURAGING, PARAPHRASING, AND SUMMARIZING

OPEN AND CLOSED QUESTIONS

ATTENDING BEHAVIOR

ETHICS, MULTICULTURAL COMPETENCE, AND WELLNESS

Conditions for creativity are to be puzzled; to concentrate; to accept conflict and tension; to be born every day; to feel a sense of self.

—*Erich Fromm*

How can confrontation help you and your clients?

CHAPTER
GOALS

Confronting client discrepancies has the capacity to activate human potential and their natural creativity. Most clients come to an interview feeling "stuck," "blocked," or "paralyzed," and seeking some sort of movement or change in their lives. They may even resist your efforts to bring about the very transformation they seek. Your task is to help them move beyond their issues and problems to create and realize their potential as human beings. An understanding of confrontation is basic to helping clients restory their lives.

Awareness, knowledge, and skills developed through the concepts of this chapter will enable you to

▲ Identify incongruity, discrepancies, or mixed messages in behavior, thought, feelings, or meanings.

▲ Increase client talk with a view toward explanation and/or resolution of conflict and discrepancies.

▲ Understand how a theory of death and dying can help us understand the client change process.

▲ Identify client change processes occurring during the interview and throughout the treatment period, using the Client Change Scale to measure creative change.

Assess your awareness, knowledge, and skills as you begin the chapter:

1. Self-Assessment Quiz: The chapter quiz will help you determine your current level of knowledge. You can take it before and after reading the chapter.
2. Portfolio of Competencies: Before you read the chapter, fill out the downloadable Self-Evaluation Checklist to assess your existing knowledge and competence in attending skills. Then, at the end of the chapter, complete the checklist again to summarize your competencies after study and practice.

8.1
DEFINING CONFRONTATION

KEY CONCEPT QUESTION

▲ What is a confrontation and how it is structured?

Cultural intentionality has been stated as a central goal for interviewers—the ability to respond to unique clients and their changing needs and issues requires us to be able to flex and have creative multiple responses in the session. Intentionality is equally important for our clients. They also need to be able to flex and meet the multiple challenges of life.

Most clients come to an interview "stuck"—having no alternatives for solving a problem or, at best, a limited range of possibilities. The task of the interviewer is to assist in freeing the client from stuckness and facilitate the development of creative thinking and expansion of choices. **Stuckness** is an inelegant but highly descriptive term coined by the Gestalt theorist Fritz Perls to describe the opposite of intentionality. Other words that represent stuckness include *immobility, blocks, repetition compulsion, inability to achieve goals, lack of understanding, limited behavioral repertoire, limited life* *script, impasse,* and *lack of motivation.* Stuckness may also be defined as an inability to resolve conflict, reconcile discrepancies, and deal with incongruity. In short, clients often come to the interview because they are stuck for a variety of reasons and seek the ability to move, expand alternatives for action, and become motivated to do something to rewrite their life stories. All of these represent the intentional client.

Confrontation, as defined below, leads clients to new ways of thinking and increased intentionality. You will find that not all clients respond to this skill; be ready with other listening skills. But if you start with careful listening and summarize the conflict accurately, your chances of success are greatly heightened.

Confrontation	Predicted Result
Supportively challenge the client: 1. Listen, observe, and note client conflict, mixed messages, and discrepancies in verbal and nonverbal behavior. 2. Point out and clarify internal and external discrepancies by feeding them back to the client, usually through the listening skills. 3. Evaluate how the client responds and whether the confrontation leads to client movement or change. If the client does not change, the interviewer flexes intentionally and tries another skill.	Clients will respond to the confrontation of discrepancies and conflict with new ideas, thoughts, feelings, and behaviors, and these will be measurable on the 5-point Client Change Scale. Again, if no change occurs, *listen.* Then try an alternative style of confrontation.

Our clients often become stuck because of internal and/or external conflict. **Internal conflict,** or **incongruity,** occurs when they have difficulty making important decisions, they feel confusion or sadness, or they have mixed feelings and thoughts about themselves or about their personal or cultural background. **External conflict** could be with others (friends, family, coworkers, or employers) or with a difficult situation, such as coping with failure, living effectively with success, having difficulty achieving an important goal financially or socially, or struggling to navigate their cultural or ethnic identities. External conflict may lead to internal feelings of incongruity. And internal conflict can lead to external conflict with others—for example, the inability to make a decision may also result in conflict with a significant other who needs to know what is going to happen. The terms most commonly used to describe these concepts in counseling are *conflict, discrepancy, incongruity,* and *mixed message;* these terms are used interchangeably in this chapter.

Effective confrontation of conflict and incongruity leads clients to creative new ways of thinking and increased intentionality. Confrontation is *not* a direct, harsh challenge but is a more gentle skill that involves listening to clients carefully and respectfully and helping them examine themselves or their situations more fully. Confrontation is not "going against" the client; it is "going with" the client, seeking clarification and the possibility of a new resolution of difficulties. Think of confrontation as a supportive challenge.

Confrontation is the word that helping professions have chosen for challenging clients to think in new ways, to examine themselves more fully, and to consider themselves and their relations with others more carefully. You may prefer to call it "a listening challenge to creativity." Confrontation is not a skill that you use frequently, but it is a powerful tool when needed. Clients often need challenges to increase their motivation to change, and a supportive confrontation helps them reach core issues of their problem or conflict more quickly and with greater precision.

Your listening skills are vital in helping the client identify conflicts, incongruities, and discrepancies. Once you have established sufficient rapport, developed a working relationship, and heard the client's story, you will identify instances of internal and external conflict. Questions, coupled with paraphrases, reflections of feeling, and summaries, will help clarify the conflict issues, but eventually you will want to provide a supportive challenge/confrontation to reach the next step of personal growth. Positive growth and development occur with the resolution of conflict and incongruity. Through this process of transformation and change, clients learn new ways to manage their lives. Use confrontation to promote creativity.

Review Case Study: Confronting a Difficult Client in Romania
Complete Weblink Critique
Complete Weblink Activity: Conflict Management and Constructive Confrontation
Complete Weblink Activity: Ohio Commission on Dispute Resolution

8.2

INSTRUCTIONAL READING: The Specific Skills of Confrontation

KEY CONCEPT QUESTIONS

▲ **How does the skill of confrontation relate to all other skills?**

▲ **What are the specific skills of confrontation?**

▲ **What are some multicultural issues related to confrontation?**

▲ **What is the Client Change Scale, and how can we evaluate client change in the here and now in the interview or over a series of sessions?**

CONFRONTATION AS A DIMENSION OF OTHER SKILLS

Confrontation is different in a major way from all the skills previously discussed, and from the influencing strategies as well. Confrontation does not stand alone as a skill but rather appears as a dimension of other skills. For example, imagine that a client talks to you about possibly leaving a job because of interpersonal conflict but also speaks of real anxiety about what this action might mean for the future. After hearing the full story, you might provide a brief summary confrontation as follows:

> Shavon, it sounds as if you really have mixed feelings about this choice. On one hand, I hear some anger with your boss and a real desire to get out and move on. But, on the other hand, I sense that you feel a bit confused and anxious about where you'd go next. Have I got that right?

In this summary, the interviewer confronts Shavon with the essence of her internal conflict and implicitly challenges her to think through what she really wants. A more forceful challenge would be "We've gone through this for some time. What do you, Shavon, really want?" If Shavon understands the challenge, she will generate her own ideas for possible resolution. The words "on one hand . . . but on the other hand . . ." are a classic way to respond to client conflict and incongruity. Amplifying the alternatives with hand and arm movements helps illustrate conflict even more clearly.

Thus, one of the most powerful influencing skills is based on careful listening. Paraphrasing is particularly useful when the conflict or incongruity involves a decision and the pluses and minuses of the decision need to be outlined. Reflection of feeling is important for emotional issues, particularly when clients have mixed feelings ("on one hand, you feel . . . , but on the other, you also feel . . ."). A summary is a good choice for bringing together many strands of thoughts, feelings, and behaviors.

▲ EXERCISE 8.1   **Identify Internal and External Conflict**

This exercise is designed to reinforce the importance of the distinctions between internal and external conflict and how the two relate to each other. The ability to identify clearly contradictions and incongruity is a useful skill to help clients eventually resolve issues. We need to understand the problem before we can resolve it. Please see page 175 for our ideas on these issues.

Identify specific areas of external conflict that are distressing Shavon. Also imagine other areas of her life that might result in external conflict. For example, perhaps she has financial challenges, difficulty with friends, and parents that are calling her frequently. Where is she stuck and lacking intentionality?

Identify areas of internal conflict that affect her emotionally with each of the different challenging problems.

How does the external become internal? How does what happens in our outside world result in emotional contradiction and decisional confusion inside our minds?

Now that you have imagined an even more complex situation for Shavon as she faces multiple issues, how might we "unwind" all of this?

CONFRONTATION, STEP 1: IDENTIFY CONFLICT BY OBSERVING MIXED MESSAGES, DISCREPANCIES, AND INCONGRUITY

This session, while abbreviated and edited, illustrates the fundamentals of confrontation—supporting while challenging.

Interviewer and Client Conversation	Process Comments
Maya: [Her body language shows excitement.] I found this wonderful friend on the Internet. We're emailing at least four times a day. It feels great. I think I'd like to meet him. But it means I may have to go out of town. [Her body language becomes more hesitant and she breaks eye contact.] I wonder what my partner would think if he found out. I'm a little bit anxious, but I really want to meet this guy.	The client demonstrates conflict or incongruity between desire and excitement about meeting the Internet friend and internal anxiety and hesitation. There is the inevitable external conflict of being involved with two people at once. Which discrepancy might you discuss first? Clients who discuss mixed feelings and conflicts usually show them nonverbally as well, through vocal tone and/or body language.

Interviewer and Client Conversation	Process Comments
Counselor: You really want to meet him, but you're a little bit anxious.	This paraphrase and reflection of feeling confront the mixed feelings in the client. This catches both verbal and nonverbal observations. Note that a confrontation always occurs as part of other microskills.
Maya: Yes, but what would happen if my partner found out? It scares me. I've got so much involved with him over the past two years. But, wow, this guy on the Internet . . .	The client responds and turns her focus to the discrepancy with her partner. There are at least two issues in this situation.

Your ability to observe verbal and nonverbal incongruities and mixed messages is fundamental to effective confrontation. It will be helpful if you review the brief exchange above and once again identify the internal and external conflicts that Maya faces.

EXAMPLES OF INTERNAL CONFLICTS

Internal conflicts can be identified in clients through mixed messages in nonverbal behavior, incongruities between two verbal statements, and discrepancies between what the person says and what he or she does. A vital part of counseling is helping clients sort out mixed and confused feelings, thoughts, and behaviors. The client's thoughts and feelings about her present partner versus the excitement of the new Internet friend clearly represents an internal conflict. The client has discrepancies within herself that need to be resolved.

EXAMPLES OF EXTERNAL CONFLICTS

Discrepancies between the client and the external world include conflicts with other individuals, ranging from friends and family to people at work or contacts in the community. Another type of external conflict is between the client and challenging situations, such as college choice, a new job, a major purchase, the decision to adopt a child, dating somebody from a different culture, or dealing with sexism or racism. Much of your counseling and interviewing work will focus on discrepancies that clients have with their external world.

For example, Maya faces what for her is a major conflict—the external attraction of the new person and the potential external conflict with her present partner. Inside, this results in mixed and torn feelings. Drawing out these emotions by using the basic listening sequence will be important in confronting the conflict. In such situations, you can see that your support while challenging is very important. The client has relationship issues, and your relationship is important to her working things through.

Discrepancies between you and the client can be challenging. Many counselor/ client relationships will have discrepancies, whether it is the counselor not fully agreeing with a client's choice or differing values or life experiences. Counseling is for the client, not you. This is a time to maintain professional behavior and accept and understand where the client is "coming from." If you sense difference or feel a conflict between you and your client, support the client by listening. If you listen carefully, most discrepancies between you and your clients will disappear as you understand how they came to think and behave as they do. Note your own or the client's discomfort with differences between you, question yourself silently, draw out the client's perceptions, and work to understand them. Keeping one's thoughts and feelings to oneself is part of being nonjudgmental. However, there are times when it is wise to use the influencing skills of feedback or self-disclosure with clients (Chapter 11). This provides a new perspective that may be useful.

As an example, Maya may be making an unwise decision that you anticipate will lead to even more difficulty in her life. Summarizing the client's point of view and then sharing your alternative thoughts via feedback or self-disclosure may help prevent problems. Furthermore, Maya may fail to see her own contributions to the problems with her present partner by blaming him. On the other hand, it is possible that Maya will take the blame herself and feel guilty while failing to see the Internet friend is the real intruder.

The distinction between internal and external conflict needs constant attention. All too often external issues result in internal conflict and anxiety for the client. For example, a woman may think that her partner's abuse is her fault and may stay in a dysfunctional relationship while blaming herself for the situation. Discrimination in the form of racism, ableism, ageism, or other forms of oppression can result in depression and "learned helplessness." Narrative therapy terms this *externalization of the problem*. When clients blame themselves for the difficult challenges they face, help them discover that it is the outside world that is the problem, not them! In these situations, we may need to take an action stance and inform clients of what is really happening. These clients often benefit from the new perspective you provide through feedback and other influencing skills.

▲ EXERCISE 8.2   **Confrontation of Abuse**

Let's assume that one of Maya's issues is her partner's abuse. If so, then working through the abusive relationship needs to take precedence over the Internet possibility. Often, a woman who has been abused may leave one relationship but soon repeat the problem by selecting another man who abuses her again. How might you confront the abuse issue and help her build awareness of what is really happening? You will likely encounter this issue in your practice. *Is the problem in the client, or in the situation?* The external issue obviously affects the internal world of the client. Please see "Comments on Exercise 8.2" on page 176 for our thoughts on this challenging issue.

CONFRONTATION, STEP 2: POINT OUT ISSUES OF INCONGRUITY AND WORK TO RESOLVE THEM

Interviewer and Client Conversation	*Process Comments*
Counselor: Could I review where we've been so far? I know you have been having some difficulties with your partner that you've detailed over the last two sessions. I also hear that you want to work things out even though you're angry with him. You have a lot of positive history together that you hate to give up. On the other hand, you've found this man on the Internet that you're excited about, and he doesn't live that far away. In the middle of all this, I sense you feel pretty conflicted. Have I got the issues right?	This summary indicates the counselor has been listening. The counselor communicates respect and a nonjudgmental attitude, both verbally and nonverbally. The counselor summarizes the major discrepancies that lead to internal and external conflict and checks out with the client to see if the listening has been accurate.
Maya: Yes, I think you've got it. I hear what you're saying; I think I've got to work a little harder on the relationship with my partner, but—wow—I sure would like to meet that guy.	Resolution of conflict and discrepancy occurs best after the situation is fully understood. Through having thoughts and feelings said back, the client starts some movement.

After 10 minutes, Maya's thoughts and feelings evolve to a new perspective.

Maya: [said with conviction] I've got so much time invested in my partner; I've really got to try harder. [The nonverbals again show hesitancy.] How am I going to work this out with my Internet friend?	Even though the conflict is moving and the client is starting to show evidence of new ways of thinking that weren't there in the first two sessions, conflict remains. But the creative process of change has started.
Counselor: [solid supportive body language and vocal tone] It looks like you really want to work it out with your partner. You sound and look very sure of yourself. Let's consider what the possibilities are with your Internet friend.	You can confront and help clients face discrepancies, incongruity, and conflict if you are able to listen and be fully supportive.

Labeling the incongruity and saying it back through nonjudgmental confrontation may be enough to resolve a situation. Focus on the elements of incongruity rather than on the person. Confrontation is too often thought of as blaming a person for his or her faults; rather, the issue is intentionally facing the incongruity and helping the client think it through. The following summarizes key dimensions of effective confrontation—supporting while challenging:

▲ *Clearly identify the incongruity or conflict in the story or comment.* Using reflective listening, summarize it for the client. The simple question "How do you put these two together?" may lead a client to self-confrontation and resolution.
▲ *Draw out the specifics of the conflict or mixed messages, using questions and other listening skills.* If necessary, share your observations. Aim for facts; avoid being

judgmental or evaluative. Address each part of the mixed message, contradiction, or conflict one at a time. If two people are involved, attempt to have the client examine both points of view.

▲ *Periodically summarize each dimension of the incongruity.* Variations of basic confrontation include "On the one hand . . . but on the other hand . . .," "You say . . . but you do . . .," "I see . . . at one time, and at another time I see . . .," and "Your words say . . . but your actions say. . . ." Follow with a checkout: "How does that sound to you?" When incongruities are pointed out, the client is confronted with facts.

Many clients are unaware of their mixed messages and discrepancies; point these out gently, but firmly.

INDIVIDUAL AND MULTICULTURAL CAUTIONS

A wide variety of attending and influencing skills may be used to follow up and elaborate on confrontations. Another very different possibility for confrontation is to listen in silence for a short time while the client struggles with internal or external contradiction. This type of confrontation is especially challenging to clients who may ask your opinion—"Don't you agree with me that my partner is wrong?" Most often such questions are answered by throwing the question back to the client (e.g., "Tell me more"). If you say nothing, clients will have to encounter your silence and may more readily find their own answers. But be aware that some could interpret your silence as disapproval.

You will find that many clients are not comfortable with confrontational and challenging approaches that are provided too quickly. Your personal support can help clients build new behaviors, thoughts, and meanings. Identify strengths and wellness qualities in clients and their relationships. When clients are aware of their wellness strengths and positive qualities, they may face difficult confrontations more easily.

The antisocial or acting-out client represents a special type of "culture" that is often very challenging to the gentler atmosphere or culture of interviewing and counseling. When confronting difficult clients, it is important to "center" your body and mind and use a stronger, firmer confronting style. Continue to listen and support, but hold your position when you believe you are on the right track. The antisocial client does not respect apparent weakness in interviewers.

Within the wide diversity that is multiculturalism, expect a considerable amount of incongruity and conflict between the individual and the external world. Directly, but with as much support as possible, confront conscious and unconscious sexism, racism, heterosexism, and discrimination of all types. For example, a woman may be depressed, but is this depression "her problem" or is her sadness the result of sexual harassment on the job? A White male client who has not been promoted may be angry, believing that his being passed over for promotion may be the result of what he terms "reverse discrimination." Your ability to understand and identify the many conflicts and discrepancies related to multiculturalism is essential. It is also critical that you, the interviewer, work through these issues in your own mind. Seek supervision when necessary.

You will find that direct, aggressive confrontations are not necessary if the client contradiction is stated kindly and with a sense of warmth and caring. Direct, blunt confrontations are likely to be culturally inappropriate for traditional Asian, Latina/Latino, and Native American clients. But even here, if good rapport and understanding exist between you and the client, confrontations can be helpful. When issues of oppression need to be confronted, your support will be especially essential.

At the same time, you should also be aware that silence is a value within much of the Native American, Dene, or Inuit tradition. You may sit with these clients for several minutes as they sort out issues. How comfortable are you with silence?

Confrontation of discrepancies can be highly challenging to any client, but particularly to one who is culturally different from you. The confrontation process may be made acceptable if the helper takes time to establish a solid relationship of trust and rapport before engaging in confrontation. If you have a fragile client or if the relationship is not solid, confrontation skills need to be used with sensitivity, ethics, and care. Each mode of confrontation must be personally authentic and meaningful for the client or it is likely to fail.

CONFRONTATION, STEP 3: EVALUATING CHANGE

The effectiveness of a confrontation is measured by how the client responds. If you observe closely in the here and now of the session, you can rate how effective your interventions have been. You will see the client change (or not change) language and behavior in the interview. When you see immediate or gradual changes, you know that your intervention has been successful. When you don't see the change you anticipate or think is needed, it is time for intentionality, flexing, and having another response, skill, or strategy available.

THE CLIENT CHANGE SCALE (CCS)*

Figure 8.1 provides a visual summary of the **Client Change Scale (CCS).** If you use the CCS as described below, you will be able to assess the outcome of your intervention. You can use that information to plan further interventions and even longer-term treatment plans. This is how you can evaluate creative change.

Client Change Scale (CCS)	*Predicted Result*
The CCS helps you evaluate where the client is in the change process. Level 1. Denial. Level 2. Partial examination. Level 3. Acceptance and recognition, but no change. Level 4. Generation of a new solution. Level 5. Transcendence.	You will be able to determine the impact of your use of skills and the creation of the *New.* Suggest new ways that you might try to clarify and support the change process though more confrontation or the use of another skill that might facilitate growth and development.

Although the progression from denial through acceptance through significant change can be linear and step by step, this is not always the case. Think of a client working through the expected death of a loved one. One possibility is for the client to move through the CCS stages one at a time. But often a client who seems to be moving forward will suddenly drop back a level or two. At one interview the client may seem acceptant of what is to come, but in the next session move back to partial examination or even denial. Then we might see a temporary jump to transcendence, followed by a return to acceptance.

---

*A paper-and-pencil measure of the Client Change Scale was developed by Heesacker and Pritchard and was later replicated by Rigazio-Digilio (cited in Ivey, Ivey, Myers, & Sweeney, 2005). A factor analytic study of more than 500 students and a second study of 1,200 revealed that the five CCS levels are identifiable and measurable.

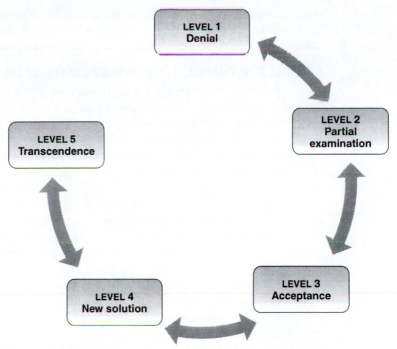

FIGURE 8.1 The Client Change Scale (CCS): The five stages of creative change may occur in step-by-step order. However, as the arrows indicate, there can be movement back and forth between the stages. In fact, the entire process can go back to the beginning as clients discover new thoughts, feelings, behaviors, and meanings.

The Client Change Scale is also useful as a broad measure of success in several types of interviews. The CCS provides a useful framework for accountability and measuring client growth. The clearest example is that of clients with substance abuse issues. Such clients may come voluntarily or be referred by the court for cocaine or alcohol abuse (often both). They may also be depressed and use the drugs to alleviate pain. Often these clients start by denying that they really have a problem (we call this Level 1 on the Client Change Scale). If we are successful in challenging, supporting, and confronting, we will see the client move to Level 2, admitting that there may be a problem. Some call this a "bargaining" stage where the client moves back and forth between denial and recognizing that something needs to be done.

Acceptance and recognition of the issue occur at Level 3, where the alcoholic admits that he or she is indeed an alcoholic and the cocaine abuser acknowledges addiction. But acceptance of the problem and recognition are not resolution. The client may be less depressed but still continue to drink and use drugs.

Change occurs when the client reaches Level 4 and actually stops the substance use, and then the depression usually lifts. While substance abuse and depression are often co-occurring mental health issues, there is some risk that depression will remain and continued treatment will be important and necessary. Even so, this is real success in counseling and therapy and is not easy to achieve, but clearly it is measurable. Level 5, transcendence and the development of new ways of being and thinking, may occur. Not all clients will achieve this level; it is represented by the user who becomes fully active in support groups and helps others move away from addiction, and continues to work on feelings, behaviors, actions, and relationships that led to the alcohol abuse and addiction. This person achieves changes in life's meaning and a much more positive view of self and the world—far more than just "getting by."

A MODEL FOR CHANGE IN THE INTERVIEW ITSELF OR OVER THE LONG TERM

Let's move the CCS to a more general level as it might appear in a counseling interview with virtually any topic. The client tells us a story and we, of course, listen. If the client is in the denial stage, the story may be distorted, others blamed unfairly, and the client's part in the story denied. In effect, the client in *denial* (Level 1) does not deal with reality. When the client is confronted effectively, the story becomes a discussion of inconsistencies and incongruity and we see Level 2 *bargaining and partial acceptance*—the story is changing. At *acceptance* (Level 3), the reality of the story is acknowledged and storytelling is more accurate and complete. Moreover, it is possible to move to *new solutions* and *transcendence* (Levels 4 and 5). When changes in thoughts, feelings, and behaviors are integrated into a new story, we see the client move into major new ways of thinking accompanied by action after the session is completed.

Virtually any problem a client presents may be assessed at one of the five levels. If your client starts with you at *denial* or *partial acceptance* (Level 1 or 2) and then moves with your help to *acceptance* and *generating new solutions* (Level 3 or 4), you have clear evidence of the effectiveness of your interviewing process. The five levels may be seen as a general way to view the change process in interviewing, counseling, and therapy. Each confrontation or other interview intervention in the here and now may lead to identifiable changes in client awareness.

Small changes in the interview will result in larger client change over a session or series of sessions. Not only can you measure these changes over time, but you can also contract with the client in a partnership that seeks to resolve conflict, integrate discrepancies, and work through issues and problems. Specifying concrete goals often helps the client deal more effectively with confrontation.

The CCS provides you with a systematic way to evaluate the effectiveness of each intervention and to track how clients change in the here and now of the interview. If you practice assessing client responses with the CCS model, eventually you will be able to make decisions automatically "on the spot" as you see how the client is responding to you. For example, if the client appears to be in denial of an issue despite your confrontation, you can intentionally shift to another microskill or approach that may be more successful.

Following is a narrative of the Client Change Scale with examples of each of the five levels as the client responds to issues around a divorce. Any time clients are working through change, they talk about their issues with varying levels of awareness. The client may be in denial one moment, the next minute talk as if he or she accepts the problem, and then return to bargaining to avoid change.

CCS Stages	Client Example
Level 1. Denial. The individual denies or fails to hear that an incongruity or mixed message exists.	"I'm not angry about the divorce. These things happen. I do feel sad and hurt, but definitely not angry."
Level 2. Partial examination. The client may work on a part of the discrepancy but fail to consider the other dimensions of the mixed message.	"Yes, I hurt and perhaps I should be angry, but I can't really feel it."

CCS Stages	Client Example
Level 3. Acceptance and recognition, but no change. The client may engage the confrontation but makes no resolution. Until the client can examine incongruity, stuckness, and mixed messages accurately, real change in thoughts, feelings, and behavior is difficult. Coming to terms with anger or some other denied emotion is an important breakthrough and often is a sufficient solution for the client.	"I guess I do have mixed feelings about it. I certainly do feel hurt about the marriage. I hurt, but now I realize how really angry I am."
Level 4. Generation of a new solution. The client moves beyond recognition of the incongruity or conflict and puts things together in a new and productive way. Needless to say, this usually does not happen immediately. It can take several interviews or sessions over a period of weeks and sometimes months.	"Yes, I've been avoiding my anger, and I think it's getting in my way. If I'm going to move on, I will accept the anger and deal with it."
Level 5. Transcendence—development of new, larger, and more inclusive constructs, patterns, or behaviors. *Many* clients will never reach this stage. A confrontation is most successful when the client recognizes the discrepancy and generates new thought patterns or behaviors to cope with and resolve the incongruity.	"You helped me see that mixed feelings and thoughts are part of every relationship. I've been expecting too much. If I expressed both my hurt and anger more effectively, perhaps I wouldn't be facing a divorce."

When you confront clients, ask them a key question, or provide any intervention, they may have a variety of responses. Ideally, they will actively generate new ideas and move forward, but much more likely they will move back and forth with varying levels of response. The idea is to note how clients respond (at what level they answer) and then intentionally provide another lead or comment that may help them grow. *Clients do not work through the five levels in a linear, straightforward pattern; they will jump from place to place and often change topic on you.* It is your task to help them move forward to significant change.

Depending on the issue, change may be slow. For some clients, movement to partial acceptance (Level 2) or partial acceptance but no change (Level 3) is a real triumph. There are a variety of issues where acceptance represents highly successful counseling and therapy. For example, the client may be in a situation where change is impossible or really difficult. Thus, accepting the situation "as it is" is a good result. A client facing death is perhaps the best example and does not always have to reach new solutions and transcendence. Simply accepting the situation as it is may be enough. There are some things that cannot be changed and need to be lived with. "Easy does it." "Life is not fair." "There is a need to accept the inevitable." For someone whose partner or parent is an alcoholic, it is a major step to realize that the situation cannot be changed; acceptance is the major breakthrough that will lead later to new solutions. The newly found solutions would facilitate the mediation process and help resolve the conflict (see Box 8.1).

BOX 8.1   Conflict Resolution and Mediation

Confrontation skills are important in the mediation process. In conflict resolution and mediation—whether between children, adolescents, or adults—the following steps are useful.

Develop a relationship with all participants and outline the structure of your session. Pay equal and neutral attention to each person. As long ago as 1992, Lane and McWhirter suggested four useful rules for children: (1) agree to solve the problem; (2) no name-calling or put-downs; (3) be honest; (4) listen—do not interrupt. You can ask adults to agree to the same rules to obtain commitment to the process of mediation. Of course, modify the wording of rules to match the age of the clients.

The King Center (2010) summarized six classic steps used by Dr. Martin Luther King for nonviolent change that are closely related to the mediation model above: (1) information gathering, (2) education, (3) personal commitment, (4) negotiations, (5) direct action, and (6) reconciliation. When you work on complex issues of institutional or community change, a review of Dr. King's everlasting model may be helpful in thinking through your approach to major challenges.

Draw out the conflicting stories nonjudgmentally. Use the basic listening sequence to clearly and *concretely* draw out the point of view of each person involved in the dispute. To avoid emotional outbursts, acknowledgment of feeling rather than reflection of feeling is recommended. Clearly summarize each person's frame of reference and carefully check out your accuracy with each one. You may ask each disputant to state the opponent's point of view. Outline and summarize the points of agreement and

disagreement, perhaps in written form if the conflict is complex.

Set goals. Use the basic listening sequence to draw out each person's wants and desires for satisfactory problem solution; focus primarily on concrete results rather than emotions and abstract or intangible ideas. This is the beginning of the negotiation process when problems and concerns may be redefined and clarified. Summarize the goals for each person, with attention to possible joint goals and points of agreement.

Create new stories and solutions. Negotiation begins in earnest; rely on your listening skills to see whether the parties can create their own satisfactory solutions. When a level of concreteness and clarity has been achieved (Steps 2 and 3), the parties involved may come close to agreement. If the parties are very conflicted, meet with each one separately as you brainstorm alternative solutions. With touchy issues, summarize them in writing. Many of the influencing skills discussed later in this section will be useful in the process of negotiation.

Contract for action and generalization to the "real world." Use the basic listening sequence; summarize the agreed-upon solution (or parts of the solution if negotiations are still in progress). Make the solution as concrete as possible, and write down touchy main issues to make sure each party understands the agreement. Obtain agreement about subsequent steps. With children, congratulate them on their hard work and ask each child to tell a friend about the resolution. Contract for follow-up with everyone.

Complete Interactive Exercise 1: Practice With the Client Change Scale
Complete Interactive Exercise 2: Practicing Confrontation
Complete Weblink Activity: Communicating Well in Conflict
Complete Weblink Activity: Mediation and Dispute Resolution Resources
Complete Weblink Activity: Elisabeth Kubler-Ross
Complete Weblink Activity: Coping With Loss—Loss in Late Life

8.3
EXAMPLE INTERVIEW: Balancing Family Responsibilities

KEY CONCEPT QUESTION

▲ How is confrontation integrated into the interview?

The integration of the three main points reviewed in this chapter is demonstrated in the example interview: (1) Listening skills are used to obtain client data. (2) Confrontations of client discrepancies are noted. (3) The effectiveness of the confrontations is considered using the Client Change Scale (CCS). You may want to study the segment carefully to fully understand how confrontation is used in the example. With more experience and practice, these concepts will be useful and important in your interviewing practice.

The following interview presents a conflict that is common to many working couples—balancing home tasks. Male attitudes and behaviors are changing, but many working women are still burdened with the responsibility for most home tasks. In this example we have both internal and external incongruity. Dominic is struggling internally with the discovery that things have changed and externally with the realization that his wife is behaving differently. Arguments may be particularly intense as two tired people come home from a hard day's work to face needy children and undone housework. Couples may blame each other rather than attributing a major portion of their issues to external causes at work. An exhausted partner or spouse often has little energy left to deal with the concrete issues at home.

Interviewer and Client Conversation	Process Comments
1. **Dominic:** I'm having a terrible time with my wife right now. She's working for the first time and we're having lots of arguments. She isn't fixing meals like she used to or watching the kids. I don't know what to do.	On the CCS, this client response is rated Level 2; he is partially aware of the problem. At the same time, he denies (Level 1) his role in the problem and is attributing the difficulties in the home to husband/wife issues, failing to see how demands at work play into the system.
2. **Ryan:** [holds out one hand to the right] So, on one hand, your wife is working outside the home, but [holds out the left hand] on the other hand you expect her to continue with all the housework, too. You don't like what's going on right now.	Confrontation presented as paraphrase and reflection of feeling. The use of hands with the words helps strengthen and clarify the conflict. Here the counselor shows imbalance or conflict concretely.
3. **Dominic:** You damn betcha she's expected to do what she's always done— I'm not confused about that.	CCS Level 1—denial.
4. **Ryan:** I see. You're not confused; you really don't like what's going on. Could you give me a specific example of what's happening—something that goes on between the two of you when you both get home?	Paraphrase, open question oriented to concreteness.

Interviewer and Client Conversation	Process Comments
5. **Dominic:** [sighs, pauses] Yeah, that's right, I don't like what's going on. Like last night, Sara was so tired that she didn't get around to fixing dinner till half an hour late. I was hungry and tired myself. We had a big argument. This type of thing has been going on for 3 weeks.	While Dominic is able to identify and talk about the conflict (CCS Level 2), he lacks awareness of how his wife feels. He seems insensitive to the fact that his wife is working and he expects her to do everything she did in the past for him (CCS Level 1).
6. **Ryan:** I hear you, Dominic; you're pretty angry. Let's change focus for a minute. When dealing with conflict, it helps to concentrate on positive things. Could we search for some positive things that have worked for you and Sara in the past?	This summary introduces an incongruity between the difficult present situation and the good things of the past through the positive asset search.
7. **Dominic:** Yes, I am angry and discouraged. Sara and I were doing pretty well until the baby came. Somehow things just got off kilter. [5-second pause and silence] . . . Well, let me try it your way. We sure had fun times together over the 3 years we've been together. We both like outdoor activity and doing things together. We never seem to have time for that now.	In couples work it is useful to remind them of positive stories from the past. This has been edited for brevity. It is important to explore positives in the relationship in much more depth than presented here.
8. **Ryan:** It's important to remember that you have a good history and have enjoyed each other. I'm wondering if part of the solution isn't finding time just to be together doing fun things. Let's make that part of our discussion later. But for the moment, let's go back to your main issues. Could you give me a specific example of what goes on between you when she gets home?	Paraphrase, suggestion, open question oriented to concreteness. Getting specifics helps clarify the situation. When counseling around couples issues, search for strengths. What brought the couple together and what maintains them now? This helps clients center themselves in positives as they struggle with the difficult negatives in a relationship. Helping couples recapture the good things in their relationship is important. We often recommend a weekly date night for couples.
9. **Dominic:** . . . [pause] Well, lately, I've had a lot of pressure at work. They're downsizing and morale is bad. I worry I'll be next. I try really hard, but when I get home, I just want to sit. Sara's got the same thing. Her new boss just wants more all the time. I guess when she comes home, she's about as exhausted and confused as I am.	It is very common for partners to get angry with each other when external stressors hit one or both individuals in the relationship. One can't argue with the boss or colleagues easily, so the partner is the scapegoat.
10. **Ryan:** I see. You both work all day and come home exhausted. Dominic, do you really think Sara can work that hard and still take care of you like she used to?	Paraphrase, closed question. Ryan's implicit confrontation is now more concrete.

Interviewer and Client Conversation	Process Comments
11. **Dominic:** I guess I hadn't thought of it that way before. If Sara is working, she isn't going to be physically able to do what she did. But where does that leave me?	For the first time, Dominic is able to see that Sara can't continue as she has in the past (CCS Level 3). Note that he is still thinking primarily of himself. To move to higher levels on the CCS, Dominic would have to be able to take Sara's perspective and articulate what she likely thinks and feels when she gets home.
12. **Ryan:** Yes, where does that leave you? Dominic, let me tell you about my experience. My wife started working and I, too, expected her to continue to do the housework, take care of the kids, do the shopping, and fix the meals. She went to work because we needed the money to make a down payment on a small house. I decided to share some of the household work with her and also shop, and I pick the kids up at day care. How does that sound to you? Do you think you want to continue to expect Sara to do it all?	Self-disclosure followed by a checkout. Ryan is speaking up directly with his ideas, but he is also allowing the client to react to them. Clearly, Ryan is trying to get Dominic's thinking and behavior to move. His last statement contains an important implicit confrontation: "On the one hand, your wife is working and if she continues, she'll need some help; on the other hand, perhaps you don't want her to work—what is your reaction to this?" This leads to Dominic starting at least to examine his own behavior.
13. **Dominic:** Uhhh . . . we need the money. We've missed the last car payment. It was my idea that Sara go to work. But isn't housework "women's work"?	Dominic is now facing up to the contradiction he is posing (CCS Level 3). He has not yet synthesized his desire for his wife to work with the need for his sharing the workload at home, but at least he is moving toward a more open attitude. As he acknowledges his part in the situation and the need for more money, he is beginning to come to a new understanding, or synthesis, of the problem— he is less incongruent and is taking beginning steps toward resolving his discrepancies.

For each developmental task completed in the interviewing process, it often seems that a new problem arises. Just as the counselor is beginning to facilitate client movement, a new obstacle ("women's work") arises. To make the progress shown took half the session. Changing the concept from "women's work" to "work in the home that must be shared if the car payments are to be met" took the rest of the session. Achieving a new and likely more lasting level of male/female cultural differences would require a major transformation in thinking and behavior (CCS Level 5). Major change may require several interviews, group sessions, and time for Dominic, or any other client, to internalize new ideas.

IN-CLASS EXERCISES: RESPONSES AND DISCUSSION

COMMENTS ON EXERCISE 8.1

Areas of external conflict distressing Shavon obviously center on her relationship with her boss. Other areas where a young woman might experience external conflict include a bill collector at the door, difficulties with friends or partner, and parents needing financial help.

Any of these conflicts could tear people apart and bring them to counseling. With the boss, Shavon experiences *anger*; on the other hand, she knows that she has to *hold in the anger* and deal with it positively for the sake of her job. With the bill collector, she likely feels some *fear* and *anxiety* over what to do and possibly some *guilt* over getting herself into this situation. She *loves* and needs the support of her partner; on the other hand, the partner is demanding and needy, which *frustrates* and *disappoints* her. Shavon *deeply cares* for her parents; on the other hand, she is *frightened* by their needs.

How does the external become internal? External pressures likely result in Shavon's feeling very confused and anxious about her ability to cope with the multiple situations. When a client faces so many external challenges, this can result in feelings of helplessness and even depression as he or she gives up. All too often our clients are being buffeted by what is occurring around them. And in the middle of all, the car may break down and need repair!

Building intentionality and finding solutions requires "unwinding" the many problems. This requires you to listen to several stories early on to get a general understanding of the complexity of Shavon's life. Listening skills and a supportive empathic attitude are central. Part of this is dealing with the many feelings underlying these issues. We suggest that you take time for a brief inventory of Shavon's strengths and personal assets that will be useful in problem resolution. Develop the emotions associated with the positive strengths. Then prioritize issues and confront the conflicts one at a time. It may be possible to develop a tentative solution for one or two issues in that first session; others can wait.

COMMENTS ON EXERCISE 8.2

We would first listen to Maya's story of excitement over her Internet friend, with special attention to the emotions Maya expressed. As we listened, we likely would hear some of the issues troubling her with her present partner. At this point, the focus of storytelling would move to uncovering what is happening currently and the nature of abuse. The basic confrontation from this is clear, but needs to be worked through carefully. "On one hand, you still have positive feelings for your partner, but on the other hand, he consistently treats you unfairly and sometimes the situation gets dangerous." Continuing, "On one hand, you realize that staying with him is likely to lead to more fear and possibly serious hurt for you, but on the other hand, where do you go?" Here we are actively rewriting the story that Maya is living with her present partner.

If we sense danger in the situation, we take an active role, confront the situation directly, and help Maya move to a new living situation. Part of this will be the positive asset search, drawing out strengths and capabilities. These can come from stories about past successes and your own feedback and observations about her strengths.

But Maya has emotional needs for a relationship. What about the Internet possibility? Here we look at the story and its possible endings. We also may need to explore Maya's family history and her past relationships with men. If she seems to have a pattern of earlier abusive relationships, this needs to be confronted. Once Maya has a better understanding of her past and present, then is the time to examine the wisdom of the Internet relationship.

"If only it were so simple!" The above is ideal, and very possibly Maya will be taking up with the Internet relationship before she fully understands what is going on. Counseling and therapy take time, and sometimes we see our clients move to repeat negative patterns. There is little we can do about it but hope that Maya will return and we can pick up the pieces and then move forward to a healthier and more positive self.

Complete Interactive Exercise 3: Identification and Classification
Complete Interactive Exercise 4: Identifying Interviewer Responses
Complete Interactive Exercise 5: Writing Confrontation Statements: Balancing Family
Responsibilities
Complete Group Practice Exercise: Group Practice With Confrontation

CHAPTER SUMMARY

| Key Points of "The Skill of Confrontation" | CourseMate and DVD Activities to Build Interviewing Competence |

8.1 Defining Confrontation

▲ Confrontation is based on effective listening and observation of client conflict, discrepancies, incongruity, and mixed feelings and thoughts.

▲ Effective confrontation leads to increased client intentionality.

▲ Confrontation is a supportive challenge to client incongruity that helps bring a clearer understanding of core issues.

1. Case Study: Confronting a Difficult Client in Romania. A good invitation to explore the skill of confrontation in a multicultural context.

2. Weblink Critique: Several links, including multicultural and grief sites, are offered to study confrontation.

3. Weblink: Conflict Management and Constructive Confrontation. Created by Gregg Walker, Department of Speech Communication, Oregon State University.

4. Weblink: Ohio Commission on Dispute Resolution and Conflict Management Programs.

8.2 Instructional Reading: The Specific Skills of Confrontation

▲ When confrontation is used clearly and concisely, it encourages clients to look at their situation from a new perspective. Confrontation is an infrequently used skill in a single session, but most interviewers and counselors use it to help clients look at themselves.

▲ There are three steps of confrontation, each using different skills: (1) identify conflict; (2) point out issues of incongruity and work to resolve them; and (3) evaluate the change.

▲ The Client Change Scale can be applied to any client statement in the interview, or you can use these concepts to evaluate client change over several sessions. The five dimensions are (1) denial; (2) partial examination; (3) acceptance and recognition (but no change); (4) generation of a new solution; and (5) development of new, larger, and more inclusive constructs, patterns, or behaviors—transcendence.

▲ Be aware of individual and cultural differences, particularly if you are culturally different from the client.

1. Interactive Exercise 1: Practice With the Client Change Scale. Test your knowledge about the CCS by classifying different clients' statements.

2. Interactive Exercise 2: Practicing Confrontation. Gain mastery of confrontation skills by writing possible responses to different clients' statements.

3. Weblink: Communicating Well in Conflict: Competence Skills and Collaboration. Created by Gregg Walker, Department of Speech Communication, Oregon State University.

4. Weblink: Mediation and Dispute Resolution Resources.

5. Weblink: Elisabeth Kubler-Ross.

6. Weblink: Coping with Loss—Loss in Late Life.

Key Points of "The Skill of Confrontation"

CourseMate and DVD Activities to Build Interviewing Competence

8.3 Example Interview: Balancing Family Responsibilities

▲ Three main points are covered in the example interview: (1) Listening skills are used to obtain client data. (2) Confrontations of client discrepancies are noted. (3) The effectiveness of the confrontations can be assessed using the Client Change Scale.

▲ Effective confrontation helps clarify issues and move toward problem resolution. Basic listening skills and observation are required to identify and clarify client issues. However, you can see that the interviewer's sharing of experience or other information can be helpful at times.

▲ Major client change is an extended process; it may require several interviews, group sessions, and some time for the client to internalize and act on new ideas.

1. Interactive Exercise 3: Identification and Classification. Review and work through a transcript and then score your responses.

2. Interactive Exercise 4: Identifying Interviewer Responses. Which interviewer response is the most helpful in producing and maintaining change in thoughts and behavior?

3. Interactive Exercise 5: Writing Confrontation Statements: Balancing Family Responsibilities. Review the CCS and write clients' responses representing all anchor points of the scale.

4. Group Practice Exercise: Group Practice With Confrontation.

Assess your awareness, knowledge, and skills as you conclude the chapter:

1. Flashcards: Use the flashcards to check your understanding of key concepts and facilitate memorization of key information.

2. Self-Assessment Quiz: Take the chapter quiz again to measure how much you have learned. The quiz will help you assess your current knowledge and prepare for course examinations.

3. Portfolio of Competencies: Evaluate your present level of competence in attending and listening skills using the downloadable Self-Evaluation Checklist. Self-assessment of your attending skills competence demonstrates what you can do in the real world.

FOCUSING THE INTERVIEW
Exploring the Story From Multiple Perspectives

9
CHAPTER

FOCUSING

CONFRONTATION

THE FIVE-STAGE INTERVIEW STRUCTURE

REFLECTION OF FEELING

ENCOURAGING, PARAPHRASING, AND SUMMARIZING

OPEN AND CLOSED QUESTIONS

ATTENDING BEHAVIOR

ETHICS, MULTICULTURAL COMPETENCE, AND WELLNESS

One very important aspect of motivation is the willingness to stop and to look at things that no one else has bothered to look at. This simple process of focusing on things that are normally taken for granted is a powerful source of creativity.

—*Edward de Bono*

How can the skill of focusing help you and your clients?

CHAPTER GOALS   Focusing is a skill that enables multiple renditions of the story and will help you and your clients think of new possibilities for restorying. Often issues are considered from only one frame of reference. Client issues are often more complex than they seem; focusing can help in seeing an issue more completely, thus ensuring more satisfying decisions and actions. For cultural competence, focusing is a crucial skill that allows and encourages us to think about broader systems and cultural issues that might affect the client.

Awareness, knowledge, and skills developed through the concepts of this chapter will enable you to

▲ Help clients tell their stories and describe their issues from multiple frames of reference, a valuable source of creative change.

▲ Increase clients' cognitive and emotional complexity, thus expanding their possibilities for restorying and resolving issues.

▲ Help clients stay on track when they need to examine themselves or other interpersonal issues in a comprehensive fashion.

▲ Operate more effectively in our diverse society by enabling clients to see themselves as selves-in-relation and persons-in-community through community and family genograms.

▲ Include advocacy, community awareness, and social change as part of your interviewing practice.

Assess your awareness, knowledge, and skills as you begin the chapter:

1. Self-Assessment Quiz: The chapter quiz will help you determine your current level of knowledge. You can take it before and after reading the chapter.

2. Portfolio of Competencies: Before you read the chapter, fill out the downloadable Self-Evaluation Checklist to assess your existing knowledge and competence in attending skills. Then, at the end of the chapter, complete the checklist again to summarize your competencies after study and practice.

9.1
DEFINING FOCUSING

KEY CONCEPT QUESTIONS

▲ **How do you focus an interview and ensure that critical information is brought up?**

▲ **What are the central dimensions of focusing?**

In one sense, focusing is a simple skill; it reminds you to change emphasis in the interview to obtain a broad array of information that may be helpful in understanding the client. Intentionality comes in when the client does not want to explore a particular area. If you use focusing as suggested below, you can move to a more comfortable focus, build further trust, and perhaps then return to the more challenging issues.

Focusing	*Predicted Result*
Use selective attention and focus the interview on the client, problem/concern, significant others (partner/spouse, family, friends), a mutual "we" focus, the interviewer, or the cultural/environmental context. You may also focus on what is going on in the here and now of the interview.	Clients will focus their conversation or story on the dimensions selected by the interviewer. As the interviewer brings in new focuses, the story is elaborated from multiple perspectives.

Focusing is an important skill that enables clients to gain a better perspective and understanding of issues affecting their lives. Often clients face multiple issues and they become stuck, unable to navigate through the many situations they face. It is the job of the counselor to help clients steer through these difficulties by helping them to focus. This is a skill that will make a difference in your practice and facilitate client cognitive/emotional development.

> Vanessa walks swiftly into the office and starts talking even before she sits down: "I'm really glad to see you. I need help. My sister and I just had an argument. She won't come home for the holidays and help me with Mom's illness. My last set of exams was a mess and I can't study. I just broke up with the guy I was going with for three years. And now I'm not even sure where I'm going to live next term. And my car wouldn't start this morning. . . ." [She continues with her list of issues and begins repeating stories almost randomly, but always with energy and considerable emotion.]

There are many clients like Vanessa, who have several issues in their lives. We, perhaps like you, often feel overwhelmed when a client spends five or more minutes telling problematic stories, jumping rapidly from topic to topic. Sometimes there is an insistence that we do "something" immediately and start solving the issues. When we fall into the trap of solving problems for clients, they often refuse to listen to us and generate more problems and difficulties. You have now added resistance to the general confusion.

So, what needs to be done here? Each client is unique, and there is no magic answer. But one rule really helps us settle down and start working with clients. *Counseling is for clients*. It is best when they use the interview to resolve their own issues. Vanessa is talking "all over the place." But we can help her focus her conversation—and usually that first focus is on the client, Vanessa, and her thoughts and feelings. We would likely gently interrupt and say something like:

> *Vanessa*, could we stop for a moment? I really hear *you* loud and clear. There are a lot of things happening right now. One of the best ways to approach these issues is to focus on how *you* are doing and feeling. Once I understand *you* a bit more, we can work on the issues that *you* describe.
>
> I get the sense that *you, Vanessa*, are hurting a lot right now and are confused about what to do next. Could *you* take a deep breath and tell me what *you* are feeling and thinking right now? What are these things doing to *Vanessa*? What's happening with *you*?

In these two paragraphs, we have used the words *Vanessa* and *you* more than ten times. The goal here is to help Vanessa focus on herself. Interviewing and counseling are first and foremost for the client who sits in front of you. Summarize the several issues, then ask her what she would like to start working on first. Explore other issues later. Once we have a better grasp of the person(s) before us, we can work more effectively toward problem resolution, but usually only one issue at a time.

Selective attention (Chapter 3) is basic to focusing—clients tend to talk about and focus on topics to which you give your primary attention. Through your attending skills (visuals, vocal tone, verbal following, and body language), you indicate to your client that you are listening. But we all tend to focus on or listen in different ways. It is important to be aware of both your conscious and unconscious patterns of selective attention; clients sometimes follow your lead rather than talk about what they really want to say. If you focus solely on individual issues, clients will talk about themselves and their frame of reference. If you focus on sexual concerns, they likely will talk about sex. Fail to include sex as a focus and the topic will soon drift away.

"Yo soy yo y mi circunstancia." The most direct translation of this quote from the Spanish philosopher José Ortega y Gasset is "I am I, and my circumstance." But direct translation into English does not always convey the full meaning. What it means to us is "I am me and my cultural/environmental/social context." Focusing on their context can help clients elaborate their story from multiple perspectives.

The individual exists in relation to others. One of the central goals of the focusing skill is to help ensure that we can see fuller client stories. The client before us became what he or she is as a result of many factors. Focusing enriches both your and your client's understanding.

With couples, families, or groups, multiple issues can occur and selective attention is important. You can be a much more effective counselor if you are able to help them focus together on one thing at a time. Often the function of group therapy is to reduce the clutter of voices and focus in on one person at a time, making sure his/her thoughts and feelings are being heard. While it can be overwhelming for the counselor, one suggestion is to have the person who is talking hold an object—a cup or a ball; when that person has finished talking, then she or he passes this object to the next person. The idea of holding something is similar to giving one person at a time the "podium" and, in essence, an audience attentively listening to what that individual has to say.

The client's problem, issue, or concern is the next most critical focus dimension. However, too many interviewers focus first on the problem and fail to understand the client and the client's perspective on issues. If a client has gone through a breakup of a significant relationship, has study difficulties, has cancer or another serious illness, we need to hear the concrete details, but in relation to how the client felt, thought, and behaved as well. Although a problem focus is essential, the client's uniqueness and background cannot be ignored.

▲ "Which issue would you like to focus on first?"
▲ "Sounds like your mother is the most important issue. Have I heard correctly?"

In addition, people live in a broader context of multiple systems. Increasingly, theorists are challenging the idea of a totally autonomous self separated from other people, friends, family, and the cultural context. We need to encourage understanding of how an individual's environment affects that person's life. Several important concepts have surfaced to help expand the idea of self. The term **self-in-relation** defines the "individual self" as really only existing in relationship to others. The term **being-in-relation,** suggested by the feminist authors Jordan, Hartling, and Walker (2004), is similar to self-in-relation but points out that all of us are beings who grow in the here-and-now moment as we interact with others.

Obonnaya (1994) stressed the **person-as-community** from an Afrocentric frame, pointing out that our family and community history live within each of us. The community genogram may be useful in helping clients gain new perspectives on themselves and their relationships to others. Clients come to the interview with many "internalized voices" affecting cognition and emotion. Bringing these to awareness through focusing is a vital but often missing part of counseling and therapy.

A **community genogram** is a visual map of the client in relation to the environment showing both stressors and assets within the person's life. You will find the community genogram a useful way to understand your client's history and a good place to identify strengths and resources.

Complete Interactive Exercise 1: Identify Focus Dimensions
Complete Weblink Critique

Clearly much individual counseling focuses on issues of conflict, incongruity, and discrepancies between the individual and family and friends. In addition, many client problems are caused by and related to issues and events in the broader context (e.g., poor schools, floods, job issues). If you help clients to see themselves and their issues as *persons-in-community*, they can learn new ways of thinking about themselves and use existing support systems more effectively. It is not the individual versus the social context; rather, social context enriches our understanding and enhances the uniqueness of each person and client. The following list offers sample comments and questions that allow the interviewer to focus the session in a specific area:

Significant Others (partner, spouse, friends, family)
▲ "Vanessa, tell me a bit more about your mother and how she is doing."
▲ "How are your friends helpful to you?"
▲ "What's going on with your sister?"
▲ "You also seem to be thinking of your breakup. Could you share a bit more about him and what he meant to you?"
▲ "Your grandmother was very helpful to you in the past. What would she say to you?"

Mutual Focus and Immediacy ("we" statements and talking about what is going on in the session *here and now*)
▲ "Vanessa, you have a lot on your plate, but *we* will work through your issues. Right now I can almost feel your hurt."
▲ "Vanessa, we've been working together for two weeks now. I sense at this moment that you felt angry at what I just said. I'm glad that you can openly express your feelings to me."

Interviewer Focus (sharing one's own experiences and reactions)
▲ "I felt really confused and worried when my mother had the same illness. I simply didn't know what to do. Is that close to the way you feel?"
▲ "I can understand your frustration with the car. It happened to me last week."

Cultural/Environmental Context (unique, personal, RESPECTFUL multicultural background [see Chapter 1] and broader issues such as the impact of the economy)
▲ "Finding a new place to stay is difficult and you think that landlords won't rent to culturally diverse renters."

▲ "What are some strengths that you gain from your spiritual orientation?"

▲ "You feel that the college is simply not supporting you at all. Could you tell me a bit more about what they are doing to make it difficult for you?"

As an interviewer, be aware of how you focus an interview and how you can broaden the session so that clients are aware of themselves more fully in relation to others and social systems. You can help them see themselves as persons-in-relation, persons-in-community. In a sense you are like an orchestra conductor, selecting which instruments (ideas, emotions, and behaviors) to focus on, thus enabling a better understanding of the whole. Some of us focus exclusively on the client and the problem, failing to recognize the impact and importance of broad contextual issues. Others may fail to give sufficient attention to the client as a person and use the interview as a chance to learn interesting details—almost as a voyeur.

 Complete Interactive Exercise 2: Community Genogram

▲ EXERCISE 9.1   **Develop a Community Genogram**

At this point, we suggest that you develop a community genogram as presented in Box 9.1. Develop a community genogram for yourself using your own style of presentation. This will help you, and later on your clients, to see how we are connected to many influences. We are persons-in-relation to others and the social context around us. Think about your own life story and how it has been affected by the many relationships in your community of origin. Usually, the family is a critical part of the community genogram. However, the **family genogram** can be a useful supplement to the community genogram and bring out additional details of family history (see Appendix II).

BOX 9.1   Developing a Community Genogram: Identifying Personal
and Multicultural Strengths

The community genogram is a "free-form" activity in which clients are encouraged to present their community of origin or present community, using their own unique style. Two visual examples of community genograms are presented here. Through the community genogram, we can better grasp the developmental history of our clients and identify client strengths for later problem solving. Clients may construct a genogram by themselves or be assisted by you through questioning and listening to the things that they include.

Step 1: Develop a Visual Representation of the Community

▲ Select the community in which you were primarily raised. The community of origin is where you tend to learn the most about culture, but any other community, past or present, may be used.

▲ Represent yourself or the client with a significant symbol. Use a large poster board or flipchart paper. Place yourself or the client in that community, either at the center or at another appropriate place. Encourage clients to be innovative and represent their communities in a format that appeals to them. Possibilities include maps, constructions, or star diagrams (used by Janet, below).

▲ It is important to place family or families, nuclear or extended, on the paper, represented by the symbol that is most relevant for you or the client.

▲ Place important, most influential groups on the community genogram, represented by

BOX 9.1   (Continued)

distinctive visual symbols. School, family, neighborhood, and spiritual groups are most often selected. For teens, the peer group is often particularly important. For adults, work groups and other special groups tend to become more central.

▲ You may wish to suggest relevant aspects of the RESPECTFUL model discussed in Chapter 1. In this way, diversity issues can be included in the genogram.

Step 2: Search for Images and Narratives of Strengths

▲ Post the community genogram on the wall during counseling sessions.

▲ Focus on one single dimension of the community or the family. Emphasize positive stories even if the client wants to start with a negative story. Do not work with the negatives until positive strengths are solidly in mind, unless the client clearly needs to tell you the difficult story.

▲ Help the client share one or more positive stories relating to the community dimension selected. If you are doing your own genogram, you may want to write it down in journal form.

▲ Develop at least two more positive images and stories from different groups within the community. It is often useful to have one positive family image, one spiritual image, and one cultural image so that several areas of wellness and support are included.

The Community Genogram: Two Visual Examples

We encourage clients to generate their own visual representations of their "community of origin" and/ or their current community support network. The examples presented here are only two of many possibilities.

1. The *star*: Janet's world during elementary school tells us a good bit about a difficult time in her life. Nonetheless, note the important support systems.

2. The *map*: The client draws a literal or metaphoric map of the community, in this case a rural setting. Note how this view of the client's background reveals a close extended family and a relatively small experiential world. The absence of friends in the map is interesting. Church is the only outside factor noted.

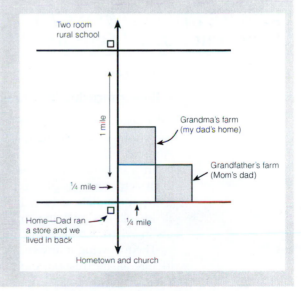

DEBRIEFING A COMMUNITY GENOGRAM

This is your chance to learn about the developmental history and cultural background of your client. It will provide you with considerable data so that you can spotlight and focus on key issues. Start by asking your client to describe the community and things that he or she considers most important in development. Obtain an overview of the client's community.

Follow this by asking for a story about each element of the genogram. Seek to obtain positive stories of fun and support, strength, courage, and survival. Bring out the facts, feelings, and thoughts within the client's story. Many of us have been raised in communities that have been challenging and sometimes even oppressive, so a positive orientation can focus on the positives and strengths that have helped the client. That platform of positives makes it possible to explore problematic issues with a greater sense of hope. (For more ideas on the community genogram, see Rigazio-DiGilio, Ivey, Grady, & Kunkler-Peck, 2005.)

Families and family stories are typically features of every community genogram. Again, search for positive family stories. In addition, you will find the family genogram in Appendix II. This is a more detailed picture of the family and often focuses on intergenerational problems. Although this information can be useful, also remember to look for survival strengths in all family members.

Complete Weblink Exercise: The Community Genogram: Identifying Strengths
Complete Weblink Exercise: Developing a Community Genogram: Identifying Personal and Multicultural Strengths
Complete Weblink Exercise: GenDraw

9.2
EXAMPLE INTERVIEW: It's All My Fault—Helping Clients Understand Self-in-Relation

KEY CONCEPT QUESTION

▲ How is focusing integrated into the interview?

Carl Rogers's person-centered counseling has had an immense and lasting influence on the way we conduct helping sessions. We can only work with the individual before us and, ultimately, it is this unique person who is most important. We live in a culture that focuses on individuality, individual responsibility, and individual achievement, thus the "I" focus. It is only natural that counseling focuses on the immediate person in the here and now. At the same time, the issues of self-in-relation and the environmental/cultural/contextual setting in which the person exists also need full attention.

The following interview with Janet focuses first on her individual issue of taking virtually all the responsibility for difficulties in her relationship with Sander. This is the second interview, and during the first session Janet completed a community genogram of the home where she grew up in Eugene, Oregon (see Box 9.1). The community genogram is used to help Janet understand how her history might affect her present behavior, feelings, and thoughts.

Interviewer and Client Conversation	Process Comments
1. **Samantha:** Nice to see you, Janet. How have things been going?	A solid relationship was established during the first session. When Samantha met Janet outside her office, Janet seemed anxious to start the session.
2. **Janet:** Well, I tried your suggestion. I think I understand things a bit better.	Last week, Samantha suggested home-work that asked Janet to spend the week listening carefully to Sander and trying to identify what he was really saying. She also advised continuing her own behavioral patterns so that she could note patterns of behavior more easily.
3. **Samantha:** You understand better? Could you tell me more?	Encourage. Focus on problem.
4. **Janet:** Well, I listened more carefully to Sander so I could understand what he really wants from me. He really wants a lot. He wants me to be a better cook, he doesn't like the way I keep house, and the more I listened, the more he wanted. I guess I haven't given him enough atten-tion. I should be doing more and doing it better.	Self-disclosure on observations. Focus on self (the client), others (Sander), and the main theme of relationship.
5. **Samantha:** Janet, it looks like you feel sad and a little guilty for not doing more. Am I hearing you right?	Reflection of feeling, focus on Janet with secondary focus on the problem.
6. **Janet:** Yes, I do feel sad and guilty; it's my responsibility to keep things together at home. He works so hard.	The key word *responsibility* appears. Janet seems to believe that she is responsible for keeping the relationship together. The acceptance of individual responsibility (and blame) is often characteristic of, but not exclusive to, women of Northern European background.
7. **Samantha:** Sounds like you're punishing yourself, when you're trying so hard and being so responsible. Could you tell me more about how Sander reacts?	Samantha makes an interpretation ("punishing yourself"). The focus is first on Janet and changes to Sander.
8. **Janet:** Yes, I get angry with myself for not doing better. It's hard with work too. But Sander wants the house to run perfectly. I try and try, but I always miss something. Then Sander blows up.	Focus on self and the problem with the close relationship in connection.
9. **Samantha:** I hear you trying very hard. Could you give me a specific example of a time that Sander blew up?	Paragraph and open question. The focus changes from Janet to Sander in a search for concreteness.

Interviewer and Client Conversation	Process Comments
10. **Janet:** Last night I made a steak dinner; I try to fix what he wants. He seemed so pleased, but I forgot to buy steak sauce and he blew up. It really shook me up . . . then, when we went to bed, he wanted to make love, but I was so tense that I couldn't. He got angry all over again.	Examples help us understand the specifics of situations. Some clients talk in vague generalities and asking for specifics can make a real difference.
11. **Samantha:** What happened then?	Open question/encourage, focus on problem. Search for more concreteness.
12. **Janet:** He went to sleep. I lay there and shook. As I calmed down, I realized that I need to do a better job and maybe he won't get so angry.	Focus on self and Sander and actions that might be taken on her part to resolve or prevent the difficulties.
13. **Samantha:** Let's see if I understand what you are saying. You are doing everything you can to make the relationship work. Sander blows up at little things, and you try harder. And then you feel the problem is your fault. Have I heard you correctly?	Samantha summarizes the situation. She recognizes it as a pattern in Janet's relationship with her husband. No matter how hard she tries, the situation escalates and Janet accepts the blame. Focus on Janet, Sander, and the main theme/problem.
14. **Janet:** It is my fault, isn't it?	Self-focus. Note the acceptance of individual responsibility for the difficulties.
15. **Samantha:** I'm not so sure. We've talked about your problem with Sander. Now let's talk about something that went right for you in the past. As you look at your community genogram, what reminds you of good times, when things were going well?	Focus shifted from interviewer and problem to cultural/environmental context and a search for positive assets and wellness strengths. Strengths and positive behaviors are helpful in understanding present situations.
16. **Janet:** [pause] Well, I loved visiting Grandma and Grandpa. They were friendly and helped me with my problems. Mom was the same way—she kept pointing out that I could do things.	Focus on family.
17. **Samantha:** Could you tell me a story about Grandma and Grandpa or your Mom when they made things better for you?	Open questions focused on positive stories in the family.
18. **Janet:** Well, Mom had to work and carry the family, and I had the most fun with Grandma and Grandpa. One time kids were teasing me at school about my braces. I thought I was funny looking because I was different. They listened to me; they told me that I was beautiful and smart. They taught me to ignore the others, just try harder, and it would all work out.	We learn about a supportive family. We also hear about a problem-solving style that focuses on ignoring underlying issues and trying harder—exactly what Janet is doing with Sander.

Interviewer and Client Conversation	Process Comments
I found that it does—if I try harder, it usually does get better.	
19. **Samantha:** So you learned in your family that ignoring things and trying harder usually helped work things out. Where did your desire to try so hard come from?	We see the emphasis on individual responsibility. Paraphrase with focus on family and client. Linking of past history with present situation.
20. **Janet:** [pauses and looks at community genogram] Dad made impossible demands of Mom. After she talked with the minister of our church, she started standing up for herself. I learned in church that it is important to care for others, but that I can't care for others unless I care for myself.	Focus on community, the family, and the church. Janet is beginning to draw on resources from the past. She is becoming aware of being a person-in-community. This helps her to start thinking of herself in new ways.
21. **Samantha:** Connections and relationships are very important to you. There is also need to care for yourself if you are to care for others. Too many women fall into the trap of always caring for others and not caring for themselves. That seems to represent the family lesson of caring. We also need to focus on your Mom starting to care for herself, how that made a difference in your family. I wonder if it would make the same difference for you now?	A brief summary followed by reframing what Janet has said with a new perspective. This is a slight extension of what Janet seemed to be already saying. Reframing and interpretation tend to be best received when they are not too far from the client's present thinking.
22. **Janet:** I think so. I so appreciate your listening to me. I wonder if I've become too much of a doormat for Sander. But I worry about what he would do if I stood up more for myself. That seems to be my family history. Thank heavens for the church.	Janet focuses on herself as a person in a family within the community. The interviewing session can now move to problem solving and restorying. Her spiritual background may be helpful.
23. **Samantha:** Could you tell me more about what the church and spirituality mean to you?	Open question with focus on the client and the cultural/environmental context in terms of the church.
24. **Janet:** Connections with friends are important, but church was a place where I could quietly meditate and work through issues. I don't seem to give myself time for that any more.	Janet is identifying herself as a self-in-relation and draws on community resources for additional strength. Meditation is a well-known and important strategy that helps build a more positive self.
25. **Samantha:** Meditation and spirituality often give us a foundation for deciding what is best. Let us explore that in more detail.	The focus turns to external sources that may help Janet build on the past to work through current issues.

Samantha helped Janet examine her situation in its broader context. The family and/or community genogram can be helpful in aiding a person who places too much responsibility on herself or himself. This can help both clients and helping professionals see the client as a person-in-relation or a person-in-community. This is a more leading approach, although the professional must continue to listen carefully to what the person has to say. As focus shifts to various dimensions of the larger client story, the client and the counselor begin to understand the multidimensionality of the issue more fully. Janet's problem is not just "Janet's problem." Rather, her issues interact with many aspects of her past and present situation. An international perspective on focusing is presented in Box 9.2.

BOX 9.2   National and International Perspectives on Counseling

Where to Focus: Individual, Family, or Culture?
WEIJUN ZHANG

Case study: Carlos Reyes, a Latino student majoring in computer science, was referred to counseling by his adviser because of his recent academic difficulties and psychosomatic symptoms. The counselor was able to discern that Carlos's major concern was his increasing dislike of computer science and growing interest in literature. He was intrigued about changing his major but felt overwhelmed by the potential consequences for his family, in which he is the oldest of four siblings. He is also the first in his family ever to attend college. Carlos has received some limited financial support from his parents and one of his younger siblings, and the family income is barely above the poverty line. The counseling was at an impasse. Carlos was reluctant to take any action and instead kept saying, "I don't know how to tell this to my folks. I'm sure they'll be mad at me."

During class discussion of this case, almost everyone argued that Carlos's problem is that he does not give priority to his personal career interests, that he should learn to think about what is good for his own mental health, and that he needs assertiveness training. I did not quite agree with my fellow students, who are all European Americans. I thought they were failing to see a decisive factor in the case: Carlos is Latino!

In traditional Hispanic culture, the extended family, rather than the individual, is the psychosocial unit of cooperation. The family is valued over the individual, and subordination of individual wants to the family needs is assumed. Also, traditional Hispanic families are hierarchical in form; parents are authority figures and children are supposed to be obedient. Given this cultural background, to encourage Carlos to make a major career decision totally by himself was impossible. Any counseling effort that did not focus on the whole family was doomed to fail.

Because it is the financial support from the family that made his college education possible, Carlos may be expected to contribute to the family when he graduates. This reciprocal relationship is a lifelong expectation in Hispanic culture, and the oldest son is especially responsible in this regard. Changing his major in his junior year does not only mean he will be postponing the date when he will be able to help his family financially, but it also means he may not be able to do so at all, for we all understand how hard it is to find a good-paying job in the field of literature. When interdependence is the norm among Hispanic Americans, how can we expect Carlos to focus entirely on his personal interests without giving more weight to his family's pressing economic needs?

If I were Carlos's counselor, rather than focusing immediately on his needs, I would first support him with his family loyalty and then help him understand that there are not just two solutions: either . . . or. . . . Together, we might brainstorm to generate some alternatives, such as having literature as his minor now and as his pastime after he graduates, changing his career when his younger siblings are off on their own, or exploring possibilities that may

BOX 9.2   (Continued)

combine the two. He could, for example, design computer programs to help schoolchildren learn literature. Each of these takes into account family needs as well as those of Carlos.

The professor praised me highly for my "different and sensitive perspective," but I shrugged it off; this is just common sense to most Third World minority people and, probably, many Italian and Jewish Americans as well. (I remember years ago, when I was trying to make major career decisions with my parents, at least ten of my relatives were involved. And these days, I am still obligated to help anyone in my extended family who is in financial need.) It took

me almost a year to realize that when I am asked, here in the United States, "How's your family?" I usually need tell only how my wife and child are faring, not my parents, grandparents, and siblings.

If the meaning of family in Hispanic culture is confusing to many counselors, the traditional extended family clan system of Native American Indians, Canadian Dene, or New Zealand Maori can be even more difficult for them to grasp. This family extension can include at times several households and even a whole village. Unless majority group counselors are aware of these differences in family structure, they may cause serious harm through their own ignorance.

 Complete Interactive Exercise 3: Focusing on Resources

Change can be measured in the session, regardless of skills used. At the beginning of this session, Janet took almost total responsibility for the problems she experienced with Sander. After review and discussion of the community genogram, she seems to have moved from denial (Level 1 on the Client Change Scale, discussed in the previous chapter) to acceptance and recognition but no change (Level 3). She clearly has a new way of thinking and is beginning to see herself as a being-in-relation. But real and lasting change will require new behavior that works successfully in the relationship. Meditation and spirituality, plus the resources of her mother's later life change, may help Janet make progress in her understanding and future relationship with Sander.

9.3

INSTRUCTIONAL READING: Multiple Contextual Perspectives on Client Concerns

KEY CONCEPT QUESTIONS

▲ **What are the important areas for focusing the interview? Why are they important?**

▲ **How does the interviewer's life history and belief system affect how he or she focuses on client topics?**

▲ **How do you help a client focus on different perspectives when working on a difficult issue?**

Although people have much in common with one another, each person we interview or counsel is totally unique. Interviewing and counseling are for the client, and learning about and focusing on that unique human being before you is the most critical and important focus. The person's name and the word *you* are central to every interview. But if we are to fully discover client uniqueness, we need to understand the broader context of the client (friends, family, community). **Contextual interviewing** strengthens the "I" focus. Multiple focusing and the family and/or community genograms provide a framework for understanding and action.

We also bring broader understanding and multiple perspectives to the session by our choice of focus on the client's life and social context. Part of what leads us to focus on certain issues is our own social context. Your developmental past is part of that context. You as interviewer can consciously or unconsciously avoid talking about certain subjects with which you are uncomfortable. You may focus on certain issues and dimensions while ignoring others. Becoming aware of your own social context and possible biases will free you to understand more fully the uniqueness of each individual and how her or his context may be similar to or different from yours.

We all work with clients who have different values and beliefs from our own. Whether the issue is the role of women, affirmative action, attitudes toward gays or lesbians, spiritual/religious beliefs, or abortion, you are going to counsel clients who think differently from you. Sharing your honest thoughts with a client on very sensitive issues can be dangerous and potentially destructive to the client. Equally or perhaps more problematic are the situations when you unconsciously direct the client in a direction that you favor. Seek supervision and consultation if you find yourself in a challenging situation in which your own thoughts get in the way of a successful counseling relationship.

▲ EXERCISE 9.2   **Focusing on Challenging Issues**

Before you continue with this section, take some time to think through your own views on a difficult and challenging issue—abortion. It will help if you take time to write your responses to the *italicized* questions below.

As an interviewer, counselor, or psychotherapist, you will encounter controversial cases and work with clients who have made different decisions than perhaps you would. Abortion is part of what is sometimes called the "culture wars." There are deeply felt beliefs and emotions around this issue. Even the language of "pro-choice" and "pro-life" beliefs can be upsetting to some. *What is your personal position on this challenging issue?*

Review the multiple dimensions of focus. What do your family and those closest to you think about abortion? Your friends? What do your community and church, both past and present, say and think? And how do your state's laws and the extensive national media coverage affect your thinking? *From a more complex, contextual point of view, spend a little time thinking about what has influenced your thinking on this issue; record what you discover.* Our individual thoughts and decisions on critical issues are deeply intertwined in our social context.

As a counselor, it is vital that you understand the situations, thoughts, and feelings of those who take varying positions on abortion or any other controversial issue, whether you agree with them or not. *Can you identify some of the thoughts and feelings of those who have a different position from your own?*

Counseling is not teaching clients how to live or what to believe. It is helping clients make their own decisions. Regardless of your personal position, you may find yourself unconsciously using the interview to further your own view. Most would agree that counselors should avoid bias in counseling. You may need to help your client understand more than one position on abortion or recognize and deal with conscious or unconscious sexism, racism, anti-Semitism, anti-Islamism, or other forms of intolerance. Effective counseling merges awareness of and respect for beliefs with unbiased probing in the interest of client self-discovery, autonomy, and growth.

▲ EXERCISE 9.3   **Focusing on the Multiple Contexts of Client Concerns**

Some school systems and agencies have written policies forbidding any discussion of abortion. Furthermore, if you are working within certain agencies (such as a faith-based agency or a pro- or anti-abortion counseling clinic), the agency may have specific policies regarding counseling about abortion. Ethically, clients should be made aware of specific agency beliefs and policies before counseling begins.

Imagine that a client comes to you who has just terminated a fetus. How would you help this client, who clearly needs to tell her story? Write your answer to this question below. There are no necessarily "right" answers to difficult issues like this.

_____

_____

_____

_____

FOCUS ON THE INDIVIDUAL AND SIGNIFICANT OTHERS

TERESA: I just had an abortion and I feel awful. The medical staff was great and the operation went smoothly. But Cordell won't have anything to do with me, and I can't talk with my parents. The people outside yelling at me as I went in scared me.

▲ What would you say to focus on Teresa as an individual?
▲ What could you say to focus on Cordell?
▲ How would you focus on the noisy and likely disrespectful crowd?
▲ How might you focus on the attitudes and possible supports from her friends?

Choosing an appropriate focus can be most challenging. Too many beginners focus only on the problem. The prompt "Tell me more about the abortion" may result in drawing out details of the abortion but little about the client's distinctive personal experience. An extremely important task is drawing out the client's story: "I'd like to hear _your_ story" or "What do _you_ want to tell me?" There are no absolute rules on where to focus, but generally, we want to hear the client's unique experience. Focusing on the individual is usually where to start—review on page 181 where the word _you_ was used frequently in the interviewer's comments.

In Teresa's case, other key figures (Cordell, the crowd, family, friends) are part of the larger picture. What are their stories? How do they relate to Teresa as a person-in-relation? You can more fully understand her situation when you draw out other stories or viewpoints. It is important to keep all significant others in mind in the process of problem examination and resolution. For a full understanding of the client's experience, all pertinent relationships eventually need to be explored.

FOCUS ON FAMILY AND SIGNIFICANT OTHERS

TERESA: My family is quite religious and they have always talked strongly against abortion; it makes me feel all the more guilty. I could never tell them.

▲ How might you focus on the family in response to her statement?
▲ How would you search for others in the family who might be helpful or supportive?

The family is where personal values and ethics are first learned. How does Teresa define "family"? There are many styles of family beyond the nuclear. African American and Hispanic clients may think of the extended family; a lesbian may see her supportive family as the gay community. Issues of single parenthood and alternative family styles continue to make the picture of the family more complex. Developing a community or family genogram may help Teresa locate resources and models that might help her. If her parents are not emotionally available, perhaps an aunt or grandmother might be.

MUTUALITY FOCUS

TERESA: I feel like everyone is just judging me. They all seem to be condemning me. I even feel a little frightened of you.

▲ How would you appropriately focus on the relationship between yourself and the client?
▲ What might you say to Teresa that focuses on the here-and-now feelings?

A mutual immediate focus often emphasizes the *we* in a here-and-now relationship. Working together in an egalitarian relationship can empower clients. Also, helping them recognize the depth of their feelings in the here and now can be valuable and powerful. "Right now at this moment, *we* have an issue." "Can *we* work together to help you?" "What are some of your thoughts and feelings about how *we* are doing?" The emphasis is on the relationship between counselor and client. Two people are working on an issue, and the interviewer accepts partial ownership of the problem.

In feminist counseling, the "we" focus may be especially appropriate: *"We* are going to solve this problem." The "we" focus provides a sharing of responsibility, which is often reassuring to the client regardless of her or his background. Many feminist counselors emphasize *we*. In some counseling theories and Western cultures, emphasizing the distinction between *you* (client focus) and *me* (interviewer focus) is more common, and *we* would be considered inappropriate.

The mutual focus often includes a *here-and-now* dimension and brings immediacy to the session. To focus on the here and now, there are several different types of responses: "Teresa, right now you are really hurting and sad about the abortion." "I sense a lot of unsaid anger right now." There is also the classic, "What are you feeling right now, at this moment?"

INTERVIEWER FOCUS

TERESA: What do you think about what I did? What should I do?

What would you say? An interviewer focus could be self-disclosure of feelings and thoughts or personal advice about the client or situation: "*I* feel concerned and sad over what happened." "Right now, *I* really hurt for you, but I know that you have what it takes to get through this." "*I* want to help." Or "*I*, too, had an abortion . . . *my* experience was. . . ." Opinions vary on the appropriateness of interviewer or counselor involvement, but the value and power of such statements are increasingly being recognized. They must not be overused; keep self-disclosures brief.

- ▲ How might you share your own thoughts and feelings appropriately?
- ▲ Would you give advice from your frame of reference? What would it be?
- ▲ What are the power issues if you share your own thoughts or the agency policies?

CULTURAL/ENVIRONMENTAL/CONTEXTUAL FOCUS

Perhaps the most complex focus dimension is the cultural/environmental context. Some topics within these broad areas are listed here, along with possible responses to the client. A key cultural/contextual issue in discussing abortion will often be religion and spiritual orientation. Whether she is conservative, liberal, Christian, Jew, Hindu, Muslim, or a nonbeliever, discussing the values issue from a spiritual perspective may be important to the client.

Given Teresa's discussion thus far, what would you say to bring in broader cultural/environmental/contextual issues?

- ▲ *The crowd:* "You said the crowd scared you. Let's talk about them some more."
- ▲ *Moral/religious issues:* "What can you draw from your spiritual background to help you?"
- ▲ *Legal issues:* "The topic of abortion brings up some legal issues in this state. How have you dealt with them?"
- ▲ *Women's issues:* "A support group for women is just starting. Would you like to attend?"
- ▲ *Economic issues:* "You said you can't pay for the operation. . . ."
- ▲ *Health issues:* "You seem to be recovering well physically, but how have you been eating and sleeping lately? Do you feel aftereffects?"
- ▲ *Educational/career issues:* "How long were you out of school/work?"
- ▲ *Ethnic/cultural issues:* "What is the meaning of abortion among people in your family/church/neighborhood?"

Any one of these issues, as well as many others, could be important to a client. With some clients all of these areas might need to be explored for satisfactory problem resolution. The counselor or interviewer who is able to conceptualize client issues broadly can introduce many valuable aspects of the problem or situation. Note that much of cultural/environmental/contextual focusing requires sensitive leading and influencing from the interviewer.

Complete Interactive Exercise 4: Writing Alternative Focusing Statements
Complete Group Practice Exercise: Group Practice With Focusing

9.4
ADVOCACY AND SOCIAL JUSTICE

All interviewing, counseling, and psychotherapy involve issues of social justice.

—*Allen Ivey*

KEY CONCEPT QUESTION

▲ **What is the interviewer's role in advocacy and social justice?**

Social justice is a moral orientation toward a just and fair world. It is a value stance that emphasizes human rights and equality. *Tikkum olim* comes from the Jewish tradition and means "repairing the world." The Roman Catholic Church stresses the life and dignity of the human person. The United Methodist Church traditionally has studied and taken positions on abortion, alcohol abuse, gambling, capital punishment, gay and lesbian rights, and inclusivity of all races, and seriously examines the meaning of military service and war (http://en.wikipedia.org/wiki/United_Methodist_Church#Social_issues). Islam forbids discrimination, cruelty, and exploitation; in addition, it pays special attention to physical health—"We are all the offspring of the same parents" (http://www.al-islam .org/lessons/3.htm). Selflessness and caring for others is one of the foundations of the Buddhist tradition.

Given these powerful definitions, we can see the interviewing and counseling constantly involve moral issues, values, and social justice. Add to the above the multiple dimensions of the RESPECTFUL diversity model, and it is clear that we cannot avoid these issues in most interviews. By becoming involved with helping others, you are already taking the route of selflessness and have an excellent start on discerning meaning in your life.

In the individual counseling session, you have the opportunity to help clients become aware of how social justice and cultural/environmental/contextual issues affect their *being-in-relation*, the fact that they are *persons-in-community*. As a very concrete example, Allen almost always has the client's community genogram available for discussion and enlightenment. A depressed client learns that he or she is not totally at fault for the depression and discovers that family and contextual factors can be addressed and the client can return to life with new meanings and purposes. Working with people from diverse cultural backgrounds and those "recovering" from oppressive religious experience, he has helped them realize that their issues and problems are not just individual issues, but deeply entwined with social history. Discovering oneself as a person-in-community broadens perspective and presents new promise for the future.

You are going to face situations in which your best counseling efforts are insufficient to help clients resolve their issues and move on with their lives. The social context of homelessness, poverty, racism, sexism, and contextual issues may leave clients in an impossible situation. The problem may be bullying on the playground, an unfair teacher, or an employer who refuses to follow fair employment practices. It is critical that we examine the societal stressors that our clients may face. Traditional approaches to interviewing and counseling may not be enough.

Advocacy is speaking out for your clients; working in the school, community, or larger setting to help clients; and working for social change. What are you going to do on a daily basis to help improve the systems within which your clients live? Advocacy counseling competencies have been defined in detail by the American Counseling Association, and we highly recommend the recent book *ACA Advocacy Competencies: A Social Justice Framework for Counselors* (Ratts, Toporek, & Lewis, 2010) for serious study and consideration.

Here are some examples when simply talking with clients about their issues may not be enough:

▲ As an elementary school counselor, you counsel a child who is being bullied on the playground.

▲ You are a high school counselor and work with a tenth grader who is teased and harassed about being gay while the classroom teacher quietly watches and says nothing.

▲ As a personnel officer, you discover systematic bias against promotion for women and minorities.

▲ Working in a community agency, you have a client who speaks of abuse in the home but fears leaving because she sees no future financial support.

▲ You are working with an African American client who has dangerous hypertension. You know there is solid evidence that racism affects blood pressure.

The elementary school counselor can work with each grade or groups of students and educate them about bullying—what it means, how it feels, and ways they can empower themselves to stop being bullied. The counselor can also work to invite parents and/or caregivers to classroom lessons on bullying so that the skills learned at school are reinforced in the home. Another course of action might be for the counselor to work with school officials to set up policies concerning bullying and harassment, actively changing the environment that allows bullying to occur. The high school counselor faces an especially challenging issue as interview confidentiality may preclude immediate classroom action. If this is not possible, then the counselor can initiate school policies and awareness programs against oppression in the classroom. The passive teacher may become more aware through training that you offer to all the teachers. You can help the African American client understand that hypertension is not just "his problem," but that his blood pressure is partially related to the constant stressors of racism in his environment. You can work to eliminate oppression in your community.

"Whistle-blowers" who name problems that others prefer to avoid can face real difficulty. They will need your support and advice on how to proceed. For example, the company may not want to have its systematic bias exposed. On the other hand, through careful consultation and data gathering, the human relations staff may be able to help managers develop promotion programs that are fairer. Again, the issue of policy becomes important. Counselors can be advocates for policy changes in their work settings. The counselor in the community agency knows that advocacy is the only possibility when abuse is apparent. For these clients, advocacy in terms of support in getting out of the home, finding new housing, and learning how to obtain a restraining order may be far more important than self-examination and understanding.

Counselors who care about their clients also become their advocates when necessary. They are willing to move out of the counseling office, meet clients in the home, connect with clients' larger network, or help clients create larger networks of support and seek social change. You may work with others on a specific cause or issue to facilitate general human development and wellness (e.g., pre-term pregnancy care, child care, fair housing, aid for the homeless, athletic fields for low-income areas). This requires you to speak out, to develop skills with the media, and to learn about legal issues. *Ethical witnessing* moves beyond working with victims of injustice to the deepest level of advocacy (Ishiyama, 2006). Counseling, social work, and human relations are inherently social justice professions, but speaking out for social concerns also needs our time and attention.

Complete Case Study Exercise: Career Counseling in Social Context: The "System" and the Client

CHAPTER SUMMARY

Key Points of "Focusing the Interview"

CourseMate and DVD Activities to Build Interviewing Competence

9.1 Defining Focusing

▲ Selective attention (Chapter 3) is basic to focusing—clients tend to talk about or focus on topics to which you give your primary attention. Through your attending skills (visuals, vocal tone, verbal following, and body language), you indicate to your client the topics that you consider most important.

▲ The interviewer, through selective attention focuses the interview on the client, problem/concern, important others (partner/spouse, family, friends, organizations such as churches, schools, clubs), or the cultural/environmental context (RESPECTFUL multicultural background, community, nation). In addition, the counselor may focus on himself or herself or on what is going on in the here and now of the interview.

1. Interactive Exercise 1: Identify Focus Dimensions. Work through the Samantha and Janet session to identify the responses using focus analysis and to score your responses.
2. Weblink Critique: Several links to counseling and therapy web sites are provided with questions to help you focus on their contents.
3. Interactive Exercise 2: Community Genogram. An exercise you can do by yourself or with a classmate to practice drawing data from a genogram.
4. Weblink: The Community Genogram: Identifying Strengths. Learn how to build community genograms. The community genogram exercise will help you understand the cultural background of your client; it is through family and community that we learn our cultural framework.
5. Weblink: Developing a Community Genogram: Identifying Personal and Multicultural Strengths. Learn how to use community genograms effectively. The community genogram is a "freeform" activity in which we encourage people to develop their own style of presenting their community of origin or present community.
6. Weblink: GenDraw, a German website with genogram drawing capability.

9.2 Example Interview: It's All My Fault—Helping Clients Understand Self-in-Relation

▲ The community genogram provided a clear visual picture of the client's background and helped the counselor understand the social context more quickly and fully.

▲ Client issues are seldom one-person issues. Focusing on the family background, the church, and spirituality enriched the session and helped the client focus on positive strengths.

1. Interactive Exercise 3: Focusing on Resources. This exercise gives you an opportunity to select the most helpful focusing response. Can you find a balance among the different possibilities?

Key Points of "Focusing the Interview"

CourseMate and DVD Activities to Build Interviewing Competence

9.3 Instructional Reading: Multiple Contextual Perspectives on Client Concerns

▲ The major focus dimensions are client, problem or issue, significant others, mutuality focus (often with immediacy), interviewer, and the broader cultural/environmental context.

▲ Counseling is for the client, and focusing on that client is central to all interviews. Secondarily, we need to focus on the client's problems, issues, and concerns within a broad contextual frame. Too often interviewing focuses just on the client and the problem, thus missing broader understanding.

▲ You as an interviewer or counselor also come from a social context, and all the focus dimensions apply to you as well. You may consciously or unconsciously focus on certain issues and dimensions while ignoring others.

1. Interactive Exercise 4: Writing Alternative Focusing Statements. A good way to practice your focusing skills.

2. Group Practice Exercise: Group Practice With Focusing.

9.4 Advocacy and Social Justice

▲ You are going to face situations when your best counseling efforts are insufficient to help your clients resolve their issues and move on with their lives. The social context of poverty, racism, sexism, and many other forms of unfairness may leave your client in an impossible situation.

▲ Those who adopt a social justice orientation and ethical witnessing move beyond understanding and take action for their clients.

1. Case Study: Career Counseling in Social Context: The "System" and the Client. A good exercise to think about the issues affecting Anne and how you would collaborate with her in facing these issues.

Assess your awareness, knowledge, and skills as you conclude the chapter:

1. Flashcards: Use the flashcards to check your understanding of key concepts and facilitate memorization of key information.

2. Self-Assessment Quiz: The quiz will help you assess your current knowledge and prepare for course examinations.

3. Portfolio of Competencies: Evaluate your present level of competence in attending and listening skills using the downloadable Self-Evaluation Checklist. Self-assessment of your attending skills competence demonstrates what you can do in the real world.

REFLECTION OF MEANING AND INTERPRETATION/REFRAMING

Helping Clients Restory Their Lives

INFLUENCING SKILLS
AND STRATEGIES

FOCUSING

CONFRONTATION

THE FIVE-STAGE INTERVIEW STRUCTURE

REFLECTION OF FEELING

ENCOURAGING, PARAPHRASING, AND SUMMARIZING

OPEN AND CLOSED QUESTIONS

ATTENDING BEHAVIOR

ETHICS, MULTICULTURAL COMPETENCE, AND WELLNESS

Then I spoke of the many opportunities of giving life a meaning. I told my comrades . . . that human life, under any circumstances has meaning. . . . I said that someone looks down on each of us in difficult hours—a friend, a wife, somebody alive or dead, or a God—and he would not expect us to disappoint him. . . . I saw the miserable figures of my friends limping toward me to thank me with tears in their eyes.

—*Viktor Frankl*

This chapter is dedicated to the memory of Viktor Frankl. The initial stimulus for the skill of reflection of meaning came from a two-hour meeting with him in Vienna shortly after we visited the German concentration camp, Auschwitz, where he had been imprisoned in World War II. He impressed on us the central value of meaning in counseling and therapy—a topic to which most theories give insufficient attention. It was his unusual ability to find positive meaning in the face of impossible trauma that impressed us most. His thoughts also affected our wellness and positive strengths orientation. His theoretical and practical approach to counseling and therapy deserves far more attention than it receives. We often recommend his short, gripping book, *Man's Search for Meaning* (1959), to clients who face serious life crises. It remains fully alive today.

200

How can the skills of reflection of meaning and interpretation/reframing help you and your clients?

CHAPTER GOALS

Viktor Frankl got through the horrors of the concentration camp at Auschwitz by focusing on possibility and meaning. Moreover, he enabled others to keep going and live by finding meaning in the most difficult of situations.

Interviewing, counseling, and therapy focus on helping clients change their thoughts, feelings, and behaviors. Underlying this basic triad is the issue of meaning and how to interpret life experience. What is the purpose or significance of it all? What sense does anything make? Two related skills are emphasized in this chapter, and both seek to enable the client to think in new and more productive ways.

Reflection of meaning is concerned with helping clients find deeper meanings underneath thoughts, feelings, and behavior. Interpretation often comes from a specific theoretical perspective, such as decisional, psychodynamic, or multicultural. In reflecting meanings, clients generate their own vision of what things mean, whereas interpretations/reframes usually come from the interviewer and provide new perspectives for the client to consider.

Awareness, knowledge, and skills developed through the concepts of this chapter will enable you to

▲ Understand the distinction and the relationship between reflection of meaning and interpretation/reframing.

▲ Assist clients, by eliciting and reflecting meaning, to explore their deeper values, visions for the future, and life mission.

▲ Examine the place of resilience in the life of your clients.

▲ Help clients, through interpretation/reframing, find an alternative way of thinking that facilitates personal development.

▲ Understand how these skills relate to and differ from other microskills.

Assess your awareness, knowledge, and skills as you begin the chapter:

1. Self-Assessment Quiz: The chapter quiz will help you determine your current level of knowledge. You can take it before and after reading the chapter.
2. Portfolio of Competencies: Before you read the chapter, fill out the downloadable Self-Evaluation Checklist to assess your existing knowledge and competence in attending skills. Then, at the end of the chapter, complete the checklist again to summarize your competencies after study and practice.

10.1
DEFINING THE SKILLS OF REFLECTING MEANING AND INTERPRETATION/REFRAMING

KEY CONCEPT QUESTIONS

▲ **How can I demonstrate the skills of eliciting and reflecting meaning and interpretation/reframing?**

▲ **How are these two skills similar, and how are they different?**

Both interpretation/reframing and reflection of meaning seek implicit issues and meanings below the surface of client conversation. If you use these skills as described below, you can predict how clients may respond.

Following are definitions and the predicted results of these related skills. Note that reflection of meaning focuses on clients' "making sense" of their experience, whereas interpretations come from the helper. Intentional use of these two skills often results in changing your helping lead from one to the other if the client finds the skills not as helpful as you thought.

Reflection of Meaning	*Predicted Result*
Meanings are close to core experiencing. Encourage clients to explore their own meanings and values in more depth from their own perspective. Questions to elicit meaning are often a vital first step. A reflection of meaning looks very much like a paraphrase, but focuses beyond what the client says. Often the words *meaning, values, vision,* and *goals* occur in the discussion.	The client discusses stories, issues, and concerns in more depth with a special emphasis on deeper meanings, values, and understandings. Clients may be enabled to discern their life goals and vision for the future.
Interpretation/Reframing	*Predicted Result*
Provide the client with a new perspective, frame of reference, or way of thinking about issues. Interpretation/reframing may come from your observations; it may be based on varying theoretical orientations to the helping field; or it may link critical ideas together.	The client may find another perspective or meaning of a story, issue, or problem. Their new perspective may have been generated by a theory used by the interviewer, from linking ideas or information, or by simply looking at the situation afresh.

The formal definition of **interpretation** is "an explanation of the meaning or significance of something"; **meaning** is defined as "what a word means" or its purpose and significance (Encarta©, 2009). From these two definitions, you can see the logic of presenting the two skills together, as they are closely related.

Reflection of meaning is related to interpretation, but the client does the meaning-making, not you. If a client voluntarily makes a statement that is clearly related to meaning issues, then it is appropriate to reflect it immediately. However, many times it is necessary to elicit meaning by asking the client some variation of the basic question, "What does . . . *mean* to you?" Effective exploration of meaning becomes a major strategy in which you bring out client stories, *past, present,* and *future.* You use all the listening, focusing, and confrontation skills to facilitate this self-examination, yet the focus remains on meaning and finding purpose in one's life. Meaning is also closely related to the discussion of advocacy and social justice presented in the preceding chapter.

The case of Charlis will serve as a way to illustrate similarities and differences between reflection of meaning and interpretation.

Charlis, a workaholic 45-year-old middle manager, has a heart attack. After several days of intensive care, she is moved to the floor where you, as the hospital social worker, work with the heart attack aftercare team. Charlis is motivated; she is following physician directives and progressing as rapidly as possible. She listens carefully to diet and exercise suggestions and seems the ideal patient with an excellent prognosis. However, she wants to return to her high-pressure job and continue moving up through the company; you observe some fear and puzzlement about what's happened.

REFLECTION OF MEANING

Charlis has had a heart attack, and you first need to hear her story with special attention to the many sad, worried, and anxious feelings that she experiences. There is a need to review what happened before the heart attack and her previous lifestyle. She will need help and support in her rehabilitation program. She will need your skills as she works through returning to her job. You prescribe and support her through an exercise program.

As you successfully complete all of these tasks, you recognize that Charlis is reevaluating the meaning of her life. She asks questions that are hard to answer—"Why me? What is the meaning of my life? What is God saying to me? Am I on the wrong track? What should I *really* be doing?" You sense that she feels that something is missing in her life, and she wants to reevaluate where she is going and what she is doing.

▲ EXERCISE 10.1   **Reflection of Meaning**

- ▲ What thoughts occur to you?
- ▲ What do you see as the key issues that relate to the meaning and purpose of Charlis's life?
- ▲ What about meaning and purpose in your own life? What are your goals and vision?

To elicit meaning, we may ask Charlis some variation of a basic meaning question: "What does the heart attack mean to you, your past and future life?" We may also ask Charlis if she would like to examine the meaning of her life through the process of discernment, a more systematic approach to meaning and purpose defined in some detail in this chapter.

As a beginning, we'd share some of the following specific questions. Sometimes these questions are enough to help clients find meaning and vision, enabling them to move to new life directions and take new actions.

- ▲ "What has given you most *satisfaction* in your job?"
- ▲ "What's been *missing* for you in your present life?"
- ▲ "What do you *value* in your life?
- ▲ "What *sense* do you make of this heart attack and the future?"
- ▲ "What things in the future will be most *meaningful* to you?"
- ▲ "What is the *purpose* of your working so hard?"
- ▲ "You've said that you wonder what *God* is saying to you with this trial. Could you share some of your thoughts?"
- ▲ "What *gift* would you like to leave the world?"

We'd ask Charlis to think of questions and issues that are particularly important to her as we work to help her discern the meaning of her life, her work, her goals, and her mission. These questions often bring out emotions, and they certainly bring out meaning in the client's thoughts and cognitions. When clients explore meaning issues, the interview becomes less precise as the client struggles with defining the almost indefinable.

Eliciting meaning often precedes reflection. Reflection of meaning as a skill looks very much like a reflection of feeling or paraphrase, but the key words *meaning, sense, deeper understanding, purpose, vision,* or some related concept will be present explicitly or implicitly. "Charlis, I sense that the heart attack has led you to question some basic understandings in your life. Is that close? If so, tell me more." Eliciting and reflecting meaning provide an opening for the client to explore issues for which there is not a final answer but rather a deeper awareness of the possibilities of life. Both reflecting meaning and interpretation/reframing are designed to help clients look deeper, first by careful listening and then by helping clients examine themselves from a new perspective.

Reflecting meaning involves *client* direction; the interpretation/reframe implies *interviewer* direction. The client provides the new and more comprehensive perspective in reflection of meaning, while an interpretation/reframe supplies the new way of being as suggested by the interviewer or counselor.

COMPARING REFLECTION OF MEANING AND INTERPRETATION/REFRAMING

Following are some brief examples of how reflection of meaning and interpretation/reframing may work for Charlis as she attempts to understand some underlying issues around her heart attack.

CHARLIS: My job has been so challenging and I really feel that pressure all the time, but I just ignored it. I'm wondering why I didn't figure out what was going on until I got this heart attack. But, I just kept going on, no matter what.

Eliciting and Reflecting Meaning

COUNSELOR: I hear you—you just kept going. Could you share what it feels like to *keep going on* and what it *means* to you? [Encourager focusing on the key words "keep going on"; open question oriented to meaning]

CHARLIS: I was raised to keep going. My mother always prided herself on doing a good job, even in the worst of times. Grandma did the same thing.

COUNSELOR: Charlis, I hear that keeping going and persistence have been a key family value that remains very important to you. [Reflection of meaning] "Hanging in" is what you are good at. [Mention of positive asset leading to wellness] Could we focus now on how that value around persistence and *keeping going on* relates to your rehab? [Open question that seeks to use the wellness dimensions to help her plan for the future]

Interpretation/Reframing

COUNSELOR: You could say that you *keep going* until you drop. How does that sound to you? [Mild reframe/interpretation followed by checkout]

CHARLIS: I was raised to keep going. My mother always prided herself on doing a good job, even in the worst of times. Grandma did the same thing.

COUNSELOR: Many of us become who we are because of family history. It sounds as if several generations have taught you to struggle and *keep going on, no matter what.* Do you want to continue that tradition? Or could you use *keeping going* on in a more positive way? [Interpretation/reframe, closed question, open question]

Both reflection of feeling and interpretation ended up in nearly the same place, but Charlis is more in control of the process with reflection of meaning. Whichever approach is used, we are closer to helping Charlis work on the difficult questions of the meaning and direction of her future life. If the client does not respond to reflective strategies, move to the more active interpretation. We need to give clients power and control of the session whenever possible. They can often generate new interpretations/reframes and new ways of thinking about their issues.

Interpretations and reframes vary with theoretical orientation. We use the joint term *interpretation/reframing* because both focus on providing a new way of thinking or a new frame of reference for the client, but **reframing** is a gentler construct. When you use influencing skills, keep in mind that interpretive statements are more directive than reflections of meaning. When we use interpretation/reframing, we are working primarily from the interviewer's frame of reference. This is neither good nor bad; rather, it is something we need to be aware of when we use influencing skills.

Linking is an important part of interpretation, although it often appears in an effective reflection of meaning as well. In linking, two or more ideas are brought together, providing the client with a new insight. The insight comes primarily from the client in reflection of meaning, but almost entirely from the interviewer in interpretation/reframing. Consider the following four examples:

Interpretation/reframe 1: Charlis, we are all reflections of our family, and it is clear that family history emphasizing success and hard work has deeply affected you, perhaps even to the point of having a heart attack. [Links family history to the heart attack. A family counselor might use this approach.]

Interpretation/reframe 2: Charlis, you seem to have a pattern of thinking that goes back a long way—we could call it an "automatic thought." You seem to have a bit of perfectionism there, and you keep saying to yourself [self-talk], "Keep going no matter what." [Links the past to the present perfectionism from a cognitive-behavioral perspective.]

Interpretation/reframe 3. It sounds as if you are using hard work as a way to avoid looking at yourself. The avoidance is similar to the way you avoid dealing with what you think you need to change in the future to keep yourself healthier. [Combines confrontation with linking to what is occurring in the interview series. This is close to a person-centered approach.]

Interpretation/reframe 4. The heart attack almost sounds like unconscious self-punishment, as if you wanted it to happen to give you time off from the job and a chance to reassess your life. [Linking interpretation from a psychodynamic perspective.]

Review Case Study: Finding Family Balance in Greece: A Universal Problem
Complete Weblink Exercises:

- ▲ International Association on Personal Meaning
- ▲ Religion and Spirituality
- ▲ Sexual Orientation: Science, Education, and Policy— University of California at Davis
- ▲ Feminist Therapy
- ▲ Sunshine for Women

- ▲ American Psychological Association (APA), Just the Facts About Sexual Orientation and Youth: A Primer for Principals, Educators, and School
- ▲ Sexual Orientation
- ▲ Heterosexuals Questionnaire
- ▲ Gender and Race—University of Iowa
- ▲ Men's Resources
- ▲ Women's Resources

10.2
EXAMPLE INTERVIEW: Travis Explores the Meaning of a Recent Divorce

KEY CONCEPT QUESTION

▲ How is reflection of meaning integrated into the interview?

In the following session, Travis is reflecting on his recent divorce. When relationships end, the thoughts, feelings, and underlying meaning of the other person and the time together often remain an unsolved mystery. Moreover, some clients are likely to repeat the same mistakes in their relationships when they meet a new person.

Both the interpretation/reframing and reflection of meaning are central skills in helping clients take a new perspective on themselves and their world. Andreas, the interviewer, seeks to help Travis think about the word *relationship* and its meaning. Note that Travis stresses the importance of connectedness with intimacy and caring. The issue of self-in-relation to others will play itself out very differently among individuals in varying cultural contexts. Many clients will focus on their need for independence.

Interviewer and Client Conversation	Process Comments
1. **Andreas:** So, Travis, you're thinking about the divorce again . . .	Encourager/restatement.
2. **Travis:** Yeah, that divorce has really thrown me for a loop. I really cared a lot about Ashley and . . . ah . . . we got along well together. But there was something missing.	
3. **Andreas:** Uh-huh . . . something missing?	Encouragers appear to be closely related to meaning. Clients often supply the meaning of their key words if you repeat them back exactly.
4. **Travis:** Uh-huh, we just never really shared something very basic. The relationship didn't have enough depth to go anywhere. We liked each other, we amused one another, but beyond that . . . I don't know . . .	Travis elaborates on the meaning of a closer, more significant relationship than he had with Ashley.

Interviewer and Client Conversation	Process Comments
5. **Andreas:** You amused each other, but you wanted more depth. What sense do you make of it?	Paraphrase using Travis's key words followed by a question to elicit meaning.
6. **Travis:** Well, in a way, it seems like the relationship was shallow. When we got married, there just wasn't enough depth for a meaningful relationship. The sex was good, but after a while, I even got bored with that. We just didn't talk much. I needed more . . .	Note that Travis's personal constructs for discussing his past relationship center on the word *shallow* and the contrast *meaningful*. This polarity is probably one of Travis's significant meanings around which he organizes much of his experience.
7. **Andreas:** Mm-hmmm . . . you seem to be talking in terms of shallow versus meaningful relationships. What does a meaningful relationship feel like to you?	Reflection of meaning followed by a question designed to elicit further exploration of meaning.
8. **Travis:** Well, I guess . . . ah . . . that's a good question. I guess for me, there has to be some real, you know, some real caring beyond just on a daily basis. It has to be something that goes right to the soul. You know, you're really connected to your partner in a very powerful way.	Connection appears to be a central dimension of meaning. We often believe that connectedness is a female construct, but many men also see it as central.
9. **Andreas:** So, connections, soul, deeper aspects strike you as really important.	Reflection of meaning. Note that this reflection is also very close to a paraphrase, and Andreas uses Travis's main words. The distinction centers on issues of meaning. A reflection of meaning could be described as a special type of paraphrase.
10. **Travis:** That's right. There has to be some reason for me to really want to stay married, and I think with her . . . ah . . . those connections and that depth were missing. We liked each other, you know, but when one of us was gone, it just didn't seem to matter whether we were here or there.	
11. **Andreas:** So there are some really good feelings about a meaningful relationship even when the other person is not there. You didn't value each other that much.	Reflection of meaning plus some reflection of feeling. Note that Andreas has added the word *value* to the discussion. In reflection of meaning it is likely that the counselor or interviewer will add words such as *meaning, understanding, sense,* and *value.* Such words lead the client to make sense of experience from the client's own frame of reference.
12. **Travis:** Uh-huh.	

Interviewer and Client Conversation	Process Comments
13. **Andreas:** Ah . . . could you fantasize how you might play out those thoughts, feelings, and meanings in another relationship?	Open question oriented to meaning.
14. **Travis:** Well, I guess it's important for me to have some independence from a person, but when we were apart, we'd still be thinking of one another. Depth and a soul mate is what I want.	Travis's meaning and desire for a relationship are now being more fully explored.
15. **Andreas:** Um-hum.	
16. **Travis:** In other words, I don't want a relationship where we always tag along together. The opposite of that is where you don't care enough whether you are together or not. That isn't intimate enough. I really want intimacy in a marriage. My fantasy is to have a very independent partner I care about and who cares about me. We can both be individuals but still have bonding and connectedness.	Connectedness is an important meaning issue for Travis. With other clients, independence and autonomy may be the issue. With still others, the meaning in a relationship may be a balance of the two.
17. **Andreas:** Let's see if I can put together what you're saying. The key words seem to be *independence* with *intimacy* and *caring*. It's these concepts that can produce bonding and connectedness, as you say, whether you are together or not.	This reflection of meaning becomes almost a summarization of meaning. Note that the key words and constructs have come from the client in response to questions about meaning and value.

Further counseling would aim to bring behavior or action into accord with thoughts. Other past or current relationships could be explored further to see how well the client's behaviors or actions illustrate or do not illustrate expressed meaning.

 Complete Interactive Exercise 1: Identifying Skills

10.3
INSTRUCTIONAL READING 1: The Specific Skills of Eliciting and Reflecting Meaning

KEY CONCEPT QUESTIONS

▲ **What are more detailed skills of eliciting and reflecting meaning?**

▲ **How does the process of discernment help clients?**

▲ **What key multicultural issues affect eliciting and reflecting meaning?**

BOX 10.1   National and International Perspectives on Counseling

What Can You Gain From Counseling in a Challenging Situation?
WEIJUN ZHANG

A good friend of mine had just started working in a hospice setting. When I asked him, "What does this mean to you?" he looked sad and started to grumble. "They are all going to die. Some of them are hard to deal with. I feel helpless." What a bleak picture he painted. No wonder many counselors prefer to work in comfortable offices and are reluctant to work with death and the dying, in a community AIDS clinic, in a kidney dialysis unit, or with the homeless.

Can counseling with issues that are so very challenging be meaningful? Certainly, there is much to learn and profit from this type of work. What my friend was missing is that there are precious moments in and important rewards from working with those who are most in need.

Years ago, I happened on an article outlining the positive aspects of working with clients who face extremely difficult futures or death.

1. We can help clients appreciate each moment. Working with people who have no choice but to live in the present can change our perspective. It can help us find a deeper meaning in life and learn to watch the world in wonder and appreciation as if this were our last day. With the homeless,

we can make each day just a bit easier and more comfortable by caring.

2. It is possible to learn something from each patient or client if we are willing to listen and be with his or her experience. Having direct contact with clients who face unthinkable pain and suffering but who still strive to live fully can help us understand the great strength that is the human spirit. This courage in the face of adversity can be very contagious.

3. Counseling challenging clients who deal daily with life's most difficult issues will enable us to witness a lot of truly unconditional love and help us to become bigger-hearted, more loving and caring persons.

I have always liked doing work with the seriously ill because of the selfless giving needed on the part of the counselor. But I have learned that it can mean a tremendous taking, learning, and satisfaction for those who help others most in need. Though this taking should never be our primary motive in doing this type of work, it is certainly rewarding to see the light when we deal with the darkness associated with major life challenges. This positive insight I wish to share with my good friends.

Meaning issues often become prominent after a person has experienced a serious illness (AIDS, cancer, heart attack, loss of sight), encountered a life-changing experience (death of a significant other, divorce, loss of a job), or gone through serious trauma (war, rape, abuse, suicide of a child). Box 10.1 presents an international perspective on counseling. Issues of meaning are also prominent among older clients who face major changes in their lives. These situations cannot be changed; they are a permanent part of the life experience.

Reflecting meaning can also help clients work through issues of daily life. In the example interview, Travis gained understanding of himself as he reflected on the meaning of divorce. Everyday issues and many typical concerns can be resolved if we turn to a serious examination of meaning, values, and life purpose. Religion and spiritual life provide a value base and can be a continuing source of strength and clarity.

ELICITING CLIENT MEANING

Understanding the client is the essential first step. Consider storytelling as a useful way to discover the background of a client's meaning-making. If a major life event is critical, illustrative stories can form the basis for exploration of meaning. Clients do not often volunteer meaning issues, even though these may be central to the clients' concerns. Critical life events such as illness, loss of a parent or loved one, accident, or divorce often force people to encounter deeper meaning issues. If spiritual issues come to the fore, draw out one or two concrete example stories of the client's religious heritage. Through the basic listening sequence and careful attending, you may observe the behaviors, thoughts, and feelings that express client meaning.

Fukuyama (1990, p. 9) outlined some useful questions for eliciting stories and client meaning systems. Adapted for this chapter, they include the following:

▲ "When in your life did you encounter existential or meaning questions? How have you resolved these issues so far?"
▲ "What significant life events have shaped your beliefs about life?"
▲ "What are your earliest childhood memories as you first identified your ethnic/cultural background? Your spirituality?"
▲ "What are your earliest memories of church, synagogue, mosque, a higher power, of discovering your parents' vital life values?"
▲ "Where are you now in your life journey? Your spiritual journey?"

REFLECTING CLIENT MEANINGS

Say back to clients their exact key meaning and value words. Reflect their own unique meaning system, not yours. Implicit meanings will become clear through your careful listening and questions designed to elicit meaning issues from the client. Using the client's key words is preferable, but occasionally you may supply the needed meaning word yourself. When you do so, carefully check that the word(s) you use feel right to the client. Simply change "You feel . . ." to "You mean. . . ." A reflection of meaning is structured similarly to a paraphrase or reflection of feeling: "You value . . . ," "You care . . . ," "Your reasons are . . . ," or "Your intention was. . . ." Distinguishing among a reflection of meaning, a paraphrase, and a reflection of feeling can be difficult. Often the skilled counselor will blend the three skills together. For practice, however, it is useful to separate out meaning responses and develop an understanding of their import and power in the interview. Noting the key words that relate to meaning (*meaning, value, reasons, intent, cause,* and the like) will help distinguish reflection of meaning from other skills.

Reflection of meaning becomes more complicated when meanings or values conflict. Here concepts of confrontation (Chapter 8) may be useful. Conflicting values, either explicit or implicit, may underlie mixed and confused feelings expressed by the client. For instance, a client may feel forced to choose between loyalty to family and loyalty to spouse. Underlying love for both may be complicated by a value of dependence fostered by the family and the independence represented by the spouse. When clients make important decisions, sorting out key meaning issues may be crucial.

For example, a young person may be experiencing a value conflict over career choice. Spiritual meanings may conflict with the work setting. The facts may be paraphrased accurately and the feelings about each choice duly noted, yet the underlying *meaning* of the choice may be most important. The counselor can ask, "What does each choice *mean* for you? What sense do you make of each?" The client's answers provide the opportunity for the counselor to reflect back the meaning, eventually leading to a decision that involves not only facts and feelings but also values and meaning. And, as in confrontation, you can evaluate client change in meaning systems using the Client Change Scale (Chapter 8).

RESILIENCE, PURPOSE, AND MEANING

Resilience is the ability to recover from a wide variety of difficulties. Two examples of resilience might be the child who is teased, but bounces back cheerfully the next day, and the adolescent who undergoes being "dumped" by a boyfriend or girlfriend, but soon gets over it and moves on. What we see here is the ability to not let bad experiences get one down.

Resilience occurs at a deeper level when the individual suffers a serious or life-threatening trauma. A child is born in poverty, experiences abuse, and somehow manages to put together a successful life. A businessman goes bankrupt, but within two years has put together a successful business. Two women lose their jobs and after both spend six months searching for employment, one goes into a deep depression, while the other continues on and eventually finds work. Two soldiers on patrol experience an ambush in which one of their comrades is killed. One soldier leaves the warfront and ends up being treated for posttraumatic stress over a period of months; the resilient soldier has a short period of mourning, grief, and recovery, but soon is back on the front lines.

Meaning and purpose are key to resilience. Viktor Frankl was a survivor of the German death camps. Many around him gave up and died, but Frankl kept his focus on the positive and looked for moments of meaning. He enjoyed a beautiful sunset, even though hungry; he thought of his beloved wife; he wrote a book in his mind while doing painful work. Frankl is the theorist who truly brought the importance of meaning to our field. His personal example and his writings remain critical for us today.

Teaching resilience results in less childhood depression (Smith et al., 2009). Martin Seligman (2009), the founder of positive psychology, has shown that teaching purpose and meaning can make a difference among children. One of his major goals was to encourage children to find purpose and positive expectations:

> One exercise involved the students' writing down three good things that happened each day for a week. Examples were: "I answered a really hard question in Spanish class," "I helped my mom shop for groceries," or "The guy I've liked for months asked me out." Next to each positive event, the students answered the following questions: "What does this mean to you?" and "How can you increase the likelihood of having more of this good thing in the future?"

"What does this mean to you?" is, of course, the basic question to elicit meaning. Meaning is then reflected and synthesized and becomes an important part of the child's cognitive/emotional processing.

High-risk African Americans who have suffered trauma, but who have a sense of purpose and meaning, experience better mental and physical health (Alim et al., 2009). At the other end of life's spectrum, older people who have a clear sense of meaning and purpose have better mental and physical health. Those with a sense of meaning and purpose are 2½ times less likely to suffer the ravages of Alzheimer's disease (Boyle, Buchman, Barnes, & Bennett, 2010). Meaning is not found just in thoughts and feelings; it also affects the body.

Purpose and meaning can and should be a center of effective mental health counseling. We have found that suggesting that clients read Frankl's *Man's Search for Meaning* (1959) can be a valuable part of treatment leading to positive life changes.

The following discussion of discernment is another way to facilitate your clients' finding deeper meanings and visions for their lives.

DISCERNMENT: IDENTIFYING LIFE MISSION AND GOALS

Listen. Listen, with intention, with love, with the "ear of the heart." Listen not only cerebrally with the intellect, but with the whole of feelings, our emotions, imaginations, and ourselves. (de Waal, 1997)

Discernment is "sifting through our interior and exterior experiences to determine their origin" (Farnham, Gill, McLean, & Ward, 1991). The word *discernment* comes from the Latin *discernere*, which means "to separate," "to determine," "to sort out." In a spiritual or religious sense, discernment means identifying when the spirit is at work in a situation—the spirit of God or some other spirit. The discernment process is important for all clients, regardless of their spiritual or religious orientation or lack thereof. Discernment has broad applications to interviewing and counseling; it describes what we do when we work with clients at deeper levels of meaning. Discernment is also a process whereby clients can focus on envisioning their future as a journey into meaning. (See Box 10.2.)

BOX 10.2   Questions Leading Toward Discernment of Life's Purpose and Meaning

Following is a systematic approach to discernment. You may find it helpful to share this list with the client before you begin the discernment process and identify together the most helpful questions to explore. Add topics and questions that occur to you and the client. Discernment is a very personal exploration of meaning. The more the client participates, the more useful it is likely to be.

You or your client may wish to begin by thinking quietly about what might give life purpose, meaning, and vision. Questions that focus on the *here and now* and intuition may facilitate deeper discovery.

Here-and-now body experience and imaging can serve as a physical foundation for intuition and discernment.

▲ Relax, explore your body, find a positive feeling of strength to serve as an anchor for your search. Build on that feeling and see where it goes.

▲ Sit quietly and allow an image (visual, auditory, kinesthetic) to build.

▲ What is your gut feeling? What are your instincts? Get in touch with your body.

▲ Discerning one's mission cannot be found solely through the intellect. What feelings and thoughts occur to you at this moment?

▲ Can you recall feelings and thoughts from your childhood that might lead to a sense of direction now?

▲ What is your felt body sense of spirituality, mission, and life goal?

Concrete questions leading to telling stories can be helpful.

▲ Tell me a story about that image above. Or a story about any of the *here-and-now* experiences listed there.

▲ Can you tell me a story that relates to your goals/vision/mission?

▲ Can you name the feelings you have in relation to your desires?

▲ What have you done in the past or are you doing presently that feels especially satisfying and close to your mission?

▲ What are some blocks and impediments to your mission? What holds you back?

BOX 10.2   (Continued)

▲ Can you tell about spiritual stories that have influenced you?

For self-reflective exploration, the following are often useful.

▲ Let's go back to that original image and/or the story that goes with it. As you reflect on that experience or story, what occurs for you?

▲ Looking back on your life, what have been some of the major satisfactions? Dissatisfactions?

▲ What have you done right?

▲ What have been the peak moments and experiences of your life?

▲ What might you change if you were to face that situation again?

▲ Do you have a sense of obligation that impels you toward this vision?

▲ Most of us have multiple emotions as we face major challenges such as this. What are some of these feelings, and what impact are they having on you?

▲ Are you motivated by love/zeal/a sense of morality?

▲ What are your life goals?

▲ What do you see as your mission in life?

▲ What does spirituality mean to you?

The following questions place the client in larger systems and relationships—the self-in-relation. They may also bring multicultural issues into the discussion of meaning.

▲ Place your previously presented experiences and images in broader context. How have various systems (family, friends, community, culture, spirituality, and significant others) related to these experiences? Think of yourself as a self-in-relation, a person-in-community.

▲ *Family.* What do you learn from your parents, grandparents, and siblings that might

be helpful in your discernment process? Are they models for you that you might want to follow, or even oppose? If you now have your own family, what do you learn from them and what is the implication of your discernment for them?

▲ *Friends.* What do you learn from friends? How important are relationships to you? Recall important developmental experiences you have had with peer groups. What do you learn from them?

▲ *Community.* What people have influenced you and perhaps served as role models? What group activities in your community may have influenced you? What would you like to do to improve your community? What important school experiences do you recall?

▲ *Cultural groupings.* What is the place of your ethnicity/race in discernment? Gender? Sexual orientation? Physical ability? Language? Socio-economic background? Age? Life experience with trauma?

▲ *Significant other(s).* Who is your significant other? What does he or she mean to you? How does this person relate to the discernment process? What occurs to you as the gifts of relationship? The challenges?

▲ *Spiritual.* How might you want to serve? How committed are you? What is your relationship to spirituality and religion? What does your holy book say to you about this process?

Discernment questions from A. Ivey, M. Ivey, J. Myers, T. & Sweeney, *Developmental Counseling and Therapy: Promoting Wellness Over the Lifespan* (Boston: Lahaska/Houghton Mifflin, 2005). Reprinted by permission.

"There is but one truly serious problem, and that is . . . judging whether or not life is worth living" (Camus, 1955, p. 3). Viktor Frankl (1978) talks about the "unheard cry for meaning." Frankl claimed that 85% of people who successfully committed suicide saw life as meaningless, and he blamed an excessive focus on self. He said that people have a need for transcendence and living beyond the self.

The vision quest, often associated with the Native American Indian, Dene, and Australian Aboriginal traditions, is oriented to helping youth and others find purpose and meaning in their lives. These individuals often undertake a serious outdoor experience to find or envision their central life goals. Meditation is used in some cultures to help members find meaning and direction.

Visioning and finding meaning may often be facilitated if issues are explored with a guide, counselor, or interviewer. This can be a spiritual or religious quest for some clients, but the discernment process will be useful for all. (See Ivey, Ivey, Myers, & Sweeney, 2005, for additional information.) In Box 10.2, the specific discernment questions lead to further examination of goals, values, and meaning. Share the list of questions and encourage the client to participate with you in deciding which questions and issues are most important.

MULTICULTURAL ISSUES AND REFLECTION OF MEANING

For practical multicultural interviewing and counseling, recall the concept of focus (Chapter 9). When helping clients make meaning, focus exploration of meaning not just on the individual but also on the broader life context. In much of Western society, we tend to assume that the individual is the person who makes meaning. But in many other cultures—for example, the traditional Muslim world—the individual will make meaning in accord with the extended family, the neighborhood, and religion. Individuals do not make meaning by themselves; *they make meaning in a multicultural context*. In truth, Western society also draws meaning from family and culture. However, individualism rather than collectivism is generally the focus.

Cultural, ethnic, religious, and gender groups all have systems of meaning that give an individual a sense of coherence and connection with others. Muslims draw on the teachings of the Qur'an. Similarly, Jewish, Buddhist, Christian, and other religious groups will draw on their writings, scriptures, and traditions. African Americans may draw on the meaning strengths of Malcolm X, Martin Luther King, Jr., or on support they receive from Black churches as they deal with difficult situations. Women may make meaning out of relationships, whereas men may focus more on issues of personal autonomy and tasks. Witness the conversations in a mixed social group. Often we find women on one side of the room talking about relationships. Men will generally be talking about sports, politics, and their accomplishments.

Viktor Frankl, a German concentration camp survivor, could not change his life situation, but he was able to draw on important strengths of his Jewish tradition to change the meaning he made of it. The Jewish tradition of serving others facilitated his survival and enabled him to help fellow sufferers. When times were particularly bad, when prisoners had been whipped and were not being given food, Frankl (1959, pp. 131–133) counseled his entire barracks, helping them reframe their terrors and difficulties, pointing out that they were developing strengths for the future.

> I quoted from Nietzsche, "That which does not kill me, makes me stronger." . . . I spoke to the future. I said that . . . the future must seem hopeless. I agreed that each of us could guess . . . how small were chances for survival. . . . I estimated my chances at about one in twenty. But I also told them that, in spite of this, I had no intention of losing hope and giving up. . . . I also mentioned the past; all its joys and how its light shone even in the present darkness. . . . Then I spoke of the many opportunities of giving life a meaning. I told my comrades . . . that human life, under any circumstances has meaning. . . . I said that someone looks down on each of us in difficult hours—a friend, a wife, somebody alive or dead, or a God—and he would not expect us to disappoint him. . . . I saw the miserable figures of my friends limping toward me to thank me with tears in their eyes.

You may counsel clients who have experienced some form of religious bias or persecution. As religion plays such an important part in many people's lives, members of dominant religions in a region or a nation may have different experiences from those who follow minority religions. For example, Schlosser (2003) talks of Christian privilege in North America, where people of Jewish and other faiths may feel uncomfortable or even unwelcome during Christian holidays. Anti-Semitism, anti-Islamism, anti–liberal Christianity, and anti–evangelical Christianity are all possible results when clients experience spiritual and/or religious intolerance. We also recall that when Christians and other religious groups find themselves in countries where they are a minority, they can suffer serious religious persecution, to the point of death.

Complete Interactive Exercise 2: Identifying/Classifying Meaning

Complete Interactive Exercise 3: Client Issues With Meaning

Complete Interactive Exercise 4: Questions Eliciting Meaning

Complete Interactive Exercise 5: Discernment Exercise: Your Future in Interviewing, Counseling, or Therapy

Complete Weblink Critique Exercise: Your Turn

Complete Group Practice Exercise 1: Group Practice With Reflection of Meaning

10.4
INSTRUCTIONAL READING 2: The Skills of Interpretation/Reframing

KEY CONCEPT QUESTIONS

▲ **What are more detailed skills of interpretation/reframing?**

▲ **How does interpretation/reframing relate to other microskills?**

▲ **How is interpretation/reframing used differently among varying theoretical approaches to counseling?**

▲ **What are some multicultural issues within interpretation/reframing?**

When you use the microskill of interpretation/reframing, you are helping the client to restory or look at the problem or concern from a new, more useful perspective. This new way of thinking is central to the restorying and action process. In the microskills hierarchy, the words *interpretation* and *reframing* are used interchangeably. Interpretation reveals new perspective and new ways of thinking beneath what a client says or does. Reframing provides another frame of reference for considering problems or issues. And eventually the client's story may be reconsidered and restored.

An interpretation/reframe generally comes from the helper rather than the client and often is presented through a specific theoretical perspective. Reflection of meaning can inspire clients to find new meanings, new ways of looking at their situation, thus enabling them to reframe/interpret their situations on their own with minimal guidance from the interviewer.

The basic skill of interpretation may be defined as follows:

▲ The counselor listens to the client story, issue, or problem and learns how the client makes sense of, thinks about, or interprets the story or issue.

▲ The counselor, drawing from personal experience or a theoretical perspective, provides an alternative meaning or interpretation of the narrative. This may include *linking* together information or ideas discussed earlier that closely relate to each other. Linking is particularly important as it integrates ideas and feelings for clients and frees them to develop new approaches to their issues.

▲ (Example based on personal experience) "You feel coming out as gay led you to lose your job, and you blame yourself for not keeping quiet. Maybe you just really needed to become who you are. You seem more confident and sure of yourself. It will take time, but I see you growing through this difficult situation." Here self-blame has been reinterpreted or reframed as a positive step in the long run.

▲ (Example based on psychoanalytic theory with multicultural awareness) "It sounds like the guy who fired you is insecure about anyone who is different from him. He sounds as if he is projecting his own insecurities on you, rather than looking at his own heterosexism or homophobia."

Consider another example of interpretation developed from the logic of the interviewer. Note how this helps the client reframe his situation.

> Allen, the client, was going through a divorce and was very angry—a common reaction for those engaged in a major breakup, particularly when finances are involved. He was telling his attorney, at some length, about what he wanted and why. Attorneys use a form of interviewing involving many questions, and it sometimes involves informal counseling. After listening for a while to Allen's issues, the attorney got out from behind the desk and stood over Allen saying: "Allen, that's your story. But I can tell you that you won't get what you want. Your wife has a story as well and what will happen is something between what you both feel you need and deserve. For your own and your children's sake, think about that." This was a rather rough and confrontational reframing of Allen's story. It also changed the focus from Allen and his problems to his wife and children. Fortunately, he heard this powerful reframe, and resolution of differences in the divorce finally began.

This story has several implications. First, even with the most effective listening, clients may still hold on to unworkable stories, ineffective thinking, and self-defeating behaviors. Clearly, they need a new perspective. Respect clients' frame of reference before interpreting or reframing their words and life in new ways. In effect, *listen before you provide your interpretation or reframe*. There will always be some clients who will need the strong, confrontational interpretation that Allen got, but recall that the attorney first listened attentively to Allen.

The value of an interpretation or reframe depends on the client's reaction to it and how he or she changes thoughts, feelings, or behaviors. Think of the Client Change Scale (CCS)—how does the client react to each interpretation? If the client denies or ignores the interpretation, you obviously are working with denial (Level 1 on the CCS). If the client explores the interpretation/reframe and makes some gain, you have moved that client to bargaining and partial understanding (Level 2). Interchangeable responses and acceptance of the interpretation (Level 3) will often be an important part of the gradual growth toward a new understanding of self and situation. If the client develops useful new ways of thinking and behaving (Level 4), movement is clearly occurring. Transcendence (Level 5) will appear only with major breakthroughs that change the direction of interviewing, counseling, and psychotherapy. But recall that movement from denial (Level 1) to partial consideration of issues (Level 2) may be a major breakthrough, beginning client improvement.

The potential power of the effective interpretation/reframe can be seen in the divorce example above. Allen was in denial about what he could "win" in the divorce and refused even to bargain. But confronted by the attorney towering over him, he moved almost immediately from denial (Level 1) to a new understanding (Level 3) by accepting the attorney's reframe. The real test of change would be whether he actually *changes his behavior as a result of his new insights*. New solutions (Level 4) are seldom reached without behavior change. Transcendence (Level 5) is rarely found in complex cases of divorce!

INTERPRETATION/REFRAMING AND OTHER MICROSKILLS

Focusing, like reflection of meaning and interpretation/reframing, is another influencing skill that greatly facilitates the generation of new client perspectives. In the story of Allen and his attorney, the focus on the wife and children was key to the successful reframe. As another example, you may work with a client who feels that he or she has been subjected to gender discrimination or sexual harassment. If you just focus on the individual, the client may blame himself or herself for the problem. By focusing on gender or other multicultural issues, you are expanding the client's perspectives. The client may then generate a new perspective, meaning, or way of solving the problem on her or his own.

Interpretation may be contrasted with the paraphrase, reflection of feeling, focusing, and reflection of meaning. In those skills, the interviewer remains in the client's own frame of reference. In interpretation, the frame of reference comes from the counselor's personal and/or theoretical constructs. The following example shows interpretation (reframing) paired with other skills.

CLIENT: [with a record of absenteeism] I'm in trouble because I missed so many days of work.

COUNSELOR: You're really troubled and worried. [reflection of feeling] You've been missing a lot of work and you know your boss doesn't like it. [paraphrase/restatement] Could you tell me what missing work means to you? [eliciting meaning; the client responds with more information] I hear that you have tended to avoid conflict of any kind for years and this relates to your avoiding work in recent months. [eliciting and reflecting meaning] As I listen to you, I sense you're angry that your friend got the promotion and you didn't. How do you react to that? [Interpretation/reframe. Note that the interviewer goes "beyond the data" and provides a new frame of reference for viewing the situation. The checkout "How do you react to that?" enables the client to deal with the interpretation in a more open fashion.]

▲ EXERCISE 10.2 **Microskills Practice**

A client has come to you for therapy and, in past interviews, has been reviewing his life as it relates to present problems in his second marriage. Last week you reviewed sexual problems in the marriage, and you noted that he tended to focus on himself with little attention to his wife. Also this week, he speaks of a dream in which he is trying to break into a room, which contains a golden icon, but is frustrated in the process.

CLIENT: [with agitation] My wife and I had another fight over sex last night. We went to a sexy movie and I was really turned on. I tried to make love and she rejected me again.

How might you:

Paraphrase what the client just said? _____

Reflect this client's feelings? _____

Interpret the client's statement? _____

Generate several possibilities. Perhaps one might relate to how this situation compares to his first marriage, another to his insensitivity to his partner, and a third to the dream. See the end of this chapter for our possibilities.

In interpretation, the interviewer or counselor adds something beyond what the client has said. Any number of interpretive responses may be made to any client statement, and they may vary according to the theory and personal experience of the counselor. How many possible interpretations can you make, if you use the focus concept as the basis? Let your imagination run free.

Interpretation has traditionally been viewed as a mystical activity in which the interviewer reaches into the depths of the client's personality to provide new insights. However, if we consider interpretation to be merely a new frame of reference, the concept becomes less formidable. Viewed in this light, the depth of a given interpretation refers to the magnitude of the discrepancy between the frame of reference from which the client is operating and the frame of reference supplied by the interviewer.

Interpretations are best given by first attending carefully to the client and listening to the story or concern, then providing the interpretation, then checking out the client's reactions to the new frame of reference ("How does that idea come across to you?"). If an interpretation is unsuccessful, the interviewer can use data obtained from the checkout to develop another, more meaningful response, most often a return to listening skills.

THEORIES OF COUNSELING AND INTERPRETATION/REFRAMING

Theoretical interpretations can be extremely valuable, as they provide the interviewer with a tested conceptual framework for thinking about the client. Each theory is itself a story—a story told about what is happening in interviewing, counseling, and therapy and what the story means. Integrative theories find that each theoretical story has some value. As you generate your own natural style, most likely you will develop your own integrative theory, drawing from those approaches that make the most sense to you.

Imagine that you have worked with Charlis for some time and she has worked through many of her fears, but still faces some real challenges. She makes the following comment:

CHARLIS: Yesterday, my manager gave me a new assignment. I sensed that something was wrong, that he wanted something from me, but wouldn't say it. It made me feel very anxious as now I'm not sure what to do. Where do I go next?

Keep in mind that the interviewer could reflect back the anxiety concerns and turn the issue back to Charlis, or even ask the meaning of the situation. In interpretation, the counselor supplies an alternative perspective. Below are several examples of how counselors with different theoretical orientations might interpret the same information. Before each of these interpretations is a brief theoretical paragraph that provides a background for the theory-oriented interpretation that follows. The following presentations are slightly exaggerated for clarity.

Decisional Theory.   A major issue in interviewing for all clients is making appropriate decisions and understanding alternatives for action. Decisions need to be made with awareness of cultural/environmental context. Interpretation/reframing helps clients find new ways of thinking about their decisions. Linking ideas together is particularly important.

COUNSELOR: Charlis, it sounds as if the manager is again giving you a double message and that causes real anxiety. It's a repetition of some of the things that led to your heart attack. We've spent some time on dealing with your tension. This seems a good place to go over breathing and relaxation again.

Person-Centered Theory.   Clients are ultimately self-actualizing. Our goal is to help them find the story that builds on their strengths and helps them find deeper meanings and purpose. Reflection of meaning helps clients find alternative ways of viewing the situation; interpretation/reframing is not used. Linking can occur through effective summarization.

COUNSELOR: Charlis, you are really feeling anxious again, and the manager seems to be giving you a difficult time again. You're wondering what all this means. [Reflection of feeling and eliciting meaning. The conversation has returned to the client.]

Brief Counseling.   Brief methods seek to help clients find quick ways to reach their central goals. The interview itself is conceived first as a goal-setting process; then methods are found to reach goals through time-efficient methods. Interpretation/reframing will be rare except for linking of key ideas.

COUNSELOR: You're facing some of the old familiar challenges since the heart attack. Let's look at this and think back on what you've done in the past that works in this situation. What comes to mind? [Linking with a mild interpretation and a focus on finding what was effective for Charlis in the past, thus reminding her of her strengths.]

Cognitive-Behavioral Theory.   The emphasis is on sequences of behavior and thinking and on what happens to the client, internally and externally, as a result. Often interpretation/reframing is useful in understanding what is going on in the client's mind and/or linking the client to how the environment affects cognition and behavior.

COUNSELOR: OK, you came in to talk and suddenly the manager seemed almost to be attacking you [antecedent], and then you become anxious [emotional consequence], and this has left you wondering what to do [behavioral consequence]. Now let's analyze the situation— let's look at what he is doing and how he seems to make you feel. Then we can develop new, alternative, and more effective ways to deal with this in the future.

Psychodynamic/Interpersonal Theory.   Individuals are dependent on unconscious forces. Interpretation/reframing is used to help link ideas and enable the client to understand how the unconscious past and long-term, deeply seated thoughts, feelings, and behaviors frame the here and now of daily client experiencing. Freudian, Adlerian, Jungian, and several other psychodynamic theories each tell different stories.

COUNSELOR: Charlis, this seems to go back to the discussions we've had in the past about your issues with authority, particularly with your father. We've even noticed that sometimes you treat me as an authority figure. We see the here and now with me, the situation with your boss, and the long-term issues with your father. What sense do you make of this possible pattern in your life?

Multicultural Counseling and Therapy (MCT).   Everyone is always situated in a cultural/environmental context, and we need to help clients interpret and reframe their issues, concerns, and problems in relation to their multicultural background (see the RESPECTFUL model). MCT is an integrative theory that uses all of the methods above, as appropriate, to help clients understand themselves and how the cultural/environmental context affects them personally. The following is from a feminist therapy frame of reference.

COUNSELOR: Sounds to me like simple sexism once again. Charlis, we've got to work on how you can deal with a work environment that seems continually to be hassling you. Perhaps a complaint is in order. But at least we have to engage in more stress management to help you, as a woman, deal with this productively so that it does not destroy your health again.

All of the above provide the client with a new, alternative way to consider the situation. In short, interpretation renames or redefines "reality" from a new point of view. Sometimes just a new way of looking at an issue is enough to produce change. Which is the correct interpretation? Depending on the situation and context, any of these interpretations could be helpful or harmful. The first two responses deal with here-and-now reality, whereas psychodynamic interpretation deals with how the past relates to the present. The feminist interpretation links the heart attack with gender discrimination on the job.

Complete Interactive Exercise 6: Interpretation/Reframe
Complete Interactive Exercise 7: Interpretation/Reframing and Avoiding Spin
Complete Group Practice Exercise 2: Group Practice With Interpretation/Reframe

POSSIBLE RESPONSES TO EXERCISE 10.2

Here are some possible responses to the client in Exercise 10.2:

You're upset and angry. [reflection of feeling]

You had a fight after the "turn-on" movie and were rejected again. [paraphrase/restatement]

As I've listened to you, I hear your deep caring for your wife, and when she rejects you, it means you fear losing her. [reflection of meaning]

Sounds like you didn't take it as slowly and easily as we talked about last week and she felt forced once again. Am I close? [interpretation #1 with checkout]

Your anger with her seems parallel to the anger you used to feel toward your first wife. I wonder what sense you make of that? [interpretation #2 with checkout]

The feelings of rejection really bother you. Those sad and angry feelings sound like the dream last week. Does that make sense? [interpretation #3 (linking) with checkout]

CHAPTER SUMMARY

Key Points of "Reflection of Meaning and Interpretation/Reframing"

CourseMate and DVD Activities to Build Interviewing Competence

10.1 Defining the Skills of Reflecting Meaning and Interpretation/Reframing

▲ A reflection of meaning looks very much like a paraphrase but focuses beyond what the client says. Often the words *meaning, values,* and *goals* will appear in the discussion. Clients are encouraged to explore their own meanings in more depth from their own perspective. Questioning and eliciting meaning are often vital as first steps.

▲ Interpretations/reframes provide the client with a new perspective, frame of reference, or way of thinking about issues. They may come from the counselor's observations, they may be based on varying theoretical orientations, or they may link critical ideas together.

▲ The two skills are similar in helping clients generate a new and potentially more helpful way of looking at things. Reflection of meaning focuses on the client's worldview and seeks to understand what motivates the client; it provides more clarity on values and deeper life meanings. An interpretation results from interviewer's observation and seeks new and more useful ways of thinking.

1. Case Study: Finding Family Balance in Greece: A Universal Problem. How would you use reflection of meaning and interpretation/reframing to help Katerina?
2. Weblink: International Association on Personal Meaning.
3. Weblink: Religion and Spirituality.
4. Weblink: Sexual Orientation: Science, Education, and Policy—University of California at Davis.
5. Weblink: Feminist Therapy.
6. Weblink: Sunshine for Women.
7. Weblink: American Psychological Association (APA), Just the Facts About Sexual Orientation and Youth: A Primer for Principals, Educators, and School.
8. Weblink: Sexual Orientation.
9. Weblink: Heterosexuals Questionnaire.
10. Weblink: Gender and Race—University of Iowa.
11. Weblink: Men's Resources.
12. Weblink: Women's Resources.

10.2 Example Interview: Travis Explores the Meaning of a Recent Divorce

▲ The counselor used listening skills and key word encouragers to focus on meaning issues.

▲ Open questions oriented to values and to meaning are often effective in eliciting client talk about meaning issues.

▲ A reflection of meaning looks very much like a paraphrase except that the focus is on implicit deeper issues, often not expressed fully in the surface language or behavior of the client.

1. Interactive Exercise 1: Identifying Skills. Practice identifying counselor responses.

10.3 Instructional Reading 1: The Specific Skills of Eliciting and Reflecting Meaning

▲ Eliciting meaning is accomplished through carefully listening to the client for meaning words. Questions related to values, meaning, life goals, and ultimate causal issues bring out client meanings. "What meaning does that have to you?" "What sense do you make of that?"

1. Interactive Exercise 2: Identifying/Classifying Meaning. Practice how to help a person clarify the meaning of his or her situation.

Key Points of "Reflection of Meaning and Interpretation/Reframing"

▲ A reflection of meaning looks much like a paraphrase or reflection of feeling except that the key words related to meaning receive special attention. "Your goals (e.g., spirituality/family/life) mean much to you as you plan your next step."

▲ Discernment is a form of listening that goes beyond our usual descriptions and could be termed "listening with the heart." Both you and the client seriously search for deeper life goals and direction. Specific discernment questions are in Box 10.2.

▲ Multicultural and family issues and stories may be key in helping clients discover personal meaning. Eliciting meaning and focusing reflection on contextual issues beyond the individual will enhance and broaden one's understanding of life's deeper concerns.

CourseMate and DVD Activities to Build Interviewing Competence

2. Interactive Exercise 3: Client Issues With Meaning. Helping a senior in college explore whether or not he wants to go on to graduate school in school counseling.

3. Interactive Exercise 4: Questions Eliciting Meaning. List questions that might be useful in bringing out the meaning of an event.

4. Interactive Exercise 5: Discernment Exercise: Your Future in Interviewing, Counseling, or Therapy. The fields of professional helping are challenging. If you have a sense of mission and purpose, this exercise may help you organize your thinking in a new way and help you get through rough times.

5. Weblink Critique Exercise: Your Turn. Create your own Weblink Critique based on your review of the following websites: Scientific Society for Logotherapy and Existential Analysis; Logotherapy; Viktor Frankl at Ninety.

6. Group Practice Exercise 1: Group Practice With Reflection of Meaning.

10.4 Instructional Reading 2: The Skills of Interpretation/Reframing

▲ First be sure that you have heard the client's story or concerns, and then draw from personal experience or a theoretical perspective to provide the client a new way of thinking and talking about issues.

▲ The effectiveness of an interpretation can be measured on the Client Change Scale. The new perspective is useful if the client moves in a positive direction.

▲ Focusing and multicultural counseling and therapy are the most certain ways to bring multicultural issues into the interview. A woman, a gay or lesbian, or a person from a minority group may be depressed over what is considered a personal failure. Helping the client see the cultural/environmental context of the issue can permit a new perspective to appear, providing a totally new and more workable meaning.

▲ Each interviewing and counseling theory provides us with a new and different story about the interview. Drawing from theory for interpretation/reframing provides a more systematic frame for considering the client. However, logic and your personal experience and observations may be as effective as a theoretically oriented reframe.

1. Interactive Exercise 6: Interpretation/Reframe. Practice interpretations to provide alternative frames of reference or perspectives for events in a client's life.

2. Interactive Exercise 7: Interpretation/Reframing and Avoiding Spin. An exercise to learn to avoid spin and use interpretation/reframing.

3. Group Practice Exercise 2: Group Practice With Interpretation/Reframe.

Assess your awareness, knowledge, and skills as you conclude the chapter:

1. Flashcards: Use the flashcards to check your understanding of key concepts and facilitate memorization of key information.
2. Self-Assessment Quiz: The quiz will help you assess your current knowledge and prepare for course examinations.
3. Portfolio of Competencies: Evaluate your present level of competence in attending and listening skills using the downloadable Self-Evaluation Checklist. Self-assessment of your attending skills competence demonstrates what you can do in the real world.

SELF-DISCLOSURE AND FEEDBACK
Bringing Immediacy
Into the Interview

INFLUENCING SKILLS AND STRATEGIES

FOCUSING

CONFRONTATION

THE FIVE-STAGE INTERVIEW STRUCTURE

REFLECTION OF FEELING

ENCOURAGING, PARAPHRASING, AND SUMMARIZING

OPEN AND CLOSED QUESTIONS

ATTENDING BEHAVIOR

ETHICS, MULTICULTURAL COMPETENCE, AND WELLNESS

"Before I open up to you (self-disclose), I want to know where you are coming from." . . . In other words, a culturally different client may not open up (self-disclose) until you, the helping professional, self-disclose first. Thus, to many minority clients, a therapist who expresses his/her thoughts and feelings may be better received in a counseling situation.

—Derald Wing Sue and Stanley Sue

How can the skills of self-disclosure and feedback help you and your clients?

CHAPTER GOALS

This chapter first reviews some key issues about interpersonal influence in the interview. Then two closely related influencing skills, most often used in humanistic/existential and person-centered theories to facilitate client's creation of new stories, are reviewed. When an interviewer effectively self-discloses or provides feedback, the here-and-now atmosphere of the interview becomes more immediate, personal, and real.

Awareness, knowledge, and skills developed through the concepts of this chapter will enable you to

- ▲ Better understand how listening skills and influencing skills facilitate client change and growth.
- ▲ Use appropriate self-disclosure, which builds a sense of equality and encourages client trust and openness. Disclosure may be about immediate here-and-now feelings, history of the interviewer, or thoughts and feelings that the counselor may have.
- ▲ Offer accurate feedback on how clients are experienced by the interviewer in the here and now of the session; how clients are progressing on issues; how others view the clients; and thoughts, feelings, and behaviors observed by the interviewer.

Assess your awareness, knowledge, and skills as you begin the chapter:

1. Self-Assessment Quiz: The chapter quiz will help you determine your current level of knowledge. You can take it before and after reading the chapter.
2. Portfolio of Competencies: Before you read the chapter, fill out the downloadable Self-Evaluation Checklist to assess your existing knowledge and competence in attending skills. Then, at the end of the chapter, complete the checklist again to summarize your competencies after study and practice.

11.1
INTERPERSONAL INFLUENCE, LISTENING SKILLS, AND INFLUENCING STRATEGIES—INTERVIEW EXAMPLE

KEY CONCEPT QUESTIONS

- ▲ **How do listening skills relate to influencing skills and strategies?**
- ▲ **What is the 1-2-3 pattern of interpersonal influence?**
- ▲ **How should we disclose to the client that we are using influencing skills?**

Influencing is part of all interviewing and counseling; even with a person-centered approach using only attending skills, the interviewer still influences what occurs in the session. Through selective attention and the topics you choose consciously or unconsciously to emphasize (or ignore), you influence what the client says. *You cannot **not** influence what happens in the interview.* Confrontation, focusing, reflection of meaning, and interpretation/reframing have all been identified as strategies of interpersonal influence with which you may have a more immediate impact on the client than with just listening alone.

Ethical practice demands respect for the client and awareness of the power relationship inherent in the interview. Interviewers and counselors, by their position, have perceived power. Use your power with awareness of client needs, and encourage client participation in the session. As you move to the direct action associated with the influencing skills, do not forget the foundation of listening and empathic understanding. Carefully developed listening skills of paraphrasing, reflecting feelings, and summarizing are sometimes lost when one masters the influencing skills and strategies. Step back and remember that counseling and interviewing are for the client, not for you. Effective use of influencing skills enables full client participation in the session.

The degree of interpersonal influence desired in the interview varies from theory to theory. The word *influence* can be upsetting to a humanistic or person-centered counselor. By way of contrast, the many proponents of cognitive-behavioral theory aim to influence directly as much client change as possible. But all theories agree that client involvement in the change process remains central.

Disclosure of what is going to happen in the session is an important part of structuring the five-stage interview. This same egalitarian disclosure is helpful when you use a specific new skill or strategy. By way of comparison, think of the effective dentist, nurse, or physician and how each one tells you ahead of time what to expect and whether it might be painful. Our clients deserve the same respect. Disclosure tends to build comfort and trust even when the next step of the interview may not be comfortable. For example, if you have focused on listening and then decide to use influencing strategies such as interpretation/reframing, self-disclosure, or feedback, or you want now to provide specific directions for the session, spend a moment acquainting the client with the change in style and the potential benefits. The general rule is to avoid surprises, although occasionally it is the very surprise that helps a client discover important new ideas.

The 1-2-3 strategic model for using influencing skills is vital to maintaining client participation in the session. Keep listening skills as your most prominent style, even though you may be using a very directive intervention.

1. *Listen.* Attend, observe, and listen to clients. Be sure that you understand where they are coming from, their issues and strengths, and whether they know what to expect from you.
2. *Assess and influence.* An influencing skill is best used after you have heard and understood the client's story. Then select a relevant influencing strategy, keep it brief, and time it appropriately to meet client needs.
3. *Check out and observe client response.* Use a checkout ("How do you respond to what I just suggested?"). Observe client response to your intervention. The client will be giving you immediate feedback on the effectiveness of your use of influencing skills. If the client ignores your lead or takes your comment in an inappropriate direction, return to listening and either try again or move to a completely different intervention.

The effectiveness of your intervention can be readily assessed using the Client Change Scale (Chapter 8). It takes some practice, but eventually, you will be able to assess client reactions to your leads in the here and now of the session. Being intentional demands that you be flexible and ready to move with the client. Is something new created with your intervention?

 Complete Interactive Exercise 1: Identifying Skills

THE CASE OF ALISIA: HOW LISTENING SKILLS CAN INFLUENCE CLIENTS

The client, Alisia, comes in with the complaint that she can't express herself and people "run all over" her. The first interview began with a short rapport phase, and permission was obtained to record the session. The following transcript is an edited version of Alisia's first two interviews, in which listening skills were used almost exclusively. This is a modified person-centered approach, in that questions are used more frequently than is typical of that theory. Effective listening skills and selective attending empower Alisia and enable her to see herself more as a self-in-relation, a person-in-community. Each of the influencing strategies presented in this and the next chapter will be applied to her issues.

The counselor spent time on Alisia's strengths and wellness assets. As part of the wellness search, the counselor encouraged Alisia to discuss several women heroes on whom Alisia would like to model herself. By the end of the interview, Alisia was still at Level 3 on the CCS but understood her issues and her strengths and resources.

Interviewer and Client Conversation	Process Comments
1. **Counselor:** Alisia, could you tell me what you'd like to talk about today?	The counselor personalizes the interview by using the client's name and the word *you* twice in the opening question.
2. **Alisia:** I simply can't express myself. I've tried many times and I can't get people to listen to me. Whether it is the boss, my partner, or the man at the garage, they all seem to run over me.	Alisia immediately identifies her central issue. She appears to start the interview with acceptance and recognition of her problems (Level 3 on the Client Change Scale, or CCS). You will find that some clients take several interviews before the problem is defined this clearly. But Alisia likely needs further understanding of this issue before change can be expected.
3. **Counselor:** Run over you?	Encourager focused on last few words.
4. **Alisia:** Yeah, I keep finding that I'm so accommodating that I'm always trying to get along. I was taught that I should please others. People like me for going along with them, but I never get what I want. I'm discouraged and disgusted with myself.	Alisia's body language is agitated and her vocal tone moves to a higher pitch, which often indicates insecurity.

Interviewer and Client Conversation	Process Comments
5. **Counselor:** Sounds as if you are really frustrated and disgusted about your inability to express yourself.	Reflection of feeling. Do you think the counselor should have changed the feeling word *discouraged* to *frustrated*?
6. **Alisia:** Right, it just goes on and on . . . I never seem to change.	Resignation in her vocal tone, almost a sound of defeat.
7. **Counselor:** Could you give me a specific example of the last time you had these feelings of discouragement? What happened? What did you say? What did they do?	Open question searching for concreteness. If you look for one specific example, you will often obtain a much clearer understanding of client style and the depth of the problem.
8. **Alisia:** Well, I was at the garage. I had called in for an early morning appointment; I had to go to a meeting at 10:00. They said come in at 8:00 and so I was there on time. When I checked at 9:30, they hadn't even started yet. The service manager just smiled and said, "Sorry, lady, we couldn't get to it." He's one of those guys who really like to demean women. But I just looked down and didn't say anything—even though I really wanted to scream. I made another appointment, but my car still has that screwy, strange sound.	As Alisia shares the concrete example, she starts speaking in a more angry tone of voice and clenches her fist. Note that her lack of assertiveness shows here in a situation where she is not fully comfortable—the garage. But she is also able to point out that the service manager was likely being unfair to other women as well. She shows awareness of cultural/contextual issues and starts to get in touch with underlying anger. This is further evidence of her functioning at the levels of acceptance and recognition, but no change on the CCS.
9. **Counselor:** And after all this the car still isn't right. As I see you now, Alisia, you also seem to be getting angry. What's happening with you as you talk to me about this?	Paraphrase, reflection of feeling, open question oriented to the here and now.
10. **Alisia:** Angry! Men!! I hurt too, deep inside. And I'm confused. [tears, but her eyes are flashing with determination] Everywhere I turn, it's there.	Listening skills also influence clients. Through sharing her story, Alisia becomes more aware of how she feels. The counselor was surprised at this explosion.
11. **Counselor:** I hear your anger and frustration. You're really angry and upset. You're tired of taking things as they are. The situation at the garage is just one instance of a pattern—something that repeats in various forms again and again. Have I heard you correctly?	Summarization with an emphasis on repeating patterns. Not all clients can see that they exhibit similar behaviors in different situations. A more concrete and less verbal client likely would be unable to realize that the situations are parallel.
12. **Alisia:** Yes, it's a pattern. The same day as the problem at the garage, my boss started leering at me again. I'm sick and tired of it. I used to think it was my fault, but now I'm wondering if men are the problem. The garage hassles me, the boss hassles me, and my partner does the same thing when he doesn't listen to me.	Notice how Alisia builds on the garage awareness to look at herself in other situations. Through effective use of listening skills, Alisia is starting to consider changes. This really is an expansion of acceptance/awareness (Level 3 on the CCS).

Listening has brought to Alisia a greater and clearer awareness of her issues. The second interview continues much the same as the first session, but near the end of the session, we hear the following.

Interviewer and Client Conversation	Process Comments
Counselor: So, Alisia, we've been talking for nearly an hour now. How do you put together all we've talked about? Have we missed something today?	The counselor could have summarized the session for Alisia, but uses the two questions as a way to involve the client in evaluation and planning for the future. The questioning process makes this version of person-centered counseling more active and influencing in style.
Alisia: I realize that much of what's been happening to me is a result of societal sexism. I learned in my family to do what "a good girl" should do and try to let the negative go and just be pleasant. But I'm not a girl; I'm a woman. I'm going to file harassment charges against my boss. And if we're going to stay together, I think I need to go with my partner for couples counseling. There! I feel better about myself right now.	She starts to see the need for behavioral change. Clients are not always so clear. This and the counselor statement above have been shortened from several client–counselor exchanges.
Counselor: You said a mouthful there, Alisia. That is a lot of things to do. Let's pick one or two of these possibilities and contract for what you might do next week as a start.	By suggesting contracting for change in behavior, the counselor has decided to move toward an active influencing approach and has presented a directive to the client. But even this directive includes Alisia in the planning of what is to happen.

Alisia is generating a new solution (Level 4 of the CCS) but still has some distance to go if her behavior is to change. Cognitively and emotionally, she is starting to touch on Level 4, but behavioral change will also be necessary to cement her newer thoughts and feelings. Drawing out her story carefully through the person-centered approach of these first two sessions provides a foundation for more active influencing later in this chapter.

 Complete Interactive Exercise 2: Using Influencing Skills in the Interview

11.2
DEFINING SELF-DISCLOSURE

KEY CONCEPT QUESTIONS

▲ **How do we define the strategy of self-disclosure?**

▲ **When is it appropriate to use this skill? When not?**

Should you share your own personal observations, experiences, and ideas with the client? When you use self-disclosure as described below, you can encourage client talk, create additional trust between you, and establish a more equal relationship in the interview.

Self-Disclosure	Predicted Result
As the interviewer, share your own related past personal life experience, here-and-now observations or feelings toward the client, or opinions about the future. Self-disclosure often starts with an "I" statement. Here-and-now feelings toward the client can be powerful and should be used carefully.	The client is encouraged to self-disclose in more depth and may develop a more egalitarian interviewing relationship with the interviewer. The client may feel more comfortable in the relationship and find a new solution relating to the counselor's self-disclosure.

Self-disclosure by the counselor or interviewer has been a highly controversial topic. Not everyone agrees that this is a wise strategy to include among the counselor's techniques. Many theorists argue against counselors' sharing themselves openly, preferring a more distant, objective persona. They express valid concerns about the counselor's monopolizing the interview or abusing the client's rights by encouraging openness too early. Self-disclosure can become therapy for the interviewer, with too much talk not really relating to the client.

Humanistically oriented and feminist counselors have demonstrated the value of appropriate self-disclosure. Multicultural theory considers self-disclosure early in the interview as key to trust building in the long run, particularly if your multicultural background is substantially different from that of your client. Box 11.1 provides an international perspective on self-disclosure. Research reveals that clients of counselors who self-disclose like their counselors more and report lower levels of symptom distress (Barrett & Berman, 2001).

▲ EXERCISE 11.1   **What Are Your Thoughts About Self-Disclosure?**

How comfortable do you feel sharing personal information? What do you think about revealing personal information to your interviewee or client? When do you think it is appropriate, and when is it not? What kind of personal information is more appropriate to share? What would be the purpose of self-disclosure? Write down your thoughts.

Now read the following online articles on self-disclosure:

Self-Disclosure & Transparency in Psychotherapy and Counseling, by Ofer Zur (2009). http://www.zurinstitute.com/selfdisclosure1.html#ethical

Ethical Aspects of Self-Disclosure in Psychotherapy: Knowing What to Disclose and What Not to Disclose, by Thomas G. Gutheil (2010). http://integral-options.blogspot.com/2010/05/thomas-g-gutheil-md-ethical-aspects-of.html

What are your thoughts about self-disclosure now, after reading these articles? What would you do differently as a consequence of reflecting on and reading about self-disclosure?

BOX 11.1 National and International Perspectives on Counseling

When Is Self-Disclosure Appropriate?
WEIJUN ZHANG

My good friend Carol, a European American, has had lots of experience counseling minority clients. She once told me that one of the first questions she asks her minority clients is "Do you have any questions to ask me?" which often results in a lot of self-disclosure on her part. She would answer questions not only about her attitudes toward racism, sexism, religion, and so forth, but also about her physical health and family problems. During the initial interview, as much as half of the time available could be spent on her self-disclosure.

"But is so much self-disclosure appropriate?" asked a fellow student in class after I mentioned Carol's experience.

"Absolutely," I replied. We know that many minority clients come to counseling with suspicion. They tend to regard the counselor as a secret agent of society and doubt whether the counselor can really help them. Some even fear that the information they disclose might be used against them. Some questions they often have in mind about counselors are "Where are you coming from?" "What makes you different from those racists I have encountered?" and "Do you really understand what it means to be a minority person in this society?" If you think about how widespread racism is, you might consider these questions legitimate and healthy. And unless these questions are properly answered, which requires a considerable amount of counselor self-disclosure, it is hard to expect most minority clients to trust and open up willingly.

Some cultural values held by minority clients necessitate self-disclosure from the counselor, too.

Asians, for example, have a long tradition of not telling personal and family matters to "strangers" or "outsiders" in order to avoid "losing face." Thus, relative to European Americans, we tend to reveal much less of ourselves in public, especially our inner experience. A mainstream counselor may well regard openness in disclosing as a criterion for judging a person's mental health, treating those who do not display this quality as "guarded," "passive," or "paranoid." However, traditional Asians believe that the more self-disclosure you make to a stranger, the less mature and wise you are. I have learned that many Hispanics and Native Americans feel the same way.

Because counseling cannot proceed without some revelation of intimate details of a client's life, what can we do about these clients who are not accustomed to self-disclosure? I have found that the most effective way is not to preach or to ask, but to model. Self-disclosure begets self-disclosure. We can't expect our minority clients to do well what we are not doing ourselves in the counseling relationship, can we?

According to the guidance found in most counseling textbooks here, excessive self-disclosure by the counselor is considered unprofessional; but if we are truly aware of the different orientation of minority clientele, it seems that some unorthodox approaches are needed.

The classmate who first questioned the practice asked with a smile, "Why are you so eloquent on this topic?" I said, "Perhaps it is because I have learned this not just from textbooks, but mainly from my own experience as both a counselor and a minority person."

You cannot expect to have experienced all that your clients bring to you. After an appropriate time for introducing the session, when a man works with a woman, for example, it may be useful to say, "Men don't always understand women's issues. The things you are talking about clearly relate to gender experience. I'll do my best, but if I miss something, let me know. Do you have any questions for me?" If you are White or African American working with a person of the other race, frank disclosure that you recognize the differences in cultures at the initiation of the interview can be helpful in developing trust. These suggestions and statements are generalizations. Imagine that you are a heterosexual Christian Asian counselor working with a conservative Christian Latina struggling with lesbian issues. Your background is very different and truly empathizing with the client may be a challenge. How would you self-disclose?

When you are multiculturally different from your client, self-disclosure requires more forethought, and it is vital that you are personally comfortable with difference. Open discussion and some self-disclosure on your part may be essential. The frank disclosure of differences can be helpful in establishing rapport. Many alcoholics are dubious about the ability of nonalcoholics to understand what is occurring for them. A client with cancer or heart disease may feel that no one can really understand if that person has not experienced this particular illness. On the other hand, if you have had a difficult life experience with alcohol, sharing your story briefly can be very helpful. It often helps to say to the client, "Please feel free to ask me questions about myself if you wish." Just remember to keep your responses brief! Be sure that what you do or do not do is authentic to your own knowledge and comfort level, but also be sure that the self-disclosure is reasonable and comfortable for the client.

Here are four dimensions of self-disclosure:

1. *Listen.* Follow the 1-2-3 pattern described earlier in the chapter. Attend to the client's story, assess the appropriateness of your self-disclosure and share it briefly, and return focus to the client, while noting how he or she receives the self-disclosure

2. *Use "I" statements.* Interviewer self-disclosure almost always involves "I" statements or self-reference using the pronouns *I, me,* and *my*—or the self-reference may be implied.

3. *Share and describe briefly your thoughts, feelings, or behaviors.* "I can imagine how much pain you feel." "I feel happy to hear you talk of that wonderful experience— it was a real change!" "My experience of divorce was hurtful." "I think your friends are taking advantage of you." "I also grew up in an alcoholic family and understand some of the confusion you feel."

4. *Use appropriate immediacy and tense.* The most powerful self-disclosures are usually made in the here and now, the present tense. ("Right now I feel _____ ." "I am hurting for you at this moment—I care.") However, variations in tense are used to strengthen or soften the power of a self-disclosure.

Making self-disclosures relevant to the client is a complex task involving the following issues, among others.

Genuineness in Self-Disclosure.   To demonstrate genuineness, the counselor must truly and honestly have the feelings, thoughts, or experiences that are shared. In addition, self-disclosure must be genuine and appropriate in relation to the client. For example, a client may have performance anxiety about a part in a school play or a presentation to a small class. The counselor may genuinely feel anxiety about giving a lecture before 100 people. The feelings may be the same or at least similar, but true genuineness demands a synchronicity with the client's world. The counselor's experience, in this case, may be too distant from that of the client.

Tense. Consider the following three self-disclosing statements.

ALISIA: I am feeling really angry about the way I'm treated by men in power positions.

COUNSELOR: (present tense) You're coming across really angry right now. I like that you finally are in touch with your feelings.

COUNSELOR: (past tense) I've had the same difficulty expressing feelings in the past. I recall when I would just sit there and take it.

COUNSELOR: (future tense) Awareness of emotions can help us all be more in touch with ourselves in the future. I know that keeping in touch with feelings will continue to aid me.

Be Careful When Clients Say, "What Would You Do If You Were in My Place?" Clients will sometimes ask you directly for opinions and advice on what you think they should do. "What do you think I should major in?" "If you were me, would you leave this relationship?" "Would you have an abortion?" Effective self-disclosure and advice can potentially be helpful, but it is *not* the first thing you need to do. Your task is to help the client make her or his own decisions. The right solution for you may not be the right solution for the client, and involving yourself too early can foster dependency and lead the client in the wrong way. Note the following exchange.

ALISIA: How do you think I ought to present the idea of couples counseling to Chris?

COUNSELOR: I'm not in your position, and I've not heard too much about Chris yet. First, could we explore your relationship in more detail?

COUNSELOR: (If you feel forced to share your thoughts when you prefer not to, keep your comments brief and ask the client for her reflections.) Alisia, my own thought would be to share with Chris how you are hurting. But I'm not you. Chris may not hear that. Does that relate to you at all?

COUNSELOR: (after drawing out more information on the relationship) From what I've heard, it sounds wise to bring up the idea of counseling when Chris is in a good mood. I think it is important to bring it up directly, and I admire the way you are thinking ahead. How does that sound?

▲ EXERCISE 11.2 **Writing Self-Disclosure Statements**

Individually or in pairs, write one effective and one ineffective self-disclosure for each of the following statements.

"I'm feeling very vulnerable talking about my stuff right now. I wonder if you're really understanding my story?"

"I have to do a class presentation, and I feel very afraid speaking in front of my classmates. I don't like this type of course assignment. What should I do?"

  Complete Interactive Exercise 3: Writing Self-Disclosure Statements

11.3
DEFINING FEEDBACK

KEY CONCEPT QUESTION

▲ How do we define feedback and identify some important considerations of using feedback?

To see ourselves as others see us,
To hear how others hear us,
And to be touched as we touch others . . .
These are the goals of effective feedback.

When you use feedback as described below, you can predict the client's response.

Feedback	Predicted Result
Present clients with clear information on how the interviewer believes they are thinking, feeling, or behaving and how significant others may view them or their performance.	Clients may improve or change their thoughts, feelings, and behaviors based on the interviewer's feedback.

Knowing how others see us is a powerful dimension in human change, and it is most helpful if the client solicits feedback. Feedback is an important influencing strategy if you have developed good rapport and enough experience with the client to know that he or she trusts you. The following feedback guidelines are critical in counseling and interviewing.

1. *The client receiving feedback should be in charge.* Listen first and use the 1-2-3 pattern to determine whether the client is ready for feedback. Feedback is more successful if the client solicits it. (Listen—provide feedback—use checkout.)
2. *Feedback should focus on strengths and/or something the client can do something about.* It is more effective to give feedback on positive qualities and build on strengths. Corrective feedback focuses on areas in which the client can improve thinking, feeling, and behaving. Corrective feedback needs to be about something the client can change, or a situation the client needs to recognize and accept as something that can't be changed.
3. *Feedback should be concrete and specific.* "You had two recent arguments with Chris that upset both of you. In each case, I hear you giving in almost immediately. You seem to have a pattern of giving up, even before you have a chance to give your own thoughts. How do you react to that?"
4. *Feedback should be relatively nonjudgmental and interactive.* Stick to the facts and specifics. Facts are friendly; judgments may or may not be. Demonstrate your nonjudgmental attitude through your vocal qualities and body language. "I do see you trying very hard. You have a real desire to accept the way Chris is and learn to live with what you can't change. How does that sound?"—not "You give in too easily, I wish you'd try harder" or the all-too-common "That was a *good* job."

5. *Feedback should be lean and precise.* Don't overwhelm the client; keep corrective feedback brief. Most of us can hear only so much and can change only one thing at a time. Select one or two things for feedback, and save the rest for later.
6. *Check out how your feedback was received.* Involve clients in feedback through the checkout. Their response indicates whether you were heard and how useful your feedback was. "How do you react to that?" "Does that sound close?" "What does that feedback mean to you?"

Positive feedback has been described as the "the breakfast of champions." Your positive, concrete feedback helps clients restory their problems and concerns. Whenever possible, find things right about your client. Even when you have to provide challenging feedback, try to include positive assets of the client. Help clients discover their wellness strengths, positive assets, and useful resources.

Corrective feedback is a delicate balance between negative feedback and positive suggestions for the future. When clients need to seriously examine themselves, corrective feedback may need to focus on things they are doing wrong or behavior that may hurt them in the future. Management settings, correctional institutions, and schools and universities often require the interviewer to provide corrective feedback in the form of reprimands and certain types of punishments. When you must give negative corrective feedback, keep your vocal tone and body language nonjudgmental and stick to the facts, even though the issues may be painful. **Praise and supportive statements** ("You can do it, and I'll be there to help") convey your positive thoughts about the client, even when you have to give troubling feedback.

Negative feedback is necessary when the client has not been willing to hear corrective feedback. For example, in cases of abuse, planned behavior that hurts self or others, and criminal behavior, negative feedback—including the negative consequences the client's actions can bring—is necessary and can be beneficial. It is our responsibility to act in these situations. But listening to the client's point of view, even if the client is a perpetrator, remains important.

COUNSELOR: Alisia, I admire your ability to hang in there with Chris and accept things as they are, but you really are giving away too much control. You need stronger boundaries. It is OK for you to be assertive and own your own space. You have strength and ability. We can work on helping you become your own person. How does that feedback sound to you?

Feedback is sometimes used when the client avoids certain topics. You may have a client who suddenly switches topic or gives you only a brief, vague response. Many clients have sensitive issues or topics that they don't want to explore. If the interview is just for one to three sessions, it is usually best to accept that behavior. But sometimes the issue really needs to be faced. Then you need to meet the client and use confrontation skills as part of the feedback. Following are some examples.

COUNSELOR: We've talked around the issue of dealing with Chris, but you never say whether you really want to stay together. On one hand, I hear you wanting to resolve issues, but on the other hand, you avoid expressing what you feel about the relationship and what you want.

COUNSELOR: On one hand, Alisia, you really do seem to want to become more assertive, but then when we start to talk seriously about how you might actually change, you avoid the issues and turn away.

COUNSELOR: Just now, I saw it again. We were starting to deal with real issues and you changed the topic. What's going on?

Following are some examples of different types of feedback.

Vague, judgmental, negative feedback	Alisia, I really don't think the way you are dealing with people who hassle you is effective. You come across as a weak person.
Concrete, nonjudgmental, positive feedback	Alisia, you have potential; I sense that you tried very hard to stand up for your rights in the garage. Now you seem more assertive than you used to be. May I suggest some specific things that might be helpful the next time you face that situation?
Corrective feedback	Your effort was in the right direction. You can do even more if we set up an assertiveness training session for you. As we do this, I can offer more suggestions to help you change behavior. (See Chapter 14 for an example of a cognitive-behavioral assertiveness training session.)

▲ EXERCISE 11.3   **Learning About Feedback From Your Own Experience**

Remember a positive and a negative experience with feedback. These may have involved feedback from friends, family, teachers, or a work supervisor. What do you notice personally about effective and ineffective feedback? Now that you have read this section of the chapter, what would you do better, or what would you avoid, when giving feedback?

Complete Interactive Exercise 4: Personal Experience With Feedback
Complete Interactive Exercise 5: Writing Feedback Statements
Complete Group Practice Exercise: Group Practice With Self-Disclosure and Feedback
Complete Weblink Exercise: What Did You Say?

CHAPTER SUMMARY

Key Points of "Self-Disclosure and Feedback"	**CourseMate and DVD Activities to Build Interviewing Competence**

11.1 Interpersonal Influence, Listening Skills, and Influencing Strategies— Interview Example

▲ Listening skills are integral to effective use of influencing skills. We need to remember that listening skills can be as or more influential than influencing skills. Help clients set their own direction as much as possible.

1. Interactive Exercise 1: Identifying Skills. Careful listening to the client's story is essential. Review the interview exchanges before working on the influencing skills.

Key Points of "Self-Disclosure and Feedback"

CourseMate and DVD Activities to Build Interviewing Competence

▲ The 1-2-3 pattern applies to all influencing skills: (1) Listen; (2) assess and influence; (3) check out and observe client response.

▲ Use the Client Change Scale (CCS) to assess the effectiveness of your leads. Listen again and be ready to flex intentionally in response to the client.

▲ Structure the interview, and disclose what is happening when you use a new strategy with the client. This helps avoid surprising the client and builds client comfort and trust.

▲ The case of Alisia again demonstrates that listening skills influence client growth and development. Listening also sets the stage for effective use of influencing skills.

2. Interactive Exercise 2: Using Influencing Skills in the Interview. What self-disclosure would be more appropriate for each segment of counselor-client interaction? Compare your responses to ours.

11.2 Defining Self-Disclosure

▲ Self-disclosures are basically "I" statements made by the interviewer to share her or his own personal experience, thoughts, or feelings. Here-and-now self-disclosures tend to be the most powerful.

▲ The self-disclosure needs to be relevant to the client's worldview. It needs to be brief, genuine and authentic, and timed appropriately to client needs. The 1-2-3 pattern is particularly important when using self-disclosure.

▲ When clients ask you what you would do in their place, your task is to help them make their own decisions. Involving yourself too early can foster dependency and potentially mislead clients. It is important to discuss multicultural differences openly and with respect.

1. Interactive Exercise 3: Writing Self-Disclosure Statements. Practice writing effective self-disclosures.

11.3 Defining Feedback

▲ Use the 1-2-3 pattern to ensure appropriateness of feedback. The client receiving feedback should be in charge. Focus on strengths, be concrete and specific, be nonjudgmental, keep feedback lean and precise, and check out with the client how the feedback was received. Involve the client as much as possible.

▲ Positive feedback is "the breakfast of champions." Use this skill relatively frequently; it will balance more challenging necessary corrective feedback.

▲ Corrective feedback will be necessary at times. It needs to focus on things that the client can actually change. Seek to avoid criticism. Be specific and clear and, where possible, supplement with client strengths that support the change.

1. Interactive Exercise 4: Personal Experience With Feedback. An exercise to help you reflect on your personal experience receiving feedback.

2. Interactive Exercise 5: Writing Feedback Statements. Practice writing positive and negative feedback statements to Alisia.

3. Group Practice Exercise: Group Practice With Self-Disclosure and Feedback.

4. Weblink: What Did You Say? The Art of Giving and Receiving Feedback—The Federal Government's Human Resources Agency, Office of Personnel Management, Performance Management.

Assess your awareness, knowledge, and skills as you conclude the chapter:

1. Flashcards: Use the flashcards to check your understanding of key concepts and facilitate memorization of key information.
2. Self-Assessment Quiz: The quiz will help you assess your current knowledge and prepare for course examinations.
3. Portfolio of Competencies: Evaluate your present level of competence in attending and listening skills using the downloadable Self-Evaluation Checklist. Self-assessment of your attending skills competence demonstrates what you can do in the real world.

LOGICAL CONSEQUENCES, INFORMATION/PSYCHOEDUCATION, AND DIRECTIVES

Helping Clients Move to Action

INFLUENCING SKILLS AND STRATEGIES

FOCUSING

CONFRONTATION

THE FIVE-STAGE INTERVIEW STRUCTURE

REFLECTION OF FEELING

ENCOURAGING, PARAPHRASING, AND SUMMARIZING

OPEN AND CLOSED QUESTIONS

ATTENDING BEHAVIOR

ETHICS, MULTICULTURAL COMPETENCE, AND WELLNESS

Blessed is the influence of one true, loving human soul on another.

—*George Eliot*

How can these influencing strategies help you and your clients?

CHAPTER
GOALS

Help clients examine new possibilities for behavioral, thought, and feeling changes by providing them with information or skill development, helping them see the logical consequences of future actions, and directly suggesting what to do in the interview or through homework assignments.

Awareness, knowledge, and skills developed through the concepts of this chapter will enable you to

▲ Help clients examine the logical consequences of alternatives and actions.
▲ Provide information and psychoeducation to clients in an appropriate and timely manner.
▲ Direct clients to try systematic change strategies such as relaxation and guided imagery and engage in specific homework activities between interviews.
▲ Connect these skills with the five-stage interview process of relationship—story and strengths—goals—restory—action.

Assess your awareness, knowledge, and skills as you begin the chapter:

1. Self-Assessment Quiz: The chapter quiz will help you determine your current level of knowledge. You can take it before and after reading the chapter.
2. Portfolio of Competencies: Before you read the chapter, fill out the downloadable Self-Evaluation Checklist to assess your existing knowledge and competence in attending skills. Then, at the end of the chapter, complete the checklist again to summarize your competencies after study and practice.

12.1
DEFINING LOGICAL CONSEQUENCES

KEY CONCEPT QUESTIONS

▲ **How do we define the skill of logical consequences?**

▲ **What are some implications of this skill, particularly in discipline situations?**

▲ **How does focusing increase the value of logical consequences?**

The strategy of logical consequences is particularly important in making decisions and is used in connection with many theoretical approaches to the interview. When you use logical consequences as described below, you can predict how your client may respond and use that response to help them explore the results of their actions in more detail.

Logical Consequences	Predicted Result
Explore specific alternatives with the client and the concrete positive and negative consequences that would logically follow from each one. "If you do this . . . , then . . ."	Clients will change thoughts, feelings, and behaviors through better anticipation of the consequences of their actions. When you explore the positives and negatives of each possibility, clients will be more involved in the process of decision making.

The skill of logical consequences is most often a gentle strategy used to help people sort through issues when a decision needs to be made. It may be used to rank alternatives when a more complex decision is faced. In interviewing, the task is to assist clients in foreseeing consequences as they sort through alternatives for action: "If you do _____ , then _____ will possibly result."

The strategy of logical consequences originated in Adlerian counseling (Sweeny, 1998), but it is also central in decisional counseling (Chapters 7 and 13). The task of the interviewer is to help individuals explore alternatives, consider consequences of alternatives, and facilitate decision making among the possibilities. For example, an individual may come to the interview aware that changing jobs offers more pay, but fearful of the effects of moving to a new city. Through further questioning and discussion, the interviewer can help the client clarify the factors involved in the decision.

For many decisions, the various alternatives are apt to have both negative and positive consequences. Potential **negative consequences** of changing jobs could include leaving a smoothly functioning and friendly workgroup, disrupting long-term friendships, moving children to a new school, and other factors that might cause problems. **Positive consequences** might be a pay raise and the opportunity for further advancement, a better school system, and money for a new home.

The interviewer or counselor may need to help clients become aware of the potential negative consequences of actions, including punishment. Consideration of possible negative consequences is important for a client who is thinking of dropping out of school, a pregnant client who has not stopped smoking, or a client who wants to "tell off" a boss, coworker, or friend. It is equally important to help clients anticipate the positive consequences, results, and rewards of specific behaviors. Finishing school will lead to a better job, the nonsmoking pregnant woman's baby is likely to be healthier, and a better alternative for handling difficult people may be simply to keep one's mouth shut for the moment. It is best when your clients can generate the likely consequences of any given action through your listening skills, questioning, and confrontation.

Emotions are critical in making decisions. The decision may be correct, even wise, but the client needs to feel satisfaction or a sense of peace to move forward with it. There are many examples of tough but wise choices, such as accepting a new job in a new city that is a good move professionally but means leaving behind friends and family, or deciding to leave an abusive partner and now facing financial hardship, or saying no to cheating at school and then being teased about it. While good decisions, these positive choices do have negative consequences that we need to prepare our clients to face. Similarly, the client may be heading toward a decision that will produce long-term problems, such as drug use, deciding to be unfaithful to a partner, or ditching school. With both wise and unwise decisions our clients make, we need to help them anticipate how they will feel *emotionally* as a consequence of the decision and work toward being able to live with the anticipated results.

It does not happen as fast as in the following example, but this is the general pattern of introducing logical consequences.

COUNSELOR: What is likely to happen if you continue smoking while pregnant?

CLIENT: I know that it isn't good, but I can't stop and I really don't want to.

COUNSELOR: Again, what are the possible negative consequences of continuing to smoke?

CLIENT: [pause] I've been told that the baby could be harmed.

COUNSELOR: Right, is that something you want? What is the benefit of stopping smoking for the baby?

CLIENT: No, I don't want to do harm. I'd be so guilty. But how can I stop smoking?

COUNSELOR: You would really miss smoking, but you have some doubts. Let's explore that feeling of guilt. Tell me more about it. . . .

For disciplinary issues or when the client has been required to come to the session, it is important to note that even more power rests with the interviewer. The school, agency, or court may ask the interviewer to recommend actions that the legal system can take. The gentle logical consequence skill becomes more powerful. Warnings are a form of logical consequences and may center on *anticipation of punishment;* if used effectively and coupled with clients' rapport and listening, warnings may reduce dangerous risk taking and produce desired behavior.

In disciplinary situations, clients need to think about the fact that they will have happier lives in the long run if they select more positive alternatives for their lives. A school disciplinary official, an attorney, or a correctional officer often needs to help the client see clearly what might be ahead. But people who hold power over others need to *follow through with the consequences* they warned about earlier, or their power to influence will be lost. The "if, then" language pattern is especially useful to summarize the situation: "If we don't make the counseling sessions work, you know what the judge will do."

It is the rare human behavior that does not have its costs and benefits. By involving the client in examining the pluses and minuses of alternatives, the interviewer gives the decision to the client, or at least shares it more openly. Consider the following suggestions for using the strategy of logical consequences.

1. *Understand the situation.* Through listening skills, make sure you understand the situation and the way your client understands it. After drawing out the situation, either you or the client can summarize what is happening.

2. *Generate alternatives.* Use questions and brainstorming to help the client generate alternatives for resolving issues. Where necessary, provide your own additional alternatives for consideration.

3. *Identify positive and negative consequences.* Work with the client to outline both the positive and negative consequences of any potential decision or action. In important cases, ask the client to generate a possible future story of what might happen if a particular choice is made. For example, "Imagine yourself two years from now. What would your life be like if you choose the alternative we just discussed?" In anticipating the future, special attention needs to be paid to likely emotional results of decisions. Emotions are often the ultimate "decider."

4. *Provide a summary.* As appropriate to the situation, provide clients with a summary of positive and negative consequences in a nonjudgmental manner, or ask them to make the summary. Pay special attention to emotional issues. With many people this step is not needed; they will have made their own judgments and decisions already. Box 12.1 discusses the issue of being or not being judgmental.

BOX 12.1 National and International Perspectives on Counseling

 Can We Be "Nonjudgmental" About Crime?
WEIJUN ZHANG

Tom, age 14, comes for help because he is scared. Several of his friends were involved in a house break-in over the weekend. Although he didn't go into the house, he was outside waiting for his friends to come out. Everyone in town is talking about the vandalism, and Tom is afraid he'll be implicated.

After reading this case in a class, the professor asked us to decide which theoretical approach was most suitable for using with Tom. Some students suggested person-centered, with a focus on listening to his story, while others preferred to use Gestalt to deal with his present feelings. We then role-played some of these ideas in class. Much to my surprise, my classmates were so good at being "nonjudgmental" that their focus was all on the boy's feelings and cognition; there wasn't the slightest hint that anyone gave any thought to the vandalism, or that being silent about vandalism might be immoral.

There was no attention to the logical consequences of his action and little attempt to provide him with ideas for building a more successful future. Listening skills, of course, are of vital importance, but to me, this is a situation in which some judgment may be helpful. I'll admit that we need to use a respectful nonjudgmental vocal tone and attitude, but judgment of right and wrong remains an issue.

I suppose a 14-year-old boy can't be expected to have very mature judgment. But given this, what effect would counseling sessions such as those we role-played have had on him? Apart from having his anxiety reduced, he might very likely have learned a moral lesson from the counselor's being nonjudgmental—namely, that there may be nothing right or wrong about break-ins, that it may be all right to not report a crime, and that one's mental health can be totally separated from social responsibility.

I am not suggesting here that counselors should act like police in dealing with the boy. I do believe, however, that the counselor should let him know clearly that the counselor personally does not approve of such law-violating actions, even though confidentiality would nevertheless be strictly observed. The point here is that when counselors believe they should remain "value free" on an issue, they should also take ethical responsibility to see that such a youngster does not get any confusing or incorrect impressions about their nonjudgmental approach.

If I were Tom's counselor, I would not feel comfortable until my duty as a responsible adult was also fulfilled in the counseling process. I would try to raise the boy's community consciousness by asking him questions like these: How would you feel if you were the victim of the break-in? What impact has this break-in had on the whole community? What difference can it make if you report the crime voluntarily? I believe that many non–European Americans, because of their strong sense of community, would agree with my orientation.

A classmate criticized me for being too judgmental, and he is probably right. After all, it is impossible to remain completely nonjudgmental about such matters. Your questions, your paraphrases, your reflection of feelings and meaning, and especially what you choose to focus on, will all reflect your judgment and have a profound impact on the youngster. The distinction I am trying to make is that by trying to be completely "nonjudgmental" with the boy, my fellow students were conveying to the boy that an individual's feelings and concerns could take precedence over a community's interests, while I suggest that we should take the good of the community into full account when we work to promote the mental health of an individual.

5. *Encourage client decision.* Encourage client decision making as much as possible. In disciplinary situations, you may have to enforce appropriate consequences if the needed decision is not made.

The following exchange with Alisia demonstrates one possible use of logical consequences. In this case, the decision is between keeping things as they are or introducing a new behavior and style.

Interviewer and Client Conversation	Process Comments
1. **Counselor:** Alisia, we know that you would like to be more assertive and speak up more for yourself. What are the likely positive consequences if you can do this?	The counselor paraphrases and then asks Alisia to identify positive consequences of change.
2. **Alisia:** Well, I've learned that if I don't speak up to the garage service manager, nothing is going to happen. I suppose that I have nothing to lose by trying to be stronger. I guess the positive result would simply be something different.	Alisia responds well and notes that she has "nothing to lose" by trying something different.
3. **Counselor:** "Something different" sounds like you'd feel better about yourself and maybe even get the car fixed.	Encourager and paraphrase. The counselor suggests another potential positive consequence of change—emotional satisfaction in feeling better about oneself.
4. **Alisia:** That would be nice, but it is a little scary.	Change is not easy. Note the fear of change. Emotions are important in making decisions.
5. **Counselor:** Change can be scary. What are the negative consequences of continuing your past behavior?	Restatement of emotion. Open question with a strong influencing dimension. Balancing the decision with negative consequences of not changing.
6. **Alisia:** I'm also embarrassed about being scared of change. Sometimes I'm just too passive and let people run over me. That makes me angry with myself.	Alisia discusses the emotional issues related to change. Unless these are recognized and dealt with, change is less likely to be made and less likely to be effective.
7. **Counselor:** I hear that conflict, Alisia. You're tired of being passive and want to do something different, but it is still a little scary. I sense you really wanted to change.	This is a confrontation focusing on Alisia's mixed feelings about taking new actions. The reflection of feeling is followed by direct feedback from counselor observations, helping enable the client to move on.
8. **Alisia:** Yes, I want to change and speak up. It is time. [said quietly with her eyes downcast]	Said tentatively, but we are seeing movement on the Client Change Scale.
9. **Counselor:** Say "It is time" again.	This is a Gestalt directive, discussed later in the chapter. Asking a client to repeat positive statements helps anchor them and leads to more certainty and positive attitudes.

Interviewer and Client Conversation	Process Comments
10. **Alisia**: It is time.	Louder and with direct eye contact at the counselor.
11. **Counselor**: Say "It is time," but louder.	Build strengths through repetition.
12. **Alisia**: IT IS TIME! [loudly and firmly]	We are now at a beginning Level 4 on the Client Change Scale, but it is only a real Level 4 if Alisia takes this experience to her daily life and becomes more assertive.
13. **Counselor:** I think you've got it, Alisia. I can see that you are acting on the assertion skills we have talked about. [pause] So, the consequences of trying a change may make something happen for the good, and you have nothing to lose. On the other hand, the consequences of staying as you are, as you say, are "not good" and embarrassing, because you know that you need to speak up.	Brief feedback followed by brief praise by the counselor. After the pause, the logical consequences of the two choices are summarized.
14. **Alisia**: Right, well I'm ready to try something new. I've decided I'm ready for a change.	The decision has been made, but it now is time for action and moving the learning to the real world.
15. **Counselor:** Let's try some practice being more assertive. A role-play will be helpful here.	The counselor turns to role-play/enactment in which Alisia will demonstrate both passive and assertive behaviors while the counselor role-plays the service manager at the garage.

In later sessions, Alisia recognized that pleasing others and allowing them to take advantage of her almost always resulted in her being unhappy. The work with the issue of the service manager in the garage could soon be generalized to examine and change her general pattern of passive behavior. As the interviews continued, Alisia realized how good it would feel to speak up for herself and accept the consequences of what it might bring. She discussed parallel issues she had with her partner and made the decision to act and speak up. Certainly being true to oneself brings more satisfaction than leaving all the decisions to others. However, change in patterns of behavior often takes several interviews; if successful, we may even reach Level 5 of the Client Change Scale.

Complete Interactive Exercise 1: Writing Logical Consequences Statements
Complete Interactive Exercise 2: Logical Consequences Using Attending Skills
Complete Group Practice Exercise: Group Practice With Logical Consequences
Complete Weblink Exercise: Natural and Logical Consequences
Complete Weblink Exercise: Positive Discipline

12.2

DEFINING INFORMATION AND PSYCHOEDUCATION

KEY CONCEPT QUESTIONS

▲ **What are the strategies of information and psychoeducation?**

▲ **What are guidelines for implementing these strategies effectively?**

Prediction of what clients will do when you provide information, give advice, or teach them skills via psychoeducation is obviously somewhat difficult as we need time to see whether the information "took" and was useful to the client. Thus, always be ready for alternatives when clients don't respond to this important skill area.

Information and Psychoeducation	*Predicted Result*
Share specific information with the client— e.g., career information, choice of major, where to go for community assistance and services. Offer advice or opinions on how to resolve issues and provide useful suggestions for personal change. Teach clients specifics that may be useful: help them develop a wellness plan, teach them how to use microskills in interpersonal relationships, educate them on multicultural issues and discrimination.	If information and ideas are given sparingly and effectively, the client will use them to act in new, more positive ways. Psychoeducation that is provided in a timely way and involves the client in the process can be a powerful motivator for change.

Giving the client information, offering your opinions and advice, or making suggestions can be an important part of interviewing and counseling. These are common strategies in career counseling, decisional counseling, and often in cognitive-behavioral counseling.

Closely allied but somewhat different is psychoeducation, in which the counselor takes on more of a teaching role and provides systematic ways for clients to increase their life skills. Psychoeducation has become prominent in cognitive-behavioral counseling in recent years, as a way to prepare the client to meet challenging situations.

INFORMATION AND ADVICE

Be aware that this area is fraught with danger. Unless the advice is actively sought, it is very difficult for even the best of information to be heard. For example, try offering teens suggestions on how they should dress, drive, and or handle alcohol. It is immensely difficult to help a person stop smoking. Children resist suggestions. Adults are told to lose or gain weight, get more exercise, and eat more fruits and vegetables, but we have real difficulty in listening to or following advice that may be critical to our physical well-being.

When listening to information the counselor supplies, the client needs to be in charge and actually want the information, perhaps even more so than when receiving feedback. Career and college counseling must provide students with career and college admissions information, and here the teen may actually listen. Students facing critical life decisions frequently want to know your opinions and advice. In family counseling or coaching, the client may be caring for an older parent and may want advice on how to handle this extremely challenging part of life.

If you seek to advise a client to develop a more relaxed lifestyle and engage in relaxation training or a meditation program, listen to the client's story and wait for a timely and appropriate opportunity to share information. Provide the client with clear information on the health gains that would result. More challenging is a student not doing well in school or the office worker who shows up late for work. Both know that change would be wise and what your advice is likely to be. It is all the more important to hear their stories and points of view before attempting any advice. Pointing out logical consequences is likely to be useful in obtaining client compliance and agreement. Change your approach when you see clients roll their eyes, slump back in the chair, or look at the ceiling.

PSYCHOEDUCATION

Psychoeducation is so closely related to information that the two often merge. The distinction is that information or advice is brief, consisting of relatively short comments in the interview, with the hope that what is said will produce action in the real world. Psychoeducation tends to be more comprehensive. For example, cognitive-behavioral counselors and therapists will often instruct clients diagnosed as depressed or borderline on the causes of their condition, the possible treatments, medications, and how to obtain a favorable outcome from counseling and therapy. This is obviously information giving, but provides a more holistic picture.

Another type of psychoeducation involves life skills training; it can involve role-playing and enactment of scenes from the client's life or anticipated for the future. This type of psychoeducation may involve several interviews, planned homework, and readings, and may also be frequently found in group work.

One clear example of the second type of psychoeducation that you can engage in now is the teaching of attending behavior and some of the listening skills to clients in the interview. The teaching of these skills as part of treatment has been basic to the microskills approach since its inception. It is particularly effective with clients who are shy or have mild depression. The method is similar to what you have encountered in your own practice sessions: (1) check out your clients' interest in learning listening skills; (2) verbally give them information on the content of the skill (the three V's + B); (3) role-play an ineffective style of listening and discuss what is unproductive; (4) role-play a positive interview using good listening skills; (5) work with the client to take action in the real world.

Think of information and psychoeducation as a continuum where you share your knowledge in useful ways with your clients. Following are some examples of information and psychoeducation that might be used with Alisia.

> *Career Information.* Alisia, it is clear that you have not allowed yourself to stretch and move ahead with the job you have now. I'd like you to explore some alternatives. I'm going to show you our career library, and we can explore other career possibilities for the future. One of the issues we need to look at is related job opportunities. You are now in computer science, but you're not using all your talents. This chart shows a projected 25% increase in information technology career opportunities coming in the next decade. How does this sound?

Advice/Information. You asked for my advice about Chris. It's not my place to tell you what to do. But I do think it is time for you to sit down with him and have a serious talk. I've got some ideas for you that might help you express your thoughts and feelings in a way that Chris might actually hear you. But first, let's summarize what you have thought of already and what makes sense to you. Then I'll share some of my thoughts.

Psychoeducation. [After the story has been told and critical feelings expressed, the client wants to continue counseling to explore issues in more depth. You need to tell the client what to expect and how the process might work.] Losing a baby in the sixth month is an awful experience, and I sense your sadness. It's good that you want to continue to talk and think toward the future. I'll be listening, but as things come up, I'll likely share some thoughts about what you say. As we continue, we will be exploring your thoughts and feelings about the loss over several interviews. And I'd like to get to know you better, particularly things that you enjoy in life. We need to talk about the anger you feel at the doctor and your relationship with your husband. Most of all, we want to look to the future and plan the next steps. As I understand it, you are very anxious about trying again. Are these the issues that you'd like to explore?

Psychoeducation: Another Example. One route in handling difficult situations is to engage in attending behavior—listening carefully to the person whom you find difficult. If you listen and observe someone carefully, you may find new ways to understand that person and act more effectively. I'll teach you the basics of effective listening. Later, we can work on a cognitive-behavioral program of assertiveness training to help you speak up. Would you like to try this?

Complete Interactive Exercise 3: Your Personal Experience With Psychoeducational Information, Advice, Opinions, and Suggestions

Complete Interactive Exercise 4: Writing Psychoeducational Information, Advice, Opinion, and Suggestion Statements

Complete Group Practice Exercise: Group Practice With Information/Psychoeducation

Complete Weblink Exercises:

- ▲ Smoking Cessation
- ▲ Cancer Support Groups
- ▲ Patient-Centered Guides
- ▲ Problem Solving
- ▲ Teaching Tolerance
- ▲ The National Center for Victims of Crime
- ▲ Sexual Assault

12.3
DEFINING DIRECTIVES

KEY CONCEPT QUESTIONS

▲ **How do we define directives?**

▲ **What are some examples of directive strategies?**

If you use directive strategies as structured below, you can predict how clients will respond. But, as with information and psychoeducation, it often takes time to see if you have made a significant impact.

Directives	Predicted Result
Direct clients to follow specific actions. Directives are important in broader strategies such as assertiveness or social skills training or specific exercises such as imagery, thought-stopping, journaling, or relaxation training. They are often important when assigning homework for the client.	Clients will make positive progress when they listen to and follow the directives and engage in new, more positive thinking, feeling, or behaving.

Directives are useful in developing a new story or thinking in new ways, and they are especially effective in helping a client move to behavioral action. A positive new story may be sufficient for some clients, but many will profit from directive strategies outlining specific behaviors and actions. Directive strategies are drawn from various counseling theories that direct the client to follow a specific sequence of events designed to produce a likely result.

THE 1-2-3 PATTERN OF INTERPERSONAL INFLUENCE APPLIED TO DIRECTIVES

Effective directives require an expansion of the 1-2-3 pattern.

1. Involve your client as co-participant in the directive strategy.   Rather than simply tell the client what to do, be sure that you have carefully heard the story, issues, and problems. Usually, directives come later in the interview or, more often, in future sessions. Inform the client what you are going to do and the likely result. Some practitioners like to use surprises (e.g., Gestalt therapy), and these can be useful in some situations. But as a general rule, we urge *working with*, rather than *working on*, your client.

Use the three V's + B (culturally appropriate visual, vocal tone, verbal following, and body language). When you use influencing skills, your attending behaviors need to be flexible in response to the needs of the client. Usually, a more forward and active behavioral style is needed. For example, when challenging an acting-out teen or an outgoing client, you may need a stronger persona with even clearer verbal and nonverbal behavior. With a quieter, more tentative client, appropriate attending may require being still and tentative yourself as you share new ways of thinking about issues.

2. Be clear and concrete in your verbal expression and time the directive to meet client needs.   Directives need to be authoritative and clear but also stated in such a way that they are in tune with the needs of the client. Compare the following:

Vague: Go out and arrange for a test.

Concrete: After you leave today, contact the testing office to take the Strong-Campbell Interest Inventory. Complete it today, and they will have the results for us to discuss during our meeting next week.

Vague: Relax.

Concrete: Sit quietly . . . feel the back of the chair on your shoulders . . . tighten your right hand . . . hold it tight . . . now let it relax slowly . . .

These examples illustrate the importance of indicating clearly to your client what you want to happen. Know what you are going to say, and say it clearly and explicitly.

3. Check out whether your directive was heard and understood.   Just because you think you are clear doesn't mean the client understands what you said. Explicitly or implicitly check to make sure your directive is understood. This is particularly important when a more complex directive has been given. For example, "Could you repeat back to me what I just asked you to do?" or "I suggested three things for you to do for homework this coming week. Would you summarize them to me to make sure I've been clear?" The Client Change Scale (CCS) can be used to determine whether the client actually changed thoughts, feelings, or behaviors as a result of your directive.

In the following example, notice that the counselor prepares Alisia ahead of time by providing an explanation, then gives a concrete directive, then follows with a checkout to see whether proceeding with the strategy is satisfactory to her.

COUNSELOR: Alisia, I've heard your story about how frustrated you feel with that service manager. I'd like to use imagery to understand the situation a bit more fully. This will require you to relax, sit back, and allow yourself to recall the situation. It often helps to visualize the specifics of the situation, what was said, and so on—almost like a movie. As we start, I'd like you to remember just one particular scene that first occurs to you. Is all this OK with you? Any questions?

▲ EXERCISE 12.1   **Testing Directives**

Test the directive strategies described in the following sections with a friend or classmate; then have the person try the same strategies with you. If you practice the details of directives, you will have a better idea of their potential and how to pace or time them. All of the directives suggested are well-known and effective strategies but must be used in full attunement with client needs and wishes. Additional directive strategies can be found in Box 12.2.

BOX 12.2   Examples of Directive Strategies Used by Counselors
of Different Theoretical Orientations

These example directive strategies are presented in very brief form. With further study and some imagination and practice—and client participation in the process—you can successfully use many of them. For more detailed presentations of these and other strategies in highly concrete form, see Ivey, D'Andrea, and Ivey's *Theories of Counseling and Psychotherapy: A Multicultural Perspective* (2012).

Specific Suggestions/	"I suggest you try . . ."
Instructions for Action	"Alisia, the next time you go to the garage, I'd like you to stand at the counter, make direct eye contact, and clearly and firmly tell the manager that

BOX 12.2   (Continued)

	you have a meeting at 10:00 and you need prompt service—now! If he says there will be a delay, get him to make a firm time commitment. Then follow up 15 minutes later."
	Detail and concreteness are very important when providing a directive.
Spiritual Images	"You say you gain strength from your spirituality and religion. Could you tell me about an image that comes to your mind related to a spiritual strength?" (Listen to the story and the feelings that go with it.)
	"Now, close your eyes and visualize (that symbol, person, experience) and allow it to enfold you completely. Just relax, focus on that image, and note what occurs in your body."
	In recent years the counseling and interviewing field has recognized the strengths and power in spirituality and religion. Many clients benefit from spiritual imagery and often find peace in their inner body and strength to move on. Forgiveness of the transgressions and omissions of others can come from spiritual imagery or your client may find new strengths to deal with a difficult illness or serious loss. A spiritual orientation even helps some clients recover from operations or serious illness.
Role-Play Enactment	"Now return to that situation and let's play it out."
	"Let's role-play it again, only change the one behavior we agreed to."
	Role-playing is an especially effective technique to make the abstract concrete. It makes the client behavior clear and specific. This is one of the most basic techniques used in assertiveness training, which you will see in Chapter 14.
Positive Reframing Combined With a Directive	"We've identified the problem and how it feels. Now feel that wellness strength in your body. Do it fully, magnify it, and take it to the problem."
	Taking real positives to attack problems through the body can be effective. If the strength is not able to meet and counteract the negative, add another resource—or just have the strength approach one part of the negative at a time.
Relaxation	"Close your eyes and focus on the moment."
	"Tighten your forearm, very tight, now let it go." (Directions for relaxation continue throughout the body parts, ending with full body relaxation.)
	Alisia may be tight and tense; once she is able to relax and be in control of her body, she will be better able to cope with stressful daily encounters. Relaxation training helps all clients loosen tight muscles and cope better with the constant stress that many of us must deal with.
Journaling	"Alisia, you like to write and think about things. How would it be if you started a journal of your work with me? You might want to reflect on each interview and its impact and how what we discuss relates to what you see happening in your life daily. You can share this with me or not as you choose."
	Keeping a journal is helpful to many clients. This helps them reflect on the interview and its impact on them during the week.

GUIDED IMAGERY FOCUSING ON A RELAXING SCENE

This is a popular technique to help clients relax and discover ways to reduce tension. All of us have past positive experiences that are important for us—maybe a lakeside or mountain scene, or a snowy setting, or a quiet, special place. The image can become a positive resource to use when we feel challenged or tense. For example, Alisia likely feels real tension in her body when she encounters conflict. If she learns to notice her internal body tension, then she can immediately and briefly focus on a relaxing scene, take a deep breath, and deal with the challenging situation more effectively. When giving a guided imagery directive, time your presentation to your observations of the client.

> Close your eyes and relax. [pause] Notice your breathing and the general feelings in your body. Focus on that place where you felt safe, comfortable, and relaxed. Allow yourself to enter that scene. What are you seeing? . . . Hearing? . . . Feeling? . . . Allow yourself to enjoy that scene in full relaxation. Notice the good feelings in your body. Enjoy it now for a moment before coming back to this room. Now, as you come back, notice your breathing [pause] and as you open your eyes, notice the room, the colors, and your surroundings. How was this experience for you?

POSITIVE IMAGES OF STRENGTH

Images of people who have meant a lot to us serve as examples of strength and give us the courage to meet difficult issues. The community genogram is a good place to find people who represent positive strengths and resources. Using the guided imagery exercise above, help the client focus on a person who was helpful in the past or who can serve as a model for the future. As you go through this exercise, add one significant dimension—ask the client to note where the feelings inspired by the image are located in her or his own body. Some note warm feelings in their stomach; others observe feelings of strength in their chest or upper arms; most of us can identify where the positive image is located in our body. This physical strength from others can be a resource to help us cope with current problems and concerns.

As you begin your work with interviewing and counseling, it is very important that you work only with positive images. Negative imaging is potentially useful in psychotherapy, and imagery can be used to go over difficult situations from the past, but this kind of exploration is highly inappropriate unless the interviewer is fully qualified and the time and situation are appropriate for the client.

ENCOURAGE PHYSICAL EXERCISE AND RELATED HEALTH ACTIVITIES

A past president of the American Psychiatric Association has stated that any physician who does not recommend exercise to patients is unethical. Interviewing, counseling, and therapy have been very weak in this area. Exercise is a preventive health activity that needs to become part of everyone's practice. Moreover, research is now showing that exercise helps clients deal with stress, which in turn helps with many difficult issues ranging from depression to Alzheimer's (Ratey, 2008).

A sound body is fundamental to mental health. Beyond encouraging clients to exercise regularly, remind them that proper eating habits and a regime of stretching and meditation can make a significant difference in their lives. Teaching clients how to nourish their bodies is becoming a standard part of counseling. We love and work more effectively if we are comfortable in our bodies.

COUNSELOR:	Alisia, you seem stressed much of the time. What's happening with exercise in your life?
ALISIA:	I simply don't have time, and when I think of it, I realize that I have some errands to run or someone calls me on the cell phone.
COUNSELOR:	Evidence is clear that tension can be relieved by exercise. I'd like you to consider the possibility. What types of exercise have you enjoyed in the past?
ALISIA:	Well, I used to run, and I did feel more "up" when I got out. But since I've moved to the city, I just don't seem to find time anymore.
COUNSELOR:	As part of dealing with Chris and your various stressors, I think it is very important that we start some sort of exercise routine. You'll feel better and will be able to deal more effectively with those challenges if you take care of yourself. What do you think?
ALISIA:	Well . . . I should consider it. I did feel better when I did run. [pause] But, how?
COUNSELOR:	Let's work on it. Tell me more about your schedule. [The session continues.]

It is obvious that we can't tell Alisia what to do. For example, "You should start exercising and running daily" simply won't work and will build client resentment. Helping clients change their behavior involves a more subtle use of directives, with the client as a full co-participant in the process. Regardless of what directive you want to provide on any topic, the client has to "buy in" and be central in the choice of action.

THOUGHT-STOPPING

This strategy has consistently been found to be one of the most effective interventions a counselor or therapist can use. If you take the time to learn and practice thought-stopping on yourself, you gain a valuable tool to increase your own self-esteem and effectiveness and you will see its potential for clients. Thought-stopping is useful for all kinds of client problems: perfectionism, excessive culture-based guilt or shame, shyness, and mild depression. This is one of our favorite strategies, and we have found it very helpful to us over the years when we get into negative thinking about ourselves. Almost everyone engages in internalized negative self-talk. Self-talk is stressful thoughts you say to yourself, perhaps several times a day. For example:

"Why did I do that?"

"I'm always too shy."

"I always foul up."

"I should have done better."

"Life is so discouraging for me."

"Nobody will listen to me."

Other negative thoughts include guilty feelings, procrastination where you "over-think" the situation, fear that things will only get worse, anger that others "never get it right," repetitively thinking about past failures, or always needing the approval of others. Negative self-talk erodes self-esteem and increases self-doubt over time.

The following is the basic process for learning and using thought-stopping.

Step 1. Learn the basic process. Relax, close your eyes, and imagine a situation in which you make the negative self-statement. Take time and let the situation evolve. When the thought comes, observe what happens and how you feel after the negative self-talk. Then tell yourself silently "STOP." If you are alone, say it loudly and firmly.

Step 2. Transfer thought-stopping to your daily life. Place a rubber band around your wrist and every time during the day that you find yourself thinking negatively, snap the rubber band and say "STOP!" This simple step almost sounds silly, but it works. (Snapping the rubber band is not a form of punishment but a way to interrupt or interfere with negative thinking. Be kind to yourself; use it with this purpose only.) The client can, of course, just say "Stop," but the rubber band adds extra reinforcement to the effort to change thought patterns.

Step 3. Add positive imaging. Once you have developed some understanding of how often you use negative self-talk, and after you say "STOP" or snap the rubber band, substitute a more positive statement about yourself immediately. You may use positive imagery, or think about an example when you had a positive experience, or use a brief broader statement emphasizing general strengths.

"I can do lots of things right."

"I am lovable and capable."

"I sometimes mess up—no one's perfect."

"I did the best I could."

HOMEWORK

The fifth stage of the interview is concerned with generalizing thoughts, feelings, and behaviors to the "real world" outside of the session. Working with the client to do something *different* or *new* during the coming week can be invaluable.

Homework assignments can include thought-stopping, using positive imaging or meditation daily, playing basketball with friends, starting a program of walking or running, or keeping a diary of foods eaten. Another type of homework may be to have the client write down the negative or faulty thought each time it occurs during the day in an *automatic thoughts chart*. The client can also record what happened just before the thought and what happened afterward. Such records can be valuable in changing faulty thinking patterns. A couple with difficulties may be asked to observe and count the number of arguments they have. They record the before, during, and after aspects of the argument to discuss with the counselor. No change is expected; the client is simply expected to observe and record what is going on. With some clients, just observing themselves leads them to change their behavior! There are endless ways to involve clients in homework following the interview.

▲ EXERCISE 12.2   **Trying Directive Strategies in Your Own Life**

By yourself, or with a friend, test out various directive strategies. If you experience them yourself, you will have a better idea of their potential and how you might want to pace or time directives with a client. Select at least one of the directive strategies presented above.

Complete Case Study: Hurricane Katrina
Complete Video Activity 1: Relaxation Demonstration
Complete Video Activity 2: Thought-Stopping Technique
Complete Weblink Exercises:

- ▲ Benson-Henry Institute for Mind Body Medicine: Eliciting the Relaxation Response
- ▲ About Relaxation
- ▲ Brief Relaxation Techniques
- ▲ Breathing Techniques
- ▲ Guided Imagery
- ▲ Self-Help Information: Worry
- ▲ Directive Counseling

CHAPTER SUMMARY

Key Points of "Logical Consequences, Information/Psychoeducation, and Directives"	CourseMate and DVD Activities to Build Interviewing Competence

12.1 Defining Logical Consequences

▲ The steps of logical consequences are (1) listen to clients and clarify issues, (2) generate alternatives with them to resolve their issues, (3) carefully sort out the positive and negative consequences of each alternative, (4) summarize the alternatives using a nonjudgmental and positive stance, and (5) encourage client decision making.

▲ "If, then" language can be helpful in clarifying key decisions. "If you do _____, then the likely result is _____." Writing down the pluses and minuses of each alternative leads to better decision making. If clients make a "wrong" decision with negative consequences, they will at least have more awareness that this was their decision.

▲ In disciplinary situations, maintain a nonjudgmental attitude, and continue to encourage client decision making. In some disciplinary situations, you may have to recommend unfortunate consequences for the client.

▲ Combine focusing with logical consequences to help clients make decisions with much more awareness of their impact on others and the cultural/environmental context.

1. Interactive Exercise 1: Writing Logical Consequences Statements. Using the five steps of the logical consequences skill, briefly indicate to the client, Alisia, what the logical consequences would be for her if she continues with her lack of assertiveness.

2. Interactive Exercise 2: Logical Consequences Using Attending Skills. Practice writing logical consequences statements for various types of clients and situations.

3. Group Practice Exercise: Group Practice With Logical Consequences.

4. Weblink Exercise: Natural and Logical Consequences.

5. Weblink Exercise: Positive Discipline.

12.2 Defining Information and Psychoeducation

▲ All influencing skills are concerned with imparting information to the client. The task is to be clear, specific, and relevant to the client's world.

1. Interactive Exercise 3: Your Personal Experience With Psychoeducational Information, Advice, Opinions, and Suggestions. Reflect on and analyze your own experience with psychoeducational information and related skills.

Key Points of "Logical Consequences, Information/Psychoeducation, and Directives"

CourseMate and DVD Activities to Build Interviewing Competence

▲ Provide clients, judiciously, with specific data that will help them in decision making; you may follow up with logical consequences. Counselor advice may be taken too seriously, so use it sparingly. Use the 1-2-3 pattern of listening when providing information or advice, and check out with the client how you were received.

2. Interactive Exercise 4: Writing Psychoeducational Information, Advice, Opinion, and Suggestion Statements.
3. Group Practice Exercise: Group Practice With Information/Psychoeducation.
4. Weblink Exercise: Smoking Cessation.
5. Weblink Exercise: Cancer Support Groups.
6. Weblink Exercise: Patient-Centered Guides.
7. Weblink Exercise: Problem Solving.
8. Weblink Exercise: Teaching Tolerance.
9. Weblink Exercise: The National Center for Victims of Crime.
10. Weblink Exercise: Sexual Assault.

12.3 Defining Directives

▲ Directive strategies are used to tell the client what to do in a timely, individually and culturally appropriate manner. Directives may range from telling the client to seek career information to using complex theoretical strategies.

▲ Used sparingly, directives can be critical in leading to client change and growth, measured on the Client Change Scale (CCS).

▲ Homework is a useful strategy to help clients generalize what they have learned in the interview to their daily life in the "real world."

1. Case Study: Hurricane Katrina
2. Video Activity 1: Relaxation Demonstration.
3. Video Activity 2: Thought-Stopping Technique.
4. Weblink Exercise: Benson-Henry Institute for Mind Body Medicine: Eliciting the Relaxation Response.
5. Weblink Exercise: About Relaxation.
6. Weblink Exercise: Brief Relaxation Techniques.
7. Weblink Exercise: Breathing Techniques.
8. Weblink Exercise: Guided Imagery.
9. Weblink Exercise: Self-Help Information: Worry.
10. Weblink Exercise: Directive Counseling.

Assess your awareness, knowledge, and skills as you conclude the chapter:

1. **Flashcards:** Use the flashcards to check your understanding of key concepts and facilitate memorization of key information.
2. **Self-Assessment Quiz:** The quiz will help you assess your current knowledge and prepare for course examinations.
3. **Portfolio of Competencies:** Evaluate your present level of competence in attending and listening skills using the downloadable Self-Evaluation Checklist. Self-assessment of your attending skills competence demonstrates what you can do in the real world.

SKILL INTEGRATION, TREATMENT PLANNING, AND USING THEORY WITH MICROSKILLS

SECTION IV

What is your style of interviewing and counseling? This section provides a framework to help you integrate the many skills and concepts of this book. Central to this process is Chapter 13, where we ask you to make a recording of a complete interview.

Chapter 13, "Decisional Counseling, Skill Integration, Treatment Plans, and Case Management" has several purposes. We have discussed decisional counseling in previous chapters and here you will see its background over the years and more specifics for practice. In addition, you will see decisional counseling in action through a complete transcript and analysis of an interview. This chapter also includes basic information on treatment planning and case management. Finally, you will be asked to record and analyze your own interview and to compare it with a previous interview to determine progress.

Chapter 14, "Applications of Microskills," shows how to use microskills with cognitive-behavioral crisis counseling and brief counseling. The chapter closes by discussing how the microskills framework can be used with many different counseling and psychotherapy methods. This chapter is a bridge to the future.

Chapter 15, "Determining Personal Style," provides an opportunity to review your work with microskills, interview structure, and various orientations to the interview. You will be asked to think about your own natural style of interviewing and plan for the future. You will be encouraged to use your knowledge and skills to build your own culturally intentional, culturally appropriate interviewing style.

We are nearly at the end of our journey through the basics of interviewing, counseling, and psychotherapy. You have skills that can be used in many settings, as these are foundational units of all communication—business, sales, law, medicine, child or peer helping, and many others throughout the nation and the world. How will you use and adapt these skills and strategies in your professional and personal life?

Decisional Counseling, Skill Integration, Treatment Plans, and Case Management

SKILL INTEGRATION
AND APPLICATIONS

INFLUENCING SKILLS
AND STRATEGIES

FOCUSING

CONFRONTATION

THE FIVE-STAGE INTERVIEW STRUCTURE

REFLECTION OF FEELING

ENCOURAGING, PARAPHRASING, AND SUMMARIZING

OPEN AND CLOSED QUESTIONS

ATTENDING BEHAVIOR

ETHICS, MULTICULTURAL COMPETENCE, AND WELLNESS

Not to decide is to decide.

—Harvey Cox

How can your intentional skill integration help you and your clients?

CHAPTER
GOALS
You and your clients will greatly benefit from a naturally flowing interview and treatment plan using a smooth integration of the skills, strategies, and concepts of intentional interviewing and counseling. This is a concept-dense chapter explaining several key issues, but the most important of these is that you conduct a complete interview and analyze what happens between you and the client.

Awareness, knowledge, and skills developed through the concepts of this chapter will enable you to

▲ Integrate the concepts, skills, and strategies learned in previous chapters.
▲ Understand the basics of decisional counseling and use this model to clarify clients' thinking, emotions, and behavior, and help them make appropriate decisions.
▲ Develop long-term treatment plans for a client, and keep systematic interview records. Examine case management as related to the treatment plan.
▲ Integrate the skills, strategies, and concepts of intentional interviewing with your understanding of other theoretical models of the helping process.
▲ Analyze your own interview, and determine the progress made.

Assess your awareness, knowledge, and skills as you begin the chapter:

1. Self-Assessment Quiz: The chapter quiz will help you determine your current level of knowledge. You can take it before and after reading the chapter.
2. Portfolio of Competencies: Before you read the chapter, fill out the downloadable Self-Evaluation Checklist to assess your existing knowledge and competence in attending skills. Then, at the end of the chapter, complete the checklist again to summarize your competencies after study and practice.

The aim of skill integration is to take the microskills, stages of the interview, and your natural expertise and then examine where you stand now. How might you want to intentionally grow? As you increase interviewing competence, the result will be increasing intentionality and the ability to flow naturally with your clients.

Skill Integration	*Predicted Result*
Integrate the microskills into a well-formed interview and generalize the skills to situations beyond the training session or classroom.	Developing interviewers and counselors will integrate skills as part of their natural style. Each of us will vary in our choices, but increasingly we will know what we are doing, how to flex when what we are doing is ineffective, and what to expect in the interview as a result of our efforts.

Skill integration will be illustrated by a complete transcript of an interview illustrating decisional counseling in action. You will see Allen demonstrate the first decisional counseling interview with Mary.

But before we examine the interview, we will first look at decisional counseling in more detail. After that discussion, we present what we think is an important new concept for interviewing planning—the checklist, a systematic way to make sure that everything necessary is remembered for that first session. The checklist has become standard in much of medical practice and will also be useful in planning for and reviewing the interview.

13.1
DECISIONAL COUNSELING: An Overview of a Practical Theory and Set of Strategies

KEY CONCEPT QUESTION

▲ **What is decisional counseling, and how does it relate to the microskills five-stage interviewing framework?**

Decisions are the stuff of life. Decisions are both our problems and our opportunities. **Decisional counseling** (Ivey & Ivey, 1987; Ivey, Ivey, & Zalaquett, 2010) may be described as a practical model that recognizes decision making and the microskills as a foundation for most—perhaps all—systems of counseling. Another term for decisional counseling is problem-solving counseling. Here you can see how decisional counseling is related to and inherent in many forms of counseling and therapy, for making decisions and solving problems are major objectives of all our work. Decisional counseling just focuses more immediately and precisely on these issues.

Creativity underlies decisions. It requires a disciplined freedom, an openness to change, and imagination. There is an old Zen fable that goes something like this, updated for today.

> A woman is hiking along a California Sierra trail along the edge of a 15-foot drop. As she rounds a bend, she sees a bear, who starts to charge. Surprised but still able, she grabs a wild vine and swings over the edge. As she thankfully hangs and looks for a safe place to jump, she sees another bear below! There are summer strawberries growing on the vine so she decides to hold on with one hand and reaches for a few berries with the other. How sweet they taste!

Your clients face bears of decisions. On one hand, one bear promises one thing, while the other may bring something else. We can help clients taste the sweetness of strawberries and the importance of the moment before they jump. Let us hope that their decisions are friendlier than bears.

Decision making is a lifelong issue. Clients are always solving problems and making decisions. Young adults must choose a college or a career or decide whether to continue a relationship, get married, or have a child. Later they will make decisions about how to succeed in a work setting, deal with difficult colleagues, and plan finances for their children's education and their own retirement. They will find that decisions don't end with retirement; the first question faced by many retirees is "What shall I do with all this time?" Finally, difficult decisions around health issues, wills, and plans for their own funeral often require counseling.

Regardless of theoretical approach, the essential issue is this: *How can you help clients work through issues and come up with new answers?* All theories of counseling, in one way or another, deal with problem solving and decisions. The person-centered counselor enables clients to make decisions for themselves through self-examination and self-reflection. The brief counselor helps clients make positive decisions during a short series of sessions. Mentoring is centrally concerned with decisions. Even the client of a psychoanalyst makes decisions.

Clients come to us with pieces of their lives literally "lying all around" without organization. The "magic" of creativity comes from the spontaneous generation of something new out of what already exists. Drawing out the story and exploring the emotional background and meaning of the story is the first key to organizing the pieces into something coherent and manageable. Focusing skills ensure that you have brought out all the important pieces necessary for a creative synthesis and effective decision. Confrontation helps the client see discrepancies, conflict, and incongruence more clearly. With just these skills, clients will often "magically" find their own solutions. But, when necessary, you can add more pieces and enable the development of new stories through the influencing skills.

The five stages of the interview (*relationship—story and strengths—goals— restory—action*) produce a solid structure for creative and intentional lives. The five-stage interview, using only listening skills and skillful confrontation, is often sufficient in itself to produce significant change.

KEY STRATEGIES OF DECISIONAL COUNSELING

Decisional counselors are open to using virtually all theories and strategies of the helping field, ranging from psychoanalytic free association through the coaching model. Nonetheless, decisional counseling has its own strategies for change. While the five-stage structure is central, the following are particularly important. Some have been mentioned before, but they are repeated here for emphasis.

Confrontation.   Also mentioned above, your ability to identify conflict and discrepancies in a clear fashion and summarize them for the client is basic to the creative process and decision making. The pieces of the puzzle need to be made clear.

Creative Brainstorming.   While decisions start with logic, generating possible alternatives for action is a creative act. Encourage clients to "loosen up" and let any thought come to their mind. You can help through your own creativity, but focus on helping clients generate their own solutions.

Logical Consequences.   This Chapter 12 skill is central to decision making and helping clients see what is likely to happen as a result of their decisions.

Cognitive and Emotional Balance Sheet.   The balance sheet is an extension of logical consequences, but here each alternative is written down in a list of gains and losses. As an example, Table 13.1 shows a balance sheet developed with a woman who has experienced abuse. Notice how powerful the arguments are for staying in an abusive relationship. This is why your support in such cases is so important.

As another example, we can help substance abusers use the **cognitive balance sheet** to look at issues around drinking or using drugs. Adding focus concepts to the balance sheet helps alcoholics see the broader implications of their drinking on the lives of others and makes possible the exploration of emotional issues. The balance

▲ **TABLE 13.1** The Cognitive and Emotional Balance Sheet

List below the positive and negative factual and emotional results for each of the possible alternatives. If there is more than one alternative, make a separate Cognitive and Emotional Balance sheet for each one. What is the decision? *What happens if I leave my abusing partner?*			
What are the possible positive gains for me?	**What are the emotional gains for me?**	**What are the possible positive gains for others?**	**What are the emotional gains for others?**
Abuse will stop and I won't get hurt.	I won't be so scared.	My Mom won't have to talk to me on the phone constantly.	Mom will be so relieved that it is over.
I'll be able to move on with my life.	Perhaps I can return to feeling good about myself.	My Mom would like to help.	She'd feel that she is important to me again.
I can be myself.	I used to feel OK, and that would be relief.		
What are the possible losses for me?	**What are the emotional losses I might face?**	**What are the possible losses for others?**	**What are the emotional losses for others?**
I'll be on my own.	This frightens me as much as staying.	None that I can think of.	Again, none that I can think of. They'll be happy to see him gone.
How can I finance things by myself?	This terrifies me.	My parents may have to support me for a while.	They aren't that well off, and they told me not to go out with him. They may be angry, even though they'll help.
I still love that man.	I'll be lonely.	My friends will be there for me.	They'll be glad for me and listen.
He might follow me, and that might make it worse.	I'll have no future and be totally alone.	My counselor is there to advise and support me.	I can sense that I'm not as alone as I might think I am. I feel supported and cared for.

sheet lists the positives—what drinking does for them—and then lists the negatives. All this is done on a balance sheet developed jointly by the helper and the client.

Emotional balancing gives special attention to how the client might feel after each possible decision. Although decision making is a cognitive activity, real satisfaction comes with feelings of pleasure and satisfaction about the result. With this activity, emotions are stressed throughout the logical consequences strategy. For example, "What do you feel and enjoy about drinking?" "What does cocaine do for you?" "Imagine yourself not drinking; how would you feel about yourself?" "What would your family feel?" Use both here-and-now emotions and those of the past and the anticipated future. Look forward to the longer-term benefits that will come with change.

Complete Interactive Exercise 1: Influence the Interview
Complete Weblink Exercise: Interviewing and Data-Gathering Techniques

13.2
PLANNING THE FIRST INTERVIEW: The Checklist

KEY CONCEPT QUESTIONS

▲ **How can a plan be developed for the first interview?**

▲ **What is a checklist, and how might it be of value?**

▲ **What if the interview does not proceed as planned?**

The decisional counseling interview presented in this chapter is a role-play conducted by Allen and Mary Ivey, based on Mary's real-life career planning. Mary role-plays a 36-year-old divorced client with two children.

She stated in her information file, completed before counseling, "I find myself bored and stymied in my present job as a physical education teacher. I think it is time to look at something new. Possibly I should think about business. Sometimes I find myself a bit depressed by it all." This initial interview illustrates that career counseling is closely related to personal counseling. The personal issues arise along with the career issues further into the session. Additionally, gender is an important multicultural issue that needs to be considered.

Given the five-stage interview structure, it is possible to develop an advance plan for the session. Before the first interview, study the client file; anticipate important issues and how you might handle them. The advance plan does not mean that you impose your views or concepts on the client; rather, you think through approaches that may enable the client to achieve whatever goal he or she may have. Equally important, just because you have a plan, don't expect things to always work out as you anticipate. Clients will bring up issues that cause you to rethink where to go next; you may even need to scrap your plan entirely. As always, intentionality—the ability to be open to alternatives and flex with the here and now—is critical.

The following plan shows Allen's assessment of his forthcoming interview with Mary. He developed the session plan from his study of Mary's file, consisting of a pre-interview questionnaire. Mary stated in her intake form, "I'd like to do something new with my career. I'm ready for something new—but what?" Note that the interview plan is oriented to help the client develop her own unique career plan and to facilitate the discussion of personal issues as well. The plan is structured to help the client achieve her objectives and make her own decisions, but remains flexible enough to adapt as the interview progresses and new issues are brought up.

Stage/Dimension Key Questions	Counselor Preparation
Relationship. Initiate the session, develop rapport/structuring. *What structure do you have for this interview? Do you plan to use a specific theory? What special issues do you anticipate with regard to rapport development?*	Mary appears to be a verbal and active person. I note she likes swimming and physical activity. I like to run, and that may be a common bond to discuss. I'll be open about structure but keep the five stages in mind. It seems she may be unhappy in her current job and want to look into another career choice. Another personal issue is her divorce. I'll need to listen to her stories and use mainly questions and reflective listening skills, and follow a decision-making model.

Stage/Dimension Key Questions	Counselor Preparation
Story and Strength. Gather data, draw out stories, concerns, problems, or issues. *What are anticipated problems? Strengths? How do you plan to define the issues with the client? Will you emphasize behavior, thoughts, feelings, meanings?*	I'll use Mary's wellness strengths early and focus on finding out what she *can* do. I'll use the basic listening sequence to bring out issues from her point of view and learn about her thoughts about her job and her thoughts about the future. I'll be interested in her personal life as well. How are things going since the divorce? What is it like to be a woman in a changing world? Mary may well bring up several issues. I'll summarize them toward the end of this phase, and we may have to list them and set priorities if there are too many issues. Mainly, however, I expect an interview on career choice.
Goals. Set goals mutually, establish outcomes. *What is the ideal outcome? How will you elicit the client's idealized self or world?*	I'll ask her what her fantasies and ideas are for an ideal resolution and follow up with the basic listening sequence. I'll end by confronting and summarizing the real and the ideal. As for outcome, I'd like to see Mary define her own direction from a range of alternatives.
Restory. Explore and create alternatives—confront client incongruities and conflict. *What theories would you probably use here? What specific incongruities have you noted or do you anticipate in the client? How will you generate alternatives?*	Working from the decisional model, I hope to begin this stage by summarizing her positive strengths and wellness assets. I'd like to see several new alternative possibilities considered. Counseling and business are indicated in her pre-interview form as two good possibilities. Are there other possibilities? The main incongruity will probably be between where she is and where she wants to go. I expect to ask her questions and develop some concrete alternatives even in the first session. I hope she will act on some of them following our first session. I think career testing may be useful.
Action. Generalize and act on new stories. *What specific plans do you have for transfer of training? What will enable you, the interviewer, to feel that the interview was worthwhile?*	I'll feel satisfied if we have generated some new possibilities and can do some exploration of career alternatives after the first session. We can plan from there. I'd like it if we could generate at least one thing Mary can do for homework before our second session.

Given the complexity of relationships, particularly professional relationships, Atul Gawande has written *The Checklist Manifesto* (2009). Focusing first on medicine, he found that a surgical checklist of basic and often obvious factors significantly reduced dangerous errors during operations. He has gone on to point out that thinking ahead about what one is going to do improves performance regardless of the field. Table 13.2 provides a checklist for the first interview. Review this list before the session, and review it afterwards checking to see if it all happened. For you, personally, what might you add or delete from this checklist? Making this checklist fit for you and your agency is essential.

▲ **TABLE 13.2** Checklist for the First Interview

Did You Complete This Step?	Significant Issues for the First Interview
	Before the Session
	Are you familiar with HIPAA, the policies of your agency, and key state laws? Are key policies posted in the agency waiting area? These need to be shared early in the session.
	If there is a file and have you read it? Do you need notes from the file for a refresher?
	Do the room and setup ensure confidentiality?
	Does the room provide adequate silence? Do you need a sound machine working outside your door?
	Do you have an inviting atmosphere where you will meet the client? Is it neutral, or do you have interesting art and objects relevant to those who may come to this office? Are chairs placed in a position where the power is relatively equalized?
	During the Session
	Relationship How did you plan to establish a relationship and connect with this unique client? Did you: 1. Discuss the client's rights and responsibilities? The interviewing, counseling, and therapy relationship works in part because of clearly defined rights and responsibilities of each person involved. 2. Provide an explanation of what might happen in the session and/or how the interview is likely to be structured? 3. Review HIPAA, agency policies, and key legal issues? If your agency requires diagnostic labels, did you explain that to the client and offer to share that diagnosis if he or she wishes? 4. Discuss confidentiality and its limits? 5. Obtain the client's permission to take notes and/or record the session? Was the client informed that these notes and the recording are available if he or she wishes to review them? 6. If working with an underage client, obtain the appropriate parental permission as required by your agency and/or state law? 7. Provide an opportunity for the client to ask you questions before you started? Were issues of multicultural differences addressed? 8. Work with the client to establish an early preliminary goal or objective for the session? 9. Come prepared if the client immediately started talking about issues and concerns? Did you listen to them carefully and return later to cover these important points?
	Story and Strengths Did you: 1. Allow and encourage the client to present the story fully? Did you reframe the word *problem* into a more positive, change-oriented perspective with words such as *issue, challenge, concern, opening for change*? 2. Bring out the key facts, thoughts, feelings, and behaviors related to the story? Did you perhaps also look for underlying deeper meanings behind the story? 3. Avoid becoming enmeshed in the client's story by becoming a voyeur (endless fascination and searching for details about the client's interesting issues) or by unconsciously putting a "negative spin" on what the client said? Are you part of the problem or part of the solution?

▲ **TABLE 13.2** (Continued)

Did You Complete This Step?	Significant Issues for the First Interview
	4. Bring out stories and concrete examples of client personal strengths and external resources? Did you search for specific images within these stories and perhaps anchor these positive images in specific areas of the body? 5. Ask the critical questions, "What else relates to what we've talked about so far?" "What else is going on in your life?" and "Is there anything else I should have asked you but didn't?"
	Goals Did you: 1. Review the early goals set by the client and revise them in accordance with new information about the story and strengths? 2. Jointly make these goals as specific and observable as possible? 3. When necessary, break down large goals into manageable step-by-step objectives that can be reached over time? Did you prioritize these goals? 4. Remind the client of the strengths and resources that he or she brings to achieve these goals?
	Restory Did you: 1. Include brainstorming without a theoretical orientation? Confront with a supportive challenge, summarizing the goal and the issue? ("On one hand the goal is _____ , but on the other hand the main challenges you face are _____. Now what occurs to you as a solution?) Often clients with your support will come up with their own unique and workable answers, often ones that you did not think of. 2. Use a variety of listening and influencing strategies to facilitate client reworking and restorying of issues? What were they? 3. Use identified positive strengths and resources to remind clients during low points of their own capabilities? 4. Did you agree on homework or personal experiments to be completed between sessions? 5. Develop a clear definition of a more workable story that can lead to action and transfer to the real world?
	Action Did you: 1. Build on the new story, or start of a new story, and work with the client to take specific action and learning from the interview to the "real world"? 2. Agree on a plan for transfer of learning that is clear and doable? 3. Contract with the client to do at least one thing differently during the week, or even tomorrow? 4. Agreed to plans to look at this homework during the next session? 5. Check how it was for client? Does he or she think they could work with you? Did you agree to work together? 6. Set a date for next session or follow-up interview?
	After the Session
	Did you: 1. Write notes that are clear to both your agency and your client? Give special attention to client strengths and resources? Sign and date the record?

(Continued)

▲ **TABLE** 13.2 Checklist for the First Interview (Continued)

Did You Complete This Step?	Significant Issues for the First Interview
	2. Review the checklist and make plans to include missing items in your next session? 3. As necessary, develop a treatment plan for the future? 4. Anything else?

Based on A. Gawande, *The Checklist Manifesto* (New York: Holt, 2009).

The first interview presents many challenges as you develop a relationship, structure the session, and get to know a new client. The interview plan and checklist may be helpful as you plan for any session, particularly that first session in which there are so many issues to cover. Experience reveals that even the most seasoned professional forgets and omits important items on this list and the treatment plan as well.

13.3
A FULL INTERVIEW TRANSCRIPT: I'd Like to Find a New Career

KEY CONCEPT QUESTIONS

▲ **How are microskills and the five-stage interview actually implemented in a real interview?**

▲ **How can you analyze the specific use of skills and their effectiveness in the session?**

▲ **How might others have responded to the same client?**

In Table 13.3, the counselor and client verbatim transcript of a career decision interview is supplemented by a skill-and-focus analysis of the session. The microskill five stages are demonstrated within the session. The Process Comments column analyzes the effectiveness of skills throughout the interview, and special attention is given to the effect of confrontations (C) on the client's developmental change. Notice that within this transcript and the process notes Allen analyzes his behavior in the session and Mary reviews and analyzes his comments. It is this type of analysis that we will ask you to do as a self-study project. This five-column format may be a suitable structure for you to use to prepare your own interview analyses.

▲ EXERCISE 13.1　**Evaluate and Assess Allen's Interviewing Style**

As you read this interview, evaluate and assess Allen's interviewing style. What responses make sense to you? Are his interventions appropriate, and what might you do differently? While it is important to define your own natural style, it is also important to look at your style and how others view it. You will find some responses and strategies are less effective than others. We all make errors; it is our ability to learn from them and change that enables us to become more effective.

▲ **TABLE 13.3** The Allen and Mary Five-Stage Decisional Interview

Skill Classifications				
Listening & Influencing	*Focus*	*C**	*Counselor and Client Conversation*	*Process Comments*
STAGE 1. RELATIONSHIP *Initiate the session, develop rapport and structuring.*				
Open question	Client		1. *Allen:* Hi, Mary. How are you today?	
	Client, interviewer		2. *Mary:* Ah . . . just fine. . . . How are you?	As Mary walked in, Allen saw her hesitate and sensed some awkwardness on her part. Note that she opens with two speech hesitations.
Information, paraphrase	Interviewer, client		3. *Allen:* Good, just fine. Nice to see you. . . . Hey, I noted in your file that you've done a lot of swimming.	
	Client, main theme		4. *Mary:* Oh, yeah, [smiling] . . . I like swimming; I enjoy swimming a lot.	Consequently, Allen decides to take a little time to develop rapport and put Mary at ease in the interview. Note that he focused on a positive aspect of Mary's past. It is often useful to build on the client's strengths even this early in the session.
Information, closed question	Main theme, interviewer, client		5. *Allen:* With this hot weather, I've been getting out. Have you been able to?	The distinction between providing information and a self-disclosure is illustrated at *Allen 5* and *Mary 6*. Allen comments that he's been getting out, whereas Mary gives information and personal feelings.
	Client		6. *Mary:* Yes, I enjoy the exercise. It's good relaxation.	
Paraphrase, reflection of feeling	Client		7. *Allen:* I also saw you won quite a few awards along the way. (*Mary:* Um-hmm.) . . . You must feel awfully good about that.	Mary's nonverbal behavior is now more relaxed. Client and counselor now have more body language symmetry.
	Client		8. *Mary:* I do. I do feel very good about that. It's been lots of fun.	
Information, closed question	Main theme		9. *Allen:* Before we begin, I'd like to ask if I can tape-record this talk. I'll need your written permission, too. Do you mind?	Obtaining permission to tape-record interviews is essential. If the request is presented in a comfortable, easy way, most clients are glad to give permission. At times it may be useful to give the tapes to clients to take home and listen to again.
Client			10. *Mary:* No, that's okay with me. [signs form permitting use of tape for *Intentional Interviewing and Counseling*]	

(Continued)

*This column will record the presence of a confrontation.

▲ **TABLE 13.3** The Allen and Mary Five-Stage Decisional Interview (Continued)

Skill Classifications				
Listening & Influencing	**Focus**	**C**	**Counselor and Client Conversation**	**Process Comments**
Information giving, self-disclosure	Client, interviewer		11. *Allen:* As we start, Mary, there are some important things to discuss. We'll have about an hour today and then we can plan for the future together. Today, I'd like to get to know you and I'll try to focus mainly on listening to your concerns. At the same time, from your file, I know that some of your issues relate to women's issues. Obviously, I'm a man, and I think it is important to bring this up so that you will be more likely to feel free to let me know if I seem to be "off-target" or misunderstand something. Feel free to ask me any questions you'd like around this or other matters. [At this point Allen discusses confidentiality, HIPPA, and other key agency policies.]	Allen provides some additional structure for the session so that Mary knows what she might expect. He introduces gender differences and provides an opportunity for Mary to react and ask questions. Note "*we* can plan." This leads to a more mutual interview.
	Client, interviewer		12. *Mary:* I feel comfortable with you already. But, a couple questions. One is that I'm interested in the counseling field as a possibility—and the other is around the issue of living with divorce and being a single parent. What can you say about those?	Mary gives the OK, but then asks two questions. She leans forward when she asks them. This question is a surprise to the interviewer and should be noted as divorce and relationship issues may show themselves to be important later in the session.
Self-disclosure, open question	Interviewer, client		13. *Allen:* Well, first I'm divorced and have one child living with me while the other is in college. Of course, I'd be glad to talk about the counseling career and share some of my thoughts. What thoughts occur to you around divorce and counseling?	Keep self-disclosures brief and return the focus to the client. But be comfortable and open in that process.
	Client, main theme		14. *Mary:* That helps. Going through my divorce was the worst thing of my life. My children are so important to me. Perhaps your experience with divorce will help you understand where I am coming from.	Mary smiles, sits back, and appears to have the information she was wondering about.

▲ **TABLE 13.3** (Continued)

Skill Classifications				
Listening & Influencing	*Focus*	*C*	*Counselor and Client Conversation*	*Process Comments*
STAGE 2. STORY AND STRENGTHS *Gather data, draw out stories, concerns, problems, issues.*				
Open question	Client		15. *Allen:* You've talked about quite a few things. Could you tell me what you'd like to start with?	In this series of leads you'll find that Allen uses the basic listening sequence of open question, encourager, paraphrase, reflection of feeling, and summary, in order. Many interviewers in different settings will use the sequence or a variation to define the client's problem.
	Client, problem/ concern, others		16. *Mary:* Well . . . ah . . . I guess there's a lot that I'd like to talk about. You know, I went through . . . ah . . . a difficult divorce and it was hard on the kids and myself and . . . ah . . . we've done pretty well. We've pulled together. The kids are doing better in school and I'm doing better. I've . . . ah . . . got a new friend. [breaks eye contact] But, you know, I've been teaching for 13 years and really feel kind of bored with it. It's the same old thing over and over every day; you know . . . parts of it are okay, but lots of it I'm bored with.	As many clients do, Mary starts the session with a "laundry list" of issues. Though the last thing in a laundry list is often what a client wants to talk about, the eye-contact break at mention of her "new friend" raises an issue that should be watched for in the interview. As the session moves along, it becomes apparent that more than the career issue needs to be looked at. Mary discusses a "pattern" of boredom. This is indicative of an abstract client who is able to reflect on herself and see patterns of behavior.
Encourage	Client		17. *Allen:* You say you're *bored* with it?	The key word *bored* is emphasized.
	Client, problem/ concern		18. *Mary:* Well, I'm bored, I guess . . . teaching field hockey and . . . ah . . . basketball and softball, certain of those team sports. There are certain things I like about it, though. You know, I like the dance, and you know, I like swimming—I like that. Ah . . . but . . . you know . . . I get tired of the same thing all the time. I guess I'd like to do some different things with my life.	Note that Mary elaborates in more detail on the word *bored*. Allen used verbal underlining and gave emphasis to that word, and Mary did as most clients would: She elaborated on the meaning of the key word to her. Many times short encouragers and restatements have the effect of encouraging client exploration of meaning and elaboration on a topic. "I'd like to do some different things" is a more positive "I" statement.

(Continued)

▲ **TABLE** 13.3 The Allen and Mary Five-Stage Decisional Interview (Continued)

Skill Classifications				
Listening & Influencing	Focus	C	Counselor and Client Conversation	Process Comments
Paraphrase	Client		19. *Allen:* So, Mary, if I hear you correctly, sounds like change and variety are important instead of doing the same thing all the time.	Note that this paraphrase has some dimensions of an interpretation in that Mary did not use the words *change* and *variety*. These words are the opposite of boredom and doing "the same things all the time." This paraphrase takes a small risk and is slightly additive to Mary's understanding. It is an example of the positive asset search, in that it would have been possible to hear only the negative "bored." Working on the positive suggests what *can* be done. Note her response.
	Client, family, problem/ concern		20. *Mary:* Yeah . . . I'd like to be able to do something different. But, you know, ah . . . teaching's a very secure field, and I have tenure. You know, I'm the sole support of my two daughters, but I think, I don't know what else I can do exactly. Do you see what I'm saying?	Mary, being heard, is able to move to a deeper discussion of her issues. Note that Mary tends to be abstract and discusses patterns and generalizations. If she were primarily concrete, she would give many more linear details and tell specific stories about her issues. She continues for most of the interview in this mode of expression. She has equated "something different" with a lack of security. As the interview progresses, you will note that she associates change with risk. It is these basic meaning constructs, already apparent in the interview, that lie under many of her issues.
Reflection of feeling, followed by checkout	Client, problem	C	21. *Allen:* Looks like the security of teaching makes you feel good, but it's the boredom you associate with that security that makes you feel uncomfortable. Is that correct?	This reflection of feeling contains elements of a confrontation as well, in that the good feelings of security are contrasted with the boredom associated with teaching.
	Client, problem/ concern		22. *Mary:* Yeah, you know, it's that security. I feel good being . . . you know . . . having a steady income and I have a place to be, but it's	Note that Mary often responds with a "Yeah" to the reflections and paraphrases before going on. Here she is wrestling with the

▲ **TABLE 13.3** (Continued)

Skill Classifications				
Listening & Influencing	**Focus**	**C**	**Counselor and Client Conversation**	**Process Comments**
			boring at the same time. You know, ah . . . I wish I knew how to go about doing something else.	confrontation of *Allen 21*. She adds new data, as well, in the last sentence. On the CCS, this would be acceptance and recognition (Level 3).
Summary, checkout	Client, family, problem/ concern	C	23. *Allen:* So, Mary, let me see if I can summarize what I've heard. Ah . . . it's been tough since the divorce, but you've gotten things together. You mentioned the kids are doing pretty well. You talked about a new relationship. *I heard you mention that.* (*Mary:* Yeah.) But the issue that you'd like to talk about now is . . . this feeling of boredom (*Mary:* Umm . . .) on the job, and yet you like the security of it. But maybe you'd like to try something new. Is that the essence of it?	This summarization concludes the first attempt at problem definition in this brief interview. Allen uses Mary's own words for the main things and attempts to distill what has been said. The positive asset search has been used briefly ("You've gotten things together . . . kids . . . doing well"). See other leads that emphasize client strength. Mary sits forward and nods with approval throughout this summary. The confrontation of the old job with "maybe you'd like to try something new" concludes the summary. Note the checkout at the end of the summary to encourage Mary to react. Nonetheless, he missed the new friend as part of his summary.
	Client, problem		24. *Mary:* That's right. That's it.	Mary again responds at Level 3 on the CCS, acceptance and recognition.
			STAGE 3. GOALS *Set goals mutually, establish outcomes.*	
Open question	Main theme		25. *Allen:* I think it might be helpful if you could specifically define what some things are that might represent a more ideal situation.	In Stage 3, find where the client wants to go in a more ideal situation. Note that the basic listening sequence is present in this stage, but it does not follow in order, as in the preceding stage.
	Client, problem, others		26. *Mary:* Ummm. I'm not sure. There are some things I like about my job. I certainly like interacting with the other professional	Mary associates interacting with people as a positive aspect of her job. When she says "enjoy working with kids," her tone changes,

(Continued)

▲ **TABLE 13.3** The Allen and Mary Five-Stage Decisional Interview (Continued)

Skill Classifications				
Listening & Influencing	Focus	C	Counselor and Client Conversation	Process Comments
			people on the staff. I enjoy working with the kids. I enjoy talking with the kids. That's kind of fun. You know, it's the stuff I have to teach I'm bored with. I have done some teaching of human sexuality and drug education.	suggesting that she doesn't enjoy it that much. But the spontaneous tone returns when she mentions "talking with them" and talks about teaching subjects other than team sports.
Paraphrase, open question	Client, main theme		27. *Allen:* So, would it be correct to say that some of the teaching, where you have worked with kids on content of interest to you, has been fun? What else have you enjoyed about your job?	The search here is for positive assets and things that Mary enjoys. Note the "what else?"
	Client, family, problem		28. *Mary:* Well, I must say I enjoy having the same summer vacations the kids have. That's a plus in the teaching field. [pause]	
Encourage			29. *Allen:* Yeah . . .	Mary found only one plus in the job. Allen probes for more data via an encourager. This type of encourager can't be classified in terms of focus.
	Client, others		30. *Mary:* You see, I like being able to . . . Oh, I know, one time I was able to do teaching of our own teachers and that was really . . . I really felt good being able to share some of my ideas with some people on the staff. I felt that was kind of neat, being able to teach other adults.	Mary brings out new data that support her earlier comment that she liked to teach when the content was of interest to her. The "I" statements here are more positive and the adjective descriptors indicate more self-assurance.
Closed question	Client, others		31. *Allen:* Do you involve yourself very much in counseling the students you have?	A closed question with a change of topic to explore other areas.
	Client, others		32. *Mary:* Well, the kids . . . you know, teaching them is a nice, comfortable environment, and kids stop in before class and after class and they talk about their boyfriends and the movies; I find I like that part . . . about their concerns.	Mary responds to the word *counseling* again with discussion of interactions with people. It seems important to Mary that she have contact with others.

▲ **TABLE 13.3** (Continued)

Skill Classifications				
Listening & Influencing	**Focus**	**C**	**Counselor and Client Conversation**	**Process Comments**
Summary, closed question	Client, problem, others		33. *Allen:* So, as we've been reviewing your current job, it's the training, the drug education, some of the teaching you've done with kids on topics other than phys. ed. (*Mary:* That's right.) And getting out and doing training and other stuff with teachers . . . ah, sharing some of your expertise there. And the counseling relationships. (*Mary:* Ummm.) Out of those things, are there fields you've thought of transferring to?	This summary attempts to bring out the main strands of the positive aspects of Mary's job. In an ongoing interview, a closed question on a relevant topic can be as facilitating as an open question. Note, however, that the interviewer still directs the flow with the closed question.
	Client, problem/ concern		34. *Mary:* Well, a lot of people in physical education go into counseling. That seems like a natural second thing. Ah . . . of course, that would require some more going to school. Umm . . . I've also thought about doing some management training for a business. Sometimes I think about moving into business . . . entirely away from education. Or even working in a college as opposed to working here in the high school. I've thought about those things, too. But I'm just not sure which one seems best for me.	Mary talks with only moderate enthusiasm about counseling. In discussing training and business, she appears more involved. Mary appears to have assets and abilities, makes many positive "I" statements, is aware of key incongruities in her life, and seems to be internally directed. She is clearly an abstract, formal-operational client. For career success, she also needs to become more concrete and action-oriented.
Paraphrase, closed question	Client, problem/ concern		35. *Allen:* So the counseling field, the training field. You've thought about staying in schools and perhaps in management as well. (*Mary:* Um-hm, um-hm.) Anything else that occurs to you?	This brief paraphrase distills Mary's ideas in her own words.
	Problem/ concern		36. *Mary:* No, I think that seems about it.	
Summary, open question, eliciting meaning	Client, problem/ concern	C	37. *Allen:* Before we go further, you've talked about teaching and the security it offers. But at the same time you talk about *boredom*. You talk with excitement about business and training. How do you put this together? What does it *mean* to you?	This summary includes confrontation and catches both content and feeling. The question at the end is directed toward issues of meaning. The word *boredom* was underlined with extra vocal emphasis.

(Continued)

▲ **TABLE 13.3** The Allen and Mary Five-Stage Decisional Interview (Continued)

Skill Classifications				
Listening & Influencing	*Focus*	*C*	*Counselor and Client Conversation*	*Process Comments*
	Client, others		38. *Mary:* Uhhh . . . Ah . . . If I stay in the same place, it's just more of the same. I see older teachers, and I don't want to be like them. Oh, a few have fun; most seem just *tired* to me. I don't want to end up like that.	Mary elaborates on the meaning and underlying structure of *why* she might want to avoid the occasional boredom of her job. When she talks about "ending up like that," we see deeper meanings. On the CCS, the client may again be rated at acceptance and recognition (Level 3). Though considerable depth of understanding and clarity are being developed, no large change has occurred. You will find that developmental movement often is slow and arduous. Nonetheless, each confrontation moves toward more complete understanding.
Encourage/ restatement	Client		39. *Allen:* You don't want to end up with that.	The key words are repeated.
	Client		40. *Mary:* Yeah, I want to do something new, more exciting. Yet my life has been so confused in the past, and it is just settling down. I'm not sure I want to risk it.	Mary moves on to talk about what she wants, and a new element—risk—is introduced. Risk may be considered Mary's opposing construct to security.
Reflection of feeling	Client		41. *Allen:* So, Mary, risk frightens you?	This reflection of feeling is tentative and said in a questioning tone. This provides an implied checkout and gives Mary room to accept it or suggest changes to clarify the feeling.
	Client, problem/ concern		42. *Mary:* Well, not really, but it does seem scary to give up all this security and stability just when I've started putting it together. It just feels strange. Yet I do want something new so that life doesn't seem so routine . . . and . . . ah . . . I think maybe I have more talent and ability than I used to think I did.	Mary responds as might be predicted with a deeper exploration of feelings of fear of change. At the same time, she draws on her personal strengths to cope with all this.
Reflection of meaning, checkout	Client, problem/ concern	C	43. *Allen:* So you've felt the meaning in this possible job change as an opportunity to use your *talent* and take risks in something new.	This reflection of meaning also confronts underlying issues that impinge on Mary's decision. It contains elements of the positive

▲ **TABLE 13.3** (Continued)

Skill Classifications				
Listening & Influencing	**Focus**	**C**	**Counselor and Client Conversation**	**Process Comments**
			This may be contrasted with the feelings of stability and certainty where you are now. But *now* means you may end up tired and burned out like some coworkers you have observed. Am I reaching the sense of things? How does that sound?	asset search or positive regard as Allen verbally stresses the word *talent*.
	Client, problem/ concern		44. *Mary:* Exactly! But I hadn't touched on it that way before. I do want stability and security, but not at the price of boredom and feeling down as I have lately. Maybe I do have what it takes to risk more.	Mary is reinterpreting her situation from a more positive frame of reference. Allen could have said the same thing via an interpretation, but reflection of meaning lets Mary come up with her own definition. This reinterpretation of Mary's meaning represents generation of a new solution (CCS Level 4). She has a new frame of reference with which to look at herself. But this newly integrated frame is *not* problem resolution; it is a *step* toward a new way of thinking and acting. Allen decides to move to Stage 4 of the interview. It would be possible to explore problem definition and detail the goals more precisely, but we can take up these matters in later interviews.
			STAGE 4. RESTORY *Explore and create alternatives, confront client incongruities and conflict.*	
Feedback	Client		45. Allen: Mary, from listening to you, I get the sense that you do have considerable ability. Specifically, you can be together in a warm, involved way with those you work with. You can describe what is important to you. You come across to me as a thoughtful, able, sensitive person. [pause]	Allen combines feedback on positive assets with some self-disclosure here and uses this lead as a transition to explore alternative actions. The emphasis here is on the positive side of Mary's experience. Allen's vocal tone communicates warmth, and he leans toward Mary in a genuine manner.

(*Continued*)

▲ **TABLE 13.3** The Allen and Mary Five-Stage Decisional Interview (Continued)

Skill Classifications				
Listening & Influencing	*Focus*	*C*	*Counselor and Client Conversation*	*Process Comments*
			46. *Mary:* Ummm . . .	During the feedback, Mary at first shows signs of surprise. She sits up, then relaxes a bit, smiles, and sits back in her chair as if to absorb what Allen is saying more completely. There are elements of praise in Allen's comment.
Directive, paraphrase, open question	Client, main theme		47. *Allen:* Other job ideas may develop as we talk . . . ah . . . I think it might be appropriate at this point to explore some alternatives you've talked about. (*Mary:* Um-hm.) The first thing you talked about was you liked teaching drug education and sexuality. What else have you taught kids?	Allen starts exploring alternatives a little more concretely and in depth. The systematic problem-solving model—define the problem, generate alternatives, and set priorities for solutions—is in his mind throughout this section. He begins with a mild directive. "What else?" keeps the discussion open.
	Client, problem/concern		48. *Mary:* Let's see . . . The general areas I liked were human sexuality, drug education, family life, and those kinds of things. Ah . . . sometimes communication skills.	
Closed question	Problem/concern		49. *Allen:* Have you attended workshops on any of these topics?	Closed questions oriented toward concreteness can be helpful in determining specific background important in decision making.
	Client, problem/concern, others		50. *Mary:* I've attended a few. I've enjoyed them . . . I really did. You know, I've gone to the university and taken workshops in values clarification and communication skills. I liked the people I met.	Note that virtually all counselor and client comments have focused on the client and the problem. It is important to consider the client in each of your responses; too heavy an emphasis on the problem may cause you to miss the unique person before you. At the same time, a broader focus might expand the issue and provide more understanding. Social work, for example, might emphasize the family and social context.
Reflection of feeling, information, checkout	Client, problem/concern		51. *Allen:* Sounds like you've really enjoyed these sessions. One of the important roles in counseling, education, and business is training—for	Allen briefly reflects her positive feelings, and then shares a short piece of occupational information. This is followed by

▲ **TABLE 13.3** (Continued)

Skill Classifications				
Listening & Influencing	**Focus**	**C**	**Counselor and Client Conversation**	**Process Comments**
			example, psychological education through teaching others skills of living and communication. How does that type of work sound to you?	a checkout returning the focus to Mary.
	Client, problem/ concern		52. *Mary:* I think I would enjoy that sort of thing. Um-hmmm . . . It sounds interesting.	
Paraphrase, open question	Client, problem/ concern		53. *Allen:* Sounds like you have also given a good deal of thought to . . . ah . . . extending that to training in general. How aware are you of the business field as a place to train and teach employees?	Mary's background and interest in a second alternative are explored.
	Client, problem/ concern, environ- mental context		54. *Mary:* I don't know that much about it. You know, I worked one summer in my dad's office, so I do have an exposure to business. That's about it. They all have been saying that a lot of teachers are moving into the business field. Teaching is not too lucrative, and with all the things happening here in California and all the cutbacks, business is a better long-term possi- bility for teachers these days. It just seems like an intriguing possibility for me to investigate or look into. The latest business cutbacks are scary, too.	Mary talks in considerably greater depth and with more enthusiasm when she talks about business. The important descriptive words she has used with teaching include *boring, security,* and *inter- personal interactions,* while *inter- est* and *excitement* were used for training and teaching psychologi- cally oriented subjects as opposed to physical education. Now she mentions cutbacks. Business has been described with more enthu- siasm and as more lucrative. We may anticipate that she will eventually associate the potential excitement of business with the negative construct of risk and the lack of summer vacations and time to be with her children.
Paraphrase, reflection of feeling	Client, problem/ concern	C	55. *Allen:* Mm-hmm, . . . so you've thought about it . . . looking into business, but you've not done too much about it yet. Neither teach- ing nor business is really promising now, and that's a little scary.	This paraphrase is somewhat subtractive. Mary did indicate that she had summer experience with her father. How much and how did she like it? Allen missed that. The paraphrase involves a confrontation between what Mary says and her lack of doing

(*Continued*)

▲ **TABLE 13.3** The Allen and Mary Five-Stage Decisional Interview (Continued)

Skill Classifications				
Listening & Influencing	Focus	C	Counselor and Client Conversation	Process Comments
				anything extensive in terms of a search. The reflection of feeling acknowledges emotion.
	Client		56. *Mary:* That's right. I've thought about it, but . . . ah . . . I've done very little about it. That's all . . .	Mary feels a little apologetic. She talks a bit more rapidly, breaks eye contact, and her body leans back a little. Mary's response is at Level 2 on the CCS. She is only partially able to work with the issues of the confrontation.
Interpretation	Problem/ concern		57. *Allen:* And, finally, you mentioned that you have considered the counseling field as an alternative. Ah . . . what about that?	Allen omitted further exploration of business. If Allen had focused on positive aspects of Mary's experience and learned more about her summer experience, the confrontation (of thinking without action) probably would have been received more easily. As this was a demonstration interview, Allen sought to move through the stages perhaps a little too fast. Also, the counseling field is an alternative, but it seems to come more from Allen than from Mary. An advantage of transcripts such as this is that one can see errors. Many of our errors arise from our own constructs and needs. This intended paraphrase is classified as an interpretation, as it comes more from Allen's frame of reference than from Mary's.
	Problem/ concern, others		58. *Mary:* Well, I've always been interested, like I said, in talking with people. People like to talk with me about all kinds of things. And *that* would be interesting . . . ah . . . I think, too.	Mary starts with some enthusiasm on this topic, but as she talks her speech rate slows and she demonstrates less energy.
Encourage			59. *Allen:* Um-hmmm.	
	Problem/ concern		60. *Mary:* You know, to explore that. [pause]	Said even more slowly.

▲ **TABLE 13.3** (Continued)

Skill Classifications				
Listening & Influencing	**Focus**	**C**	**Counselor and Client Conversation**	**Process Comments**
Encourage			61. *Allen:* Um-hmmm. [pause]	Allen senses her change of enthusiasm, is a bit puzzled, and sits silently, encouraging her to *talk more*. When you have made an error and the client doesn't respond as you expect, return to attending skills.
	Problem/ concern		62. *Mary:* But . . . I'd have to take some *courses* . . . if I really wanted to get into it.	One reason for Mary's hesitation appears.
Interpreta- tion/ reframe	Client, problem/ concern		63. *Allen:* So putting those three things together, it seems that you want people-oriented occupations. They are particularly interesting to you.	This is a mild interpretation, as it labels common elements in the three jobs. It could be classified also as a paraphrase. Not all skill distinctions are clear.
	Client		64. *Mary:* Definitely . . . and that's where I am most happy.	Mary has returned to a Level 3 on the CCS.
Feedback	Client, problem/ concern	C	65. *Allen:* And, Mary, as I talk I see you . . . ah . . . coming across with a lot of enthusiasm and interest as we talk about these alternatives. I do feel you are a little less enthu- siastic about returning to school. (*Mary:* Right!) I might contrast your enthusiasm about the pos- sibilities of business and training with your feelings about educa- tion. There you talk a little more slowly and almost seem bored as you talk about it. You seem lively when you talk about business possibilities.	Allen gives Mary specific and concrete feedback about how she comes across in the inter- view. There is a confrontation as he contrasts her behavior when discussing two topics. Confrontation—the presentation of discrepancies or incongruity— may appear with virtually all skills of the interview. It may be used to summarize past conversation and stimulate further discussion, leading toward a resolution of the incongruity.
	Client, problem/ concern		66. *Mary:* Well, they sound kind of exciting to me, Allen. But I just don't know how to go about getting into those fields or what my next steps might be. They sound very exciting to me, and I think I may have some talents in those areas I haven't even discovered yet.	Mary talks rapidly, her face flushes slightly, and she gestures with enthusiasm. She meets the confrontation and seems to be willing to risk more. This, how- ever, may still be considered a Level 3 on the CCS, although there may be movement ahead.

(Continued)

▲ **TABLE 13.3** The Allen and Mary Five-Stage Decisional Interview (Continued)

Skill Classifications				
Listening & Influencing	**Focus**	**C**	**Counselor and Client Conversation**	**Process Comments**
Feedback, information, logical consequences	Client, problem/ concern	C	67. *Allen:* Um-hmmm. Well, Mary, I can say one thing. Your enthusiasm and ability to be open will be helpful to you in your search. Ah . . . at the same time, business and schools represent different types of lifestyles. I think I should give you a warning that if you go into the business area you're going to lose those summer vacations.	This statement combines mild feedback with logical consequences. A warning about the consequences of client action or inaction is spelled out. Mary is also confronted with some consequences of choice.
	Client, problem/ concern others		68. *Mary:* Yeah, I know that . . . and you know, that special friend in my life—he's in education—I don't think he would like it if I was, you know, working all summer long. But business does pay a lot more, and it might have some interesting possibilities. (*Allen:* Um-hmm.) . . . It's a difficult situation.	Confrontations often result in clients' presenting new important concepts and facts that have not been discussed previously. A new problem has emerged that may need definition and exploration. Mary is still responding at Level 3 on the CCS, but Allen is obtaining a more complete picture of the problem and of the client.
Encourage/ restatement	Problem		69. *Allen:* A difficult situation?	Again, the encourager is used to find deeper meanings and more information.
	Client, problem, others		70. *Mary:* Um-hmm. I guess I'm saying that . . . I'm . . . ah . . . you know, my friend . . . I don't think he would approve or like the idea of me having two weeks' vacation. (*Allen:* Uh-huh.) He wants me to stay in some field where I have the same vacation time I have now so we can spend that time together.	Mary has more speech hesitations and difficulties in completing a sentence here than she has anywhere in the interview. This suggests that her relationship is important to her, and her friend's attitude may be important in the final career decision. Much career counseling involves personal issues as well as career choice. Both require resolution for true client satisfaction.
Interpretation/ reframe, open question	Client, others, cultural/ environmental context	C	71. *Allen:* I hear you saying that your friend has a lot to say about your future. How does that strike you as an independent woman who has been on your own successfully for quite a while?	Here we see the introduction of gender relations as a cultural/environmental/contextual issue. Allen's reframing of the situation offers Mary a chance to explore her relationship with her friend from a different contextual perspective.

▲ **TABLE 13.3** (Continued)

Skill Classifications				
Listening & Influencing	**Focus**	**C**	**Counselor and Client Conversation**	**Process Comments**
	Others		72. *Mary:* It really is . . . well, Bo's a special person . . .	Mary's eyes brighten.
Interpretation	Client, others		73. *Allen:* And, I sense you have some reactions to his . . .	Allen interrupts, perhaps unnecessarily. It might have been wise to allow Mary to talk about her positive feelings toward Bo.
	Client, problem/ concern		74. *Mary:* Yeah, I'd like to be able to explore some of my own potential without having those restraints put on me right from the beginning.	Mary talks slowly and deliberately, with some sadness in her voice. Feelings are often expressed through intonation. Here we see the beginning of a critical gender issue. Women often feel constraints in career or personal choices, and men in this culture often place implicit or explicit restraints on critical decisions. Feminist counseling theorists argue that a male helper may be less effective with these types of problems. What are your thoughts on this issue?
Interpretation/ reframe, checkout	Client, problem/ concern, others, cultural/ environmental context	C	75. *Allen:* Um-hmm . . . In a sense he's almost placing similar constraints on you that you feel in the job in physical education. There are certain things you have to do. Is that right?	This interpretation relates the construct of boredom and the implicit constraint of being held down to the constraints of Bo. The interpretation clearly comes from Allen's frame of reference. With interpretations or helping leads from your frame of reference, the checkout of client reactions is even more important. The drawing of parallels is abstract, formal-operational in nature.
	Client, problem/ concern, others		76. *Mary:* Yes, probably so. He's putting some limits on me . . . setting limits on the fields I can explore and the job possibilities I can possibly have. Setting some limits so that my schedule matches his schedule.	Mary answers quickly. It seems the interpretation was relatively accurate and helpful. One measure of the function and value of a skill is what the client does with it. Mary changes the word *constraints* to the more powerful word *limits*. Mary remains at Level 3 on the CCS, as she is still expanding on aspects of the problem.

(Continued)

▲ **TABLE 13.3** The Allen and Mary Five-Stage Decisional Interview (Continued)

Skill Classifications				
Listening & Influencing	**Focus**	**C**	**Counselor and Client Conversation**	**Process Comments**
Open question, oriented to feeling	Client		77. *Allen:* In response to that you feel . . . ? [deliberate pause, waiting for Mary to supply the feeling]	Research shows that *some* use of questions facilitates emotional expression.
	Client, problem/ concern		78. *Mary:* Ah . . . I feel I'm not at a point where I want to *limit things*. I want to see what's open, and I would like to keep things open and see what all the alternatives are. I don't want to shut off any possibility that might be really exciting for me. (*Allen:* Um-hmm.) A total lifetime of careers.	Mary determinedly emphasizes that she does not want limits.
Reflection of feeling, paraphrase	Client, problem/ concern	C	79. *Allen:* So you'd like to have a life of exciting opportunity, and you sense some limiting . . .	A brief, but important, confrontation of Bo versus career.
	Client, problem/ concern, others, cultural/ environmental context		80. *Mary:* He reminds me of my relationship with my first husband. You know, I think the reason that all fell apart was my going back to work. You know, assuming a more nontraditional role as a woman and exploring my potential as a woman rather than staying home with the children . . . ah . . . you know, sort of a similar thing happened there.	Again, the confrontation brings out important new data about Mary's present and past. Is she repeating old relationship patterns in this new relationship? The counselor should consider issues of cultural sexism as an environmental aspect of Mary's planning. This does not appear in this interview, but a broader focus on issues in the next session seems imperative. Other focus issues of possible importance include Mary's parental models, others in her life, a women's support group, the present economic climate, the attitudes of the counselor, and "we"—the immediate relationship of Mary and Allen. So far he has assumed a typical Western "I" form of counseling in which the emphasis is on the client. With the development of new, more integrated data, this could move toward a more inclusive construct (Level 5 response on the CCS).

▲ **TABLE 13.3** (Continued)

Skill Classifications				
Listening & Influencing	*Focus*	*C*	*Counselor and Client Conversation*	*Process Comments*
Summary	Client, problem/ concern, others, cultural/ environ- mental context	C	81. *Allen:* There really are a variety of issues that . . . you're look- ing at. One of these is the whole business of a job. Another is your relationship with Bo and your desire to find your own space as an inde- pendent woman.	The interview time is waning, and Allen must plan a smooth ending and plan for the next session. He catches the incongruity that Mary faces between work and relation- ship. Allen fails to pick up fully on the cultural/environmental context focus. Many of Mary's issues relate to women's issues in a sometimes sexist world.
			82. *Mary:* [slowly] Um-hmmm . . .	Mary looks down, relaxes, and seems to go into herself.
Reflection of feeling	Client		83. *Allen:* You look a little sad as I say that.	This reflection of feeling comes from nonverbal observations and picks up on her facial reactions.
	Problem/ concern		84. *Mary:* It would be nice if the two would mesh together, but it seems difficult to have both things fit together nicely.	Mary is describing her ideal reso- lution. Here the interview could recycle back to Stages 2 and 3, with more careful delineation of the problem between job and personal relationships and defining the ideal resolution more fully. *A problem exists only if there is a difference between what is actually happening and what you desire to have happen.* This sentence illustrates the impor- tance of problem definition and goal setting. Mary's response to the confrontation is 4 on the CCS. We have an important new insight, but insight is not action. She also needs to act on this awareness.
Information, directive	Problem		85. *Allen:* Well, that's something we can explore a little bit further. It seems this is an important part of the puzzle. Let's work on that next week. Would that be okay? I see our time is about up now. But it might be useful if we can think of some actions we can take between now and the next time we get together.	Many clients bring up central issues just as the interview is about to end. Allen makes the decision, difficult though it is, to stop for now and plan for more discussion later. Note that Mary is still talking about her relationship mainly from an abstract, formal-operational orientation. Clients often bring up central issues late in the session.

(Continued)

▲ **TABLE 13.3** The Allen and Mary Five-Stage Decisional Interview (Continued)

Skill Classifications				
Listening & Influencing	*Focus*	*C*	*Counselor and Client Conversation*	*Process Comments*
			STAGE 5. ACTION *Generalize and act on new stories.*	
Summary, open question	Client, problem/ concern		86. *Allen:* We have come up so far with three things that seem to be logical: business, counseling, and training. I think it would be useful, though, if you were to take a set of career tests. (*Mary:* Uh-huh.) That will give us some additional things to check out to see if there are any additional alternatives for us to consider. How do you feel about taking tests?	Allen continues his statement and moves to Stage 5. He summarizes the career alternatives generated so far and raises the possibility of taking a test. Note that he provides a checkout to give Mary an opportunity to make her own decision about testing.
	Client, problem/ concern		87. *Mary:* I think that's a good idea. I'm at the stage where I want to check all alternatives. I don't want *anything* to be limited. I want to think about a lot of alternatives at this stage. And I think it would be good to take some tests.	Mary approves of testing and views this as a chance to open alternatives. She verbally emphasizes the word *anything*, which may be coupled with her desire to avoid limits to her potential. Some women would argue that a female counselor is needed at this stage. A male counselor may not be sufficiently aware of women's needs to grow. Allen could unconsciously respond to Mary in the same ways she views Bo as responding to her.
Information	Client, problem/ concern, interviewer		88. *Allen:* Then another thing we can do . . . ah . . . is helpful. I have a friend at a local firm who originally used to be a coach. She's moved into personnel and training at Jones. (*Mary:* Ummm.) I can arrange an appointment for you to see her. Would you like to go down and look at the possibilities there?	Allen suggests a concrete and specific alternative for action. Mary is predominantly abstract formal-operational; she has tended to talk about issues and avoid action. This avoidance of action is also indicative of Level 3 on the confrontation impact scale. Until Mary takes some form of concrete action or resolves the issue in her mind, she will remain at Level 2 or 3 on the CCS. If some action is taken on the issue during the coming week, then she will have moved at least partially to Level 4 on the CCS.

▲ **TABLE 13.3** (Continued)

Skill Classifications				
Listening & Influencing	**Focus**	**C**	**Counselor and Client Conversation**	**Process Comments**
	Client, problem/ concern		89. *Mary:* Oh, I would like to do that. I'd get kind of a feel for what it's like being in a business world. I think talking with someone would be a good way to check it out.	Stated with enthusiasm. The proof of the helpfulness of the suggestion will be determined by whether she does indeed have an interview with the friend at Jones and finds it helpful in her thinking.
Feedback, open question	Client, problem		90. *Allen:* You're a person with a lot of assets. I don't have to tell you all the things that might be helpful. What other ideas do you think you might want to try during the week?	Allen recognizes he may be taking charge too much and pulls back a little. Although he is pushing Mary for action, he is now using her ideas. Too much direction and advice can make a client resistant to your efforts.
	Client, problem		91. *Mary:* What about checking into the university and ah . . . advanced degree programs? I have a bachelor's degree, but . . . maybe I should check into school and look into what it means to do more coursework.	Mary, on her own, decides to look into the university alternative. This is particularly important, as earlier indications were that she was not all that interested. Note that real generalization is usually concrete *action*.
Summary	Client, problem, cultural/ environmental context		92. *Allen:* Okay, that's something else you could look into as well. (*Mary:* Uh-huh.) So let's arrange for you then to follow up on that. I'd like to see you doing that. (*Mary:* Um-hmmmm.) And . . . ah . . . we can get together and talk again next week. You did express some concern about your relationship with your friend, Bo, ah . . . would you like to talk about that as well next week? And, as I look back on this session, one theme we haven't discussed yet is how being a woman with family responsibilities relates to all this. Maybe this is something to be explored next week as well?	Allen is preparing to terminate the interview. Fortunately, he does consider the women's issue. Probably this should have been done sooner in the session. Is this an issue with which he can help, or would you recommend referral?
	Client, cultural/ environmental context		93. *Mary:* I think so, they sort of all . . . one decision influences another. You know. It all sort of needs to be discussed. And thanks	An important insight at the end. Mary realizes her career issue is more complex than she originally believed. If you were Allen's

▲ **TABLE** 13.3 The Allen and Mary Five-Stage Decisional Interview (Continued)

Skill Classifications				
Listening & Influencing	Focus	C	Counselor and Client Conversation	Process Comments
			for bringing up the women's issue and my children. That's important to me.	supervisor, would *you* recommend a primary emphasis on career counseling or on personal counseling in the next session? Or perhaps some combination of them both? What else would you advise him to do?
Self-disclosure	Client, interviewer		94. *Allen:* Okay. I look forward to seeing you next week, then.	
			95. *Mary:* Thank you.	

Our use of listening skills influences what a client says next and the general direction of the interview. Allen, the interviewer, constructed the session addressing primarily career counseling and focused on pertinent, important decisions. However, a person-centered counselor might have responded rather quickly to personal issues, with a greater emphasis on reflection of feeling and reflection of meaning (see page 219). A brief solution-focused counselor would focus much sooner on client goals and use many more questions (see page 219). Nonetheless, the systematic decisional microskills counseling model underlies most, perhaps all, theories of helping and provides a good foundation for mastering other theories, including your own unique approach to the helping field.

Complete Interactive Exercise 2: Interview Transcript: Allen and Mary's Decisional Counseling

Complete Weblink Exercise: Career Resources

13.4
TRANSCRIPT ANALYSIS

KEY CONCEPT QUESTION

▲ What is the pattern of skills used by the interviewer in the decisional counseling transcript?

Through the way you listen and the topics you select to reinforce by attending, you influence what happens in the session. Effective listening will increase the control clients have and allow them to become partners or "co-constructors" of what happens in the session. Examine your behavior in the interview and become aware of your impact on your client. If you plan a career session, you most likely will have a career session. If you decide to "let the client talk and see what happens," the interview may lack direction, but "what happens will happen."

Of course, any interview will not completely follow your plan, but your personal decisions influence what happens. This is a critical reason to use the checklist, develop interview plans, examine your notes, and reexamine your own style. We recommend a detailed analysis of your interviewing style and behavior continuously throughout your career. We also suggest appropriate sharing of your thoughts and analysis with your clients. Seek colleagues and supervisors to review your work, enabling you to constantly grow and improve.

SKILLS AND THEIR IMPACT ON THE CLIENT

Let us turn to a microskill analysis of the interview. Table 13.4 presents a skill summary of Allen's interview with Mary. You will see that each stage of the interview involved different patterns of microskill usage.

▲ EXERCISE 13.2   **Your Approach to the Interview**

What do you see as strengths of this interview? What do you think should have been done differently? How would you have approached a client such as Mary? What advice would you give Allen for the future?

Stage 1: Relationship. The interview began with Allen using both listening and influencing skills. We see open questions, a combined paraphrase and reflection of feeling, information giving, and self-disclosure. He focused immediately on Mary's wellness strengths in swimming, obtained permission to record the session, and offered the opportunity for Mary to ask him questions. Observation skills helped Allen decide when it was time to move on with the session.

Stage 2: Story and Strengths. Only listening skills were used to draw out Mary's story and concerns. The primary focus was on changing careers from physical education to either counseling or business.

Stage 3: Goals. Again, only listening skills were used as Mary spoke about her goals in more detail. The issue of security in teaching versus risk in business was an important issue, as revealed in the reflection of meaning with a confrontation (*Allen* 43).

Stage 4: Restory. In this "working" phase of the interview, Allen used both influencing skills and confrontation of incongruity and discrepancies extensively. The cultural/contextual issue of gender is explored though interpretation/reframing, listening skills (71), and an important summary (81).

Stage 5: Action. The interview closes with specific plans for generalization and homework for Mary. This stage begins with a summary of the interview and ends with a summary of plans for the future.

In terms of the overall balance of skill usage, Allen used a ratio of approximately two attending skills for every influencing skill. When you look at competence levels, Allen is able to identify and classify the several skills and stages of the interview. He is able to identify the impact of his skills on the client. Allen also demonstrates his ability to use the basic listening sequence to structure an interview in five stages and

▲ **TABLE 13.4** Skill Summary of Allen and Mary Interview Over Five Stages

	Listening/Attending							Focus							Influencing						C
	Open question	Closed question	Enc./restatement	Paraphrase	Reflection of feeling	Reflection of meaning	Summary	Client	Problem/concern	Significant others	Family	Mutuality "we"	Interviewer	C/E/C	Interpretation/reframe	Logical consequences	Self-disclosure	Feedback	Info/adv./etc.	Directive	Confrontation
STAGE 1. *RELATIONSHIP:* Initiating the session 6 attending skills 3 influencing skills	2	2		2	1			6	2				3				1		4		
STAGE 2. *STORY AND STRENGTHS:* Gathering data 4 attending skills 0 influencing skills 2 confrontation skills	1			1			1	5	2		1										2
STAGE 3. *GOAL:* Mutual goal setting 14 attending skills 0 influencing skills 2 confrontation skills	3	3	2	2	1	1	2	8	6	2											2
STAGE 4. *RESTORY:* Working 16 attending skills 15 influencing skills 7 confrontation skills	4	1	3	4	3		1	15	14	4				3	6	1		3	3	2	7
STAGE 5. *ACTION:* Terminating 5 attending skills 3 influencing skills	2						3	5	4				2	1			1	1	1		
Total: 45 attending skills 21 influencing skill 11 confrontation skills	12	6	5	9	5	1	7	39	28	6	1	0	5	4	6	1	2	4	8	2	11

Skill Classifications

to employ intentional interviewing skills. Note that Allen focused primarily on the client in the earlier phases of the interview and only in the later portions increased emphasis on the career issues and challenges. This demonstrates that he can balance focus between the person and client concerns and problems. An ineffective interviewer might have focused early on the problem and missed Mary as a unique person.

In terms of focus dimensions, Allen's focus remained primarily on the client, although the majority of his focus dimensions were dual, combining focus on Mary with focus on the issues and concerns Mary brought to the session. Focus analysis points out that Allen did not focus extensively on others and the family (on Bo or on Mary's children, for example). The relationship with Bo appeared with greater clarity later in the interview. Allen brought in the cultural/environmental/contextual focus (71, 75, and 81), enabling a beginning discussion of gender issues that clearly need further work.

Mary appeared to move a little deeper into personal insights concerning her present and future life following each of the 11 confrontations in Stage 4. She responded primarily at Level 3 on the Client Change Scale (CCS) each time. Note that Mary was led into the important area of her personal life and relationship with Bo (71). At 79, Allen comments on Mary's desire to have a life of "exciting opportunity," but she senses that Bo is putting limits on her. Mary moves easily here to discuss her own personal wishes in more depth. Allen included a checkout at the end of a slightly inaccurate confrontation (75), and Mary was able to introduce her important construct, substituting the word *limits* for Allen's *constraints*.

As the session ended, Mary appeared ready and willing to take action. On the Client Change Scale, she has moved to Level 4, or a new way of thinking about her issues. The CCS (presented in Chapter 8) is a systematic way for you to monitor the progress of your clients toward their goals both inside and outside the session. The real proof of the success of the interview, however, will have to wait until the next meeting so that we can determine whether the generalization plan was indeed acted on. Thoughts, feelings, and behaviors need to change for true generalization. *The work that clients do after the interview is as important as or more important than what they do in the session.* You can also evaluate the effectiveness of an interview in terms of the number of choices available to Mary. "If you don't have at least three possibilities, you don't have a choice." Mary appears to have achieved that objective in the interview. In addition, the issue of her relationship with Bo has been unearthed, and this topic may open her to further counseling possibilities. The question must be raised whether Allen, as a man, is the most appropriate interviewer for Mary to see. Answers to that question will vary with your personal worldview. Again, what are your evaluations? What would you do differently?

    Complete Interactive Exercise 3: Interview Summary

13.5

ADDITIONAL CONSIDERATIONS: Referral, Treatment Planning, and Case Management

KEY CONCEPT QUESTIONS

▲ **What are some key issues and client implications in referral, treatment planning, and case management?**

▲ **How can we ensure that action happens in the "real world"?**

Whether you are involved with decisional, person-centered, crisis, or brief counseling, you need to consider what to do next. The following section addresses three key issues in planning for the future.

REFERRAL

The word **referral** appears in the interview process notes. No interviewer has all the answers, and in the case of Mary, Allen thought referral to a women's group might be helpful as many of her issues are common to women looking for career change. In addition, his notes indicate the need for a referral to the university financial aid office. An important part of individual counseling is helping your clients find community resources that may facilitate their growth and development. The community genogram (presented in Chapter 9) helps interviewers and clients think more broadly and consider appropriate referral sources.

Sometimes the client/interviewer relationship simply doesn't work as well as we all would like. When you sense the relationship isn't doing well, avoid blaming either the client or yourself. Focus on client goals, and seek to hear the story completely and accurately. Ask the client for feedback on how you might be more helpful. Seek consultation and supervision, and most often these "difficult patches" can be resolved to the benefit of all. When an appropriate referral needs to be arranged, we do not want to leave our clients "hanging" with no sense of direction or fearful that their problems are too difficult. Maintain contact with the client as the referral process evolves, sometimes even continuing for a session or two until arrangements are complete.

Another key referral issue is whether this is a case where interviewer expertise and experience are sufficient to help the client. Even if the counselor thinks that he or she is working effectively, this may not be enough; it may be a case in which supervision and case conferences can be helpful. Opening up your work to others' opinion is an important part of professional practice. Clients, of course, should be made aware that you as counselor or therapist are being supervised. It is critical that the client never feel rejected by you. Your support during the referral process is essential.

Your first counseling interview with a client is always unique and gives you an opportunity to learn about diversity and the complexity of the world. Box 13.1 (page 296) presents an international view of the work with clients and recommendation for practice.

TREATMENT PLANNING

Treatment planning, in the form of specific written goals and objectives, is increasingly standard and often required by agencies and insurance companies. When possible, negotiate the specific goals with the client and write them down for joint evaluation. They should be as concrete and clear as possible, and indicators of behavioral change and emotional satisfaction are important. The more structured counseling theories, such as cognitive-behavioral, strongly urge interview and treatment plans with specific goals developed for each issue. Their interview and treatment plans are often more specific than those presented in the interview example here.

Less structured counseling theories (Gestalt, psychodynamic, person-centered) tend to give less emphasis to treatment plans, preferring to work in the moment with the client. In short-term counseling and interviewing, the interview plan serves as the treatment plan. As you move toward longer-term counseling (5 to 10 sessions), a more detailed treatment plan with specific goals is often required. Many agencies also use case management as part of the treatment plan.

Following is Allen's interview plan for the second session with Mary. The plan is developed from information gained in the first interview and organizes the central issues of the case, allowing for new input from Mary as the session progresses.

Stage/Dimension Key Questions	Counselor Assessment and Plan
Relationship. Initiate the session, develop rapport and structuring. *What structure do you have for this interview? Do you plan to use a specific theory? What special issues do you anticipate with regard to rapport development?*	Mary and I have reasonable rapport. As I look at the first session, I note I did not focus enough on Mary's context nor did I attend to other things that might be going on in her life. It may be helpful to plan some time for general exploration *after* I follow up on the testing and her interviews with people during the week. Mary indicated an interest in talking about Bo. Two issues need to be considered at this session in addition to general exploration of her present state. I'll introduce the tests and follow that with discussion of Bo. For Bo, I think a person-centered method emphasizing listening skills may be helpful.
Story and Strengths. Gather data, draw out stories, concerns, problems, or issues. *What are anticipated problems? Strengths? How do you plan to define the issues with the client? Will you emphasize behavior, thoughts, feelings, meanings?*	1. Check how Mary sees her career concerns defined now. Use basic listening sequence. 2. Later, and as appropriate, open up the issue of Bo with a question, then follow through with reflective listening skills. Be alert to a woman's perspective. 3. Mary has many assets. She is bright, verbal, and successful in her job. She has good insight and is willing to take reasonable risks and explore new alternatives. These assets should be noted in our future interviews. 4. Explore women's issues with her.
Goals. Set goals mutually, establish outcomes. *What is the ideal outcome? How will you elicit the client's idealized self or world?*	We have already discussed her career goals, but they may need to be reconsidered in light of the tests, further discussion of Bo, and so on. It is possible that late in this interview or in a following session we may need to define a new outcome in which careers and her relationships are both satisfied.
Restory. Explore and create alternatives, confront client incongruities and conflict, restory. *What theories would you probably use here? What specific incongruities have you noted or do you anticipate in the client? How will you generate alternatives?*	1. Check on results of tests and report them to Mary. 2. Explore her reactions and consider alternative occupations. 3. Use person-centered, Rogerian counseling and explore her issues with Bo. 4. Relate careers to the relationship with Bo. Give special attention to confronting the differences between her needs

Stage/Dimension Key Questions	Counselor Assessment and Plan
	as a "person" and Bo's needs for her. Note and consider the issue of women in a changing world. Does Mary need referral to a woman or to a women's group for additional guidance? Would assertiveness training be useful?
Action. Generalize and act on new stories. *What specific plans do you have for transfer of training? What will enable you, the interviewer, to personally feel that the interview was worthwhile?*	At the moment it seems clear that further exploration of careers outside the interview is needed. We will have to explore the relationship with Bo and determine her objectives more precisely.

Those working in community and hospital clinics are often required to have precise goals in their treatment plans. Stating the goals specifically on a treatment plan is both for the good of the client and to fulfill the requirements of many insurance companies. A brief outline of what many agencies consider important on a treatment plan is shown in Figure 13.1. Note the emphasis on concrete goals, specificity of interventions, and a planned date for evaluation of goal achievement. Increasingly, you will find that you will be working with some variation of this goal-oriented form, regardless of setting.

CASE MANAGEMENT AS RELATED TO TREATMENT PLANNING

Although this book focuses on the interview and counseling skills, treatment planning often needs to be extended to case management. For human service professionals, social workers, and school counselors, case management will be as or more important than treatment planning. Case management requires the professional helper to coordinate community services for the benefit of the client and, very often, the client's family as well. Let's look at the complexity of case management with an example.

> A single mother and her 10-year-old boy are referred to a social worker in a family services agency. The family physician thinks that the child's social interaction issues may be a result of Asperger's syndrome. The child has few friends, but is doing satisfactorily in school. The social worker interviews the mother and reports the child's social and academic situation at school. The mother says that she has financial problems and difficulty in finding work. Meeting with the child a few days later, the worker notes good cognitive and language competence, but that the child is unhappy and demonstrates some repetitive, almost compulsive behavioral patterns.

The agency staff meets and starts to initiate a case management treatment plan. It is clear that the mother needs counseling, and that the child needs psychological analysis and likely treatment as well. At the staff meeting, the following plan is developed: The child is to be referred to a psychologist for evaluation and, based on that evaluation, recommendations for treatment are likely to be made. The social worker is assigned to do supportive counseling with the mother and to take overall responsibility for case management. The social worker will eventually need a treatment plan for the mother and a case management plan for the family.

COMMUNITY CLINIC
Behavioral Treatment Plan

This form will be reviewed again in no less than two months, and progress toward goals will be noted. Changes in interventions or goals should be noted immediately.

Patient's Name, Address, Phone, Email: _____

_____

Clinic Record Number _____   Insurance _____

Diagnosis: Summary of Patient's Original Concerns:

Axis I _____   _____

Axis II _____   _____

Axis III _____   _____

Axis IV _____   _____

Axis V _____   _____

Identified Patient Strengths and Resources (to be added to throughout therapy):

_____

_____

_____

_____

Interview Progress Narrative

Problem/Concern #1		
Goal	Interventions	Progress Toward Goal

Problem/Concern #2		

Problem/Concern #3		

Signature _____   Date _____

Patient signature _____   Date _____

If patient is a child:

Name of child _____   Age _____

Parent signature _____   Date _____

FIGURE 13.1 Treatment Plan Example

BOX 13.1   National and International Perspectives on Counseling

What's Happening With Your Client While You Are Counseling?
ROBERT MANTHEI, CHRISTCHURCH UNIVERSITY, NEW ZEALAND

There is more going on in interviewing beyond what we see happening during the session. Clients are good observers of what you are doing, and they may not always tell you what they think and feel. Research shows that clients expect counseling to be shorter than do most counselors and therapists. Clients see counselors as more directive than counselors see themselves. And what the counselor sees as a good session may be seen otherwise by clients, and vice versa. Counselors and clients may vary in their perceptions of interviewing effectiveness.

I conducted a study of client and counselor experience of counseling. Among the major findings are the following:

Clients often have sought help before. Most people don't come for counseling immediately. Talking with friends and family and trying to work it out on their own were usually tried first. Reading self-help books, prayer, and alcohol and drugs are among other things tried. Some deny that they have problems until these become more serious.

Implications for practice. Ask clients what they have tried before they came to you, and find out what aspects of prior efforts seemed to have helped. You may want to build on past successes. This is an axiom of brief solution-oriented counseling (see Chapter 14).

First impressions are important. That first interview sets the stage for the future, and the familiar words "relationship and rapport" are central. I found that clients generally had favorable impressions of the first sessions and viewed what happened even more positively than counselors. Sometimes sharing experience helps. One client who did not feel positive about the

first session commented, "Maybe if the counselor had gone through a similar experience of divorce and children, it would have helped."

Implications for practice. Obviously, be ready for that first session. Cover the critical issues of confidentiality and legal issues in a comfortable way. Structuring and letting the client know what to expect seem important. Some personal sharing, used carefully, can help. And empathic listening always remains central.

Counseling helps, but so do events outside of the interview. Resolution of their issues was attributed to counseling by 69% of clients, while 31% believed events outside the session made the difference. Among things that helped were talking and socializing more with family and friends, taking up new activities, learning relaxation, and involvement with church.

Implications for practice. The interview is important, but generalization of behavior and thought to daily life is central. Homework and specific ideas for using what is learned in the session are important.

Things that clients liked. Relationship variables such as warmth, understanding, and trust were important. Clients liked being listened to and being involved in making decisions about the course of counseling. Reframes and interpretations helped them see their situations in a new way; also valued were new skills such as imagery, relaxation training, and thought-stopping to eliminate negative self-talk.

All of the above speaks to respecting the client's ability to participate in the change process. I think it is vital that we tell clients what we are doing, but also ask them to share their perceptions of the session(s) with us. We can learn from and learn with more client participation.

School performance is good, but the social worker contacts the elementary school counselor and finds that the counselor already has the child in counseling. In fact, the counselor has been working with the teacher to help the client develop better classroom and playground relationships. The school counselor and social worker discuss the situation and realize that the boy is often left alone without much to do. Staying after school for an extended day would likely be helpful. The school counselor, with the social worker's backing, contacts a source for funding through a local men's club.

This is but a beginning for many cases—case management involves multiple dimensions beyond this basic scenario. Counseling and therapy are an important part of case management, but only a part. The Division of Youth Services or Family and Children Services may need to be called to review the situation if the counselor, or another mandated reporter, suspects there are signs of abuse or neglect; the mother may need short-term financial assistance and career counseling. If the father has not been making child support payments, legal services may have to be called in.

Through all of the above, the social worker maintains awareness of all aspects of the case. Through individual counseling with the mother, the worker is in constant touch with all that is going on and remains in contact with key figures working toward the positive treatment of the child. But the child can only be successfully treated in the family, school, and community situation.

Advocacy action is important throughout, and it does not happen by chance. Throughout this process, the social worker will constantly advocate for the mother and the child by establishing connections with various agencies. The school counselor who goes to the local men's club for funding has to advocate for the child; in addition, encouraging a teacher to change teaching style to meet individual child needs requires advocacy. Elementary counselors spend a good deal of time in various activities advocating for students individually, and often advocating for fair school policies as well.

Social justice issues may appear. In many communities and agencies, all of the above services may not exist—and equally or more likely, they do not work together effectively. This may require organizing to produce change. The child may be bullied, but the school has no policy to prevent bullying. Social justice and fairness demand that each child be safe. Discrimination against those who are poor or are from minoriy backgrounds may exist. Individual and group education or actual forced change may be necessary.

13.6
MAINTAINING CHANGE: Relapse Prevention

KEY CONCEPT QUESTION

▲ How can we reduce relapses or slips?

One of the approaches used to reduce slips of relapses is the Relapse Plan. This approach, originally developed for use in alcohol or drug abuse treatment, is now used in a variety of settings. Relapse prevention is now a standard part of most cognitive-behavioral counseling, and the concepts presented here will enrich all theories of helping described in this chapter.

The Maintaining Change Worksheet: Self-Management Strategies for Skill Retention form (Box 13.2) helps the client plan to avoid relapse or slips into the old behavior. The counselor hands the client the worksheet and they work through it together, giving special emphasis to things that may come up to prevent treatment success. Research and clinical experience in counseling both reveal that this may be the most important thing you can do with clients—help them ensure that they actually *do* something different as a result of their experience in the interview.

BOX 13.2   Maintaining Change Worksheet: Self-Management Strategies for Skill Retention

---

I. Choose an Appropriate Behavior, Thought, Feeling, or Skill to Increase or Change

Describe in detail what you intend to increase or change.

Why is it important for you to reach this goal?

What will you do specifically to make it happen?

II. Relapse Prevention Strategies

A. Strategies to help you anticipate and monitor potential difficulties: regulating stimuli

Strategy	Assessing Your Situation
1. Do you understand that a temporary slip may occur but it need not mean total failure?	
2. What are the differences between learning the behavioral skill or thought and using it in a difficult situation?	
3. Support network: Who can help you maintain the skill?	
4. High-risk situations: What kinds of people, places, or things will make retention or change especially difficult?	
B. Strategies to increase rational thinking: regulating thoughts and feelings	
5. What might be an unreasonable emotional response to a temporary slip or relapse?	
6. What can you do to think more effectively in tempting situations or after a relapse?	
C. Strategies to diagnose and practice related support skills: regulating behaviors	
7. What additional support skills do you need to retain the skill? Assertiveness? Relaxation? Microskills?	

BOX 13.2   (Continued)

D. Strategies to provide appropriate outcomes for behaviors: regulating consequences	
8. Can you identify some probable outcomes of succeeding with your new behavior?	
9. How can you reward yourself for a job well done? Generate specific rewards and satisfactions.	

III. Predicting the Circumstance of the First Possible Failure (Lapse)

Describe the details of how the first lapse might occur; include people, places, times, and emotional states.

Permis sion to use this adaptation of the Relapse Prevention Worksheet was given by Robert Marx.

Remember that every client is unique and the relapse prevention plan needs to be tailored to the individual client and the characteristics of the problem. Every person's prevention plan should look somewhat different.

Include in the plan a list of people who can be counted on for support, the triggers to the unwanted behaviors, possible responses to those triggers, and rewarding alternatives client can do instead of engaging in the negative behavior.

People you can contact via telephone, e-mail, or texting: Develop a list of three or more people who can provide immediate support during times of potential relapse or slip.

Triggers to undesired behavior, and possible responses for those triggers: Discuss and write down the biggest temptations or triggers. What are the things or situations that make you want to engage in the negative behavior (frequenting some people, visiting a particular place, the sight, smell, or sound of . . .)? What can you plan to do in response to each of these triggers?

Alternative and rewarding activities that do not involve the unwanted behavior: Inactivity and boredom are some of the greatest threats to change. Working with your client, develop a plan of activities that he or she enjoys and wants to do to prevent voids of time that can facilitate negative behaviors.

Rewards the client can give her- or himself for meeting change targets: What would the client do when reaching a week of progress, two weeks, and longer periods of change?

Remember that change takes time and effort. Maintaining an intentional effort to change will pay off at the end.

13.7
YOUR TURN: An Important Audio or Video Exercise

KEY CONCEPT QUESTIONS

▲ **How have you grown and progressed through this learning process?**

▲ **What is your current skill level, and has your personal preferred style been clarified or changed?**

You have engaged in the systematic study of the interviewing process. By now, you have experienced many ideas for analyzing your interviewing style and skill usage. Furthermore, the microskills learned through this text and the practice exercises have provided you with alternatives for intentional responses. These responses must be genuinely your own. If you use a skill or strategy simply because it is recommended, it could be ineffective for both you and your client. Not all parts of the microskills framework are appropriate for everyone. We hope that you will draw on the ideas presented here, but ultimately it is *you* who will put the science together in your own art form.

You have practiced varying patterns of helping skills with diverse clientele. We hope that you have developed awareness and knowledge of individual and multi-cultural differences. It is important to learn how to "flex" and be intentional when you encounter diversity among clients. Developing trust and a working alliance takes more time with some clients than with others. For example, you may be more comfortable with teenagers than with children or adults, or you may have difficulties with elders.

Now that you are finishing this chapter, it is your turn to complete another audio or video interview. Follow the guidelines provided here (these are the same guidelines that were presented in Box 1.3, page 22).

Guidelines for Audio or Video Recording
1. *Find a volunteer client* willing to role-play a concern, problem, opportunity, or issue.
2. *Interview the volunteer client* for at least 15 minutes. Avoid sensitive topics. Feel free to go further to gain a sense of completion.
3. *Use your own natural communication style.*
4. *Ask the volunteer client,* "May I record this interview?"
5. *Inform the client* that the tape recorder may be turned off any time he or she wishes.

6. *Select a topic*. You and the client may choose interpersonal conflict, a specific issue, or one of the elements from the RESPECTFUL model.
7. *Follow the ethical guidelines* from Chapter 2. Common sense demands ethical practice and respect for the client.
8. *Obtain feedback*. You will find it very helpful to get immediate feedback from your client. As you practice the microskills, use the Client Feedback Form (Box 1.4, page 23).

Transcribe the interview (see Box 13.3), and use what you have learned so far to fully analyze your work. A careful analysis of your behavior in the session will aid in identifying your natural style and its special qualities as well as your skill level. Use the ideas presented in this chapter to further examine and analyze your own interview, referral and consultation, and case management and treatment planning.

As you review your own work, pay special attention to your understanding and use of cultural/environmental/contextual issues. Examine your interview from the perspective of someone from a cultural group and gender that are different from your own. How would he or she consider and evaluate your work? This will give you an indication of your current natural style and expertise. It is invaluable to identify your personal style and current skill level after your systematic training.

Finally, go back to the transcript or recording of the interview you recorded in Chapter 1, and note how your style has changed and evolved since then. What particular strengths do you note in your own work?

Remember that it is *you* who will integrate what you have learned here into your own practice.

BOX 13.3   Transcribing an Interview

Type the transcript in a format similar to the transcripts presented in this text. An interview transcript may be useful in obtaining a job, demonstrating your competence, and so on. Consider the transcript a permanent part of your developing professional life and your portfolio.

Check off the following points to make sure you have included all the necessary information in the transcript:

_____ Describe the client briefly. Do not use the client's real name.

_____ Outline your interview plan *before* the session begins.

_____ Be sure you obtain the client's permission before recording the interview, and include a summary of this agreement in the transcript. The client should be free to withdraw at any time, and even at the end of the interview the client may say, "Do not use this session." If that happens, then find another client for your recording. It is critical to protect the rights of the client.

_____ Number all interactions during the interview, and be sure to indicate who is speaking at the beginning of each interaction.

_____ Mark the focus of each interaction and note your use of attending and influencing skills.

_____ If you confront, mark a C in the confrontation column, and note how the client responds, using the Confrontation Impact Scale.

_____ Comment on your interactions when you feel that it is appropriate. Discuss what you feel was good or poor in your interviewing skills, and describe *why* you feel that way. If you feel that you used a skill inappropriately, describe what you feel would have been a better approach. Note also what skills worked well! Include comments on your *personal constructs* (from observation).

(Continued)

BOX 13.3 Transcribing an Interview (Continued)

Noting how you personally influenced the direction of the interview from the constructs is an important part of the transcript.

_____ Indicate when you feel that you have reached the end of a stage. Do not feel that you must cover all stages; in some cases, you may cycle back to an earlier stage or forward to a later stage. Just be aware of the stages you cover in the interview.

_____ Write a commentary on the interview that summarizes what happened.

_____ Summarize your use of skills through a skill count.

_____ Assess your competence levels. What skills have you mastered, and what do you need to do next? This is also a summary of your

strengths and the areas that need further development. What did you like and not like about your work? Your ability to understand and process "where you are" and discuss yourself is important.

_____ End the transcript with a treatment plan for a hypothetical future series of interviews.

For the exercise in this section, you don't have to transcribe a full interview in typed form, although a full transcript likely will be most beneficial to you. Twenty minutes of transcript from within a longer session is enough. But if you do such an excerpt, be sure to indicate what happened in the rest of the interview session so that the context of the transcript is clear.

Complete Interactive Exercise 4: A Full Interview Transcript
Complete Group Practice Exercise: Your Turn: An Important Audio or Video Exercise

CHAPTER SUMMARY

Key Points of "Decisional Counseling, Skill Integration, Treatment Plans, and Case Management"

 CourseMate and DVD Activities to Build Interviewing Competence

13.1 Decisional Counseling: An Overview of a Practical Theory or Set of Strategies

▲ Decisional counseling is a practical model that helps clients work through issues and come up with new answers.

▲ Also known as _problem solving,_ decisional counseling helps clients make a wide array of decisions from simple ones in daily life to complex problems.

▲ Decision making and the microskills are a foundation for most systems of counseling.

▲ All counseling theories can be organized using the five-stage interviewing model, even though the content and practice of each theory may vary extensively.

▲ The microskills five-stage interview is basic to the process of creative decision making.

▲ Five strategies that may be helpful in decisional counseling are creative brainstorming, logical consequences, the balance sheet, emotional balancing, and future imaging.

1. Interactive Exercise 1: Influence the Interview. This exercise provides a good review of the skills learned.
2. Weblink Exercise: Interviewing and Data-Gathering Techniques—University of Missouri, St. Louis, College of Business Administration.

| **Key Points of "Decisional Counseling, Skill Integration, Treatment Plans, and Case Management"** | **CourseMate and DVD Activities to Build Interviewing Competence** |

13.2 Planning the First Interview: The Checklist

▲ Systematically plan for an initial interview using the five-stage structure of the interview.

▲ It is important to be intentionally flexible and ready to change your plan if events in the session suggest that another approach is needed. The five stages provide a useful checklist to ensure covering all points even if the interview does not go as expected.

▲ The first interview presents many challenges; using the first interview checklist helps you plan for the session and reduce the chances of missing the basics.

13.3 A Full Interview Transcript: I'd Like to Find a New Career

▲ The microskills can be integrated into a well-formed interview that prepares the interviewer to focus on clients' presenting issues but still allows room to flex to meet emerging needs. Counselors with varying theoretical orientations will approach the case in different ways.

▲ A person-centered counselor might focus more on Mary and her feelings, thoughts, and meanings about herself.

▲ A cognitive-behavioral counselor would be more interested in specific behavioral descriptions and aim for change in actions.

▲ The interview transcript provides a systematic analysis through process comments and behavioral counts of microskill usage, examines the five-stage structure of the interview, and evaluates client movement on the Client Change Scale.

1. Interactive Exercise 2: Interview Transcript: Allen and Mary's Decisional Counseling. This exercise is a good way to learn how to analyze the interview and rate counselor's responses.

2. Weblink Exercise: Career Resources. This web page provides additional tools to students, career specialists, and individuals interested in career resources.

13.4 Transcript Analysis

▲ Self-analysis of one's own interviewing style and its impact on the client, microsupervision, and consultation with teachers and expert colleagues, both short and long term, are recommended as a continuing part of professional practice.

▲ The five-stage structure of the interview is clinically useful for short- and long-term interviewing planning. Specifying goals and outcomes is particularly important.

1. Interactive Exercise 3: Interview Summary. This exercise is based on the Interview Transcript of Allen and Mary's Decisional Counseling. To achieve basic competence, we suggest that you write an Interview Summary of that session, using the Interview Summary form provided.

Key Points of "Decisional Counseling, Skill Integration, Treatment Plans, and Case Management"

CourseMate and DVD Activities to Build Interviewing Competence

13.5 Additional Considerations: Referral, Treatment Planning, and Case Management

▲ Interview notes and treatment plan notes should be made available to the client if he or she wishes to see them.

▲ When referral is necessary, provide sufficient support during the transfer process.

▲ Case management involves coordinating a broad treatment plan for a client and may involve many professionals.

13.6 Maintaining Change: Relapse Prevention

▲ An effective approach to reduce slips of relapses is the Relapse Plan.

▲ The Maintaining Change Worksheet: Self-Management Strategies for Skill Retention Form helps clients plan to avoid relapses or slips.

13.7 Your Turn: An Important Audio or Video Exercise

▲ Conduct a full audio or video recorded interview using the guidelines for recording.

▲ Transcribe and analyze your work. Pay attention to your natural style and skill level; note your capacity to flex and work with your client.

▲ Compare and contrast your current interview with the one completed in Chapter 1.

1. Interactive Exercise 4: A Full Interview Transcript. Prepare a paper in which you demonstrate your interviewing style, classify your behavior, and comment on your own skill development as an interviewer.

2. Group Practice Exercise: Your Turn: An Important Audio or Video Exercise.

Assess your awareness, knowledge, and skills as you conclude the chapter:

1. Flashcards: Use the flashcards to check your understanding of key concepts and facilitate memorization of key information.

2. Self-Assessment Quiz: The quiz will help you assess your current knowledge and prepare for course examinations.

3. Portfolio of Competencies: Evaluate your present level of competence in attending and listening skills using the downloadable Self-Evaluation Checklist. Self-assessment of your attending skills competence demonstrates what you can do in the real world.

APPLICATIONS OF MICROSKILLS
Cognitive-Behavioral Crisis Counseling and Brief Interviewing and Counseling

DETERMINING PERSONAL STYLE AND THEORY

SKILL INTEGRATION AND APPLICATIONS

INFLUENCING SKILLS AND STRATEGIES

FOCUSING

CONFRONTATION

THE FIVE-STAGE INTERVIEW STRUCTURE

REFLECTION OF FEELING

ENCOURAGING, PARAPHRASING, AND SUMMARIZING

OPEN AND CLOSED QUESTIONS

ATTENDING BEHAVIOR

ETHICS, MULTICULTURAL COMPETENCE, AND WELLNESS

The interview is not about doing something *to* people; rather it is doing something *with* people.

—*Adapted from Kenneth Blanchard*

How are the microskills used in multiple settings?

CHAPTER
GOALS

You can use microskills with many different styles and theories of interviewing, counseling, and psychotherapy. Intentional competence in the microskills and the five-stage interview enables you to understand and master multiple approaches for facilitating your clients' growth.

This chapter focuses on two key applications of microskills, crisis counseling and brief counseling, which you will find highly useful in many ways. Equipped with microskills, you are prepared for competence in many aspects of the helping field.

Awareness, knowledge, and skills developed through the concepts of this chapter will enable you to

▲ Engage in the ABC crisis intervention using cognitive behavioral theory.
▲ Engage in the basics of brief solution-oriented interviewing and counseling.
▲ Briefly examine theories of counseling and psychotherapy as possible future applications of the microskills models of interviewing.

Assess your awareness, knowledge, and skills as you begin the chapter:

1. Self-Assessment Quiz: The chapter quiz will help you determine your current level of knowledge. You can take it before and after reading the chapter.
2. Portfolio of Competencies: Before you read the chapter, fill out the downloadable Self-Evaluation Checklist to assess your existing knowledge and competence in attending skills. Then, at the end of the chapter, complete the checklist again to summarize your competencies after study and practice.

14.1

INTENTIONALITY, MICROSKILLS, AND APPLICATIONS WITH MULTIPLE APPROACHES TO THE INTERVIEW

KEY CONCEPT QUESTION

▲ How can we intentionally apply the microskills beyond this book?

In *Essentials of Intentional Interviewing*, we have stressed that the interviewer needs to be ready for surprises and able to intentionally flex and find another comment, lead, or approach that may help clients grow and develop. Part of that flexibility is becoming fully competent in the skills and strategies that have been presented. Mastery and competence are required for full intentionality.

But there are also purpose, value, and meaning issues that need to be considered. Why are we helping? Are we going to work *with* or *on* our clients? We have suggested that an egalitarian relationship can be most effective. For example, in dealing with clients who had experienced sexual or childhood assault, it was found that those who had a say in their treatment were the ones who improved the most. This was true whether they chose medication or a therapeutic strategy in which they gradually encountered the origins of their trauma. The therapeutic strategy was slightly more effective than medication, but most important was whether these clients were able to choose how they were treated (Keller, Zoeliner, & Feeny, 2010). "Clients liked being listened to and being involved in making decisions about the course of counseling" (Manthei, 2008). In effect, we can be most intentional if we also draw on the intentionality and creativity of our clients.

Thus, using microskills and the structure of the interview in multiple ways and applications makes sense and may even be required for long-term effectiveness. Although microskills can be applied to many different theoretical and practical approaches, in this chapter we will present two key systems of helping that you can draw on to act even more intentionally as you work with clients. First, we will explore crisis intervention. If you do interviewing for any period of time, you will encounter clients in crisis. The structure discussed here provides a useful beginning for crisis counseling. Second, we will look at brief intervention. You will sometimes find that you only have one to three interviews available to work with a client. For these situations, we have provided the most basic ideas of brief counseling and therapy.

With two approaches to the interview, you can change your style if your first system does not seem compatible with your client. Moreover, you can start involving the client in treatment choices. As you develop further competence in the field, there are many theories of helping available for your consideration. We recommend that you move beyond your favorite theory and involve your clients in choice of strategies and in treatment planning. You are much more likely to find them complying and taking action in the real world, outside of the session.

Complete Group Practice Exercise: Reflect, Identify, Share, and Plan for Advancement
Complete Weblink Exercise: Counseling Approaches
Complete Weblink Exercise: Examples of Treatment Modalities

14.2
COGNITIVE-BEHAVIORAL CRISIS COUNSELING:
The ABC Model and Example Interview

by Kristi Kanel, Ph.D.

KEY CONCEPT QUESTION

▲ **What are some of the basics and specifics for the use of the ABC model of crisis counseling?**

The microskills of this book work well in crisis counseling and provide a foundation that will enable you to handle the first sessions in crisis intervention more comfortably. We strongly suggest that you practice crisis counseling, rather than just read about it. That will make the difference.

It is critical that you develop a relationship and working alliance with the client, and many times this must be done quickly. Nonetheless, try to make the client as comfortable as possible. Even a harried crisis counselor needs to focus on rapport immediately, or the rest of the session may not be successful. You want to be seen as a supportive person.

The ABC model of crisis intervention is a three-stage system for conducting brief sessions with clients whose level of functioning has significantly diminished following a crisis or severe psychosocial stressor. It is a problem-focused approach and is most effectively applied immediately or within 4–6 weeks of the stressor or traumatic event. The number of sessions typically ranges from two to four. We seek to identify the client's *cognition* (thinking, perception, beliefs) related to the precipitating stressor. The main goal is to diminish the power of unmanageable feelings as early as possible by modifying negative cognitions and helping the client organize a plan that enables her or him to cope more effectively. This cognitive-behavioral model of crisis intervention is based on experience with multiple types of trauma and a review of research findings. For a more detailed description of the model, see Kanel (2011); Kanel (2008) also provides a live demonstration on DVD of the crisis model in action.

The ABC model is based on the idea that people's thoughts and beliefs about the crisis stressors lead to negative emotions. Severely distressing emotions, in turn, lead to impairment in cognitive functioning and ineffective or confused decision making. These thoughts and beliefs are termed *schemas*—stories or maps of how clients understand what they are facing. The crisis counselor seeks to help organize thinking and deal with the negative emotions. Out of this comes a new schema (or story) and an action plan in which thoughts and feelings become more suitable to face very challenging situations. Crisis counseling seeks to help clients "remap" their view of life's most challenging issues and return to normal functioning as soon as possible.

The ABC stages are not always practiced in a linear format, but are best understood as interacting dimensions that are used in the moment to meet immediate client needs—new schema, stories, or maps that lead to active coping. The outline presented below should be tailored to best meet the immediate client needs.

A: DEVELOPING AND MAINTAINING CONTACT

Trauma victims need a listening ear and a shoulder to lean on. Relationship and interpersonal contact are essential at the beginning and continue throughout the interview, along with appropriate structuring. Clients in crisis need solid contact with a person who can help them gain a sense of security and calm, organize their thoughts, and decide what needs to be done next. This stage is virtually identical with the relationship phase presented throughout this book, but it is implemented in a brief period of time. Some clients want immediate answers, and these may be necessary to establish a relationship. Thus Step B, or even C, may sometimes come first.

A central issue in crisis work relationship development is multicultural difference. Under conditions of crisis (for example, Hurricane Katrina in New Orleans or the major BP oil spill in the Gulf), many clients already distrust "the system," and if you are different in race, ethnicity, or gender, establishing a working alliance may be more challenging. Here are three major ways that you can work on this issue. First, never defend "the system." Listen to clients' complaints and concerns, and validate these when appropriate. Indicate that you also have challenges with administration, but indicate that, nonetheless, you can move forward together as a team. Second, encourage the client to ask you questions, and acknowledge differences frankly. Third, and perhaps most important, focus on the clients' goals and specify clearly what you actually can do to help them reach what they need today and in the near future. Of course, don't overreach with your promises.

B: IDENTIFYING THE PROBLEM AND PROVIDING THERAPEUTIC INTERACTION

Those in a trauma state need to tell their stories, describe what happened to them, and vent their emotions. There likely are many issues and challenges to be met and resolved. Prioritization of which issues need to come first is essential. The first issue is safety, then housing and food for some populations. In the case of an abused woman, the story needs to be told and options for the future considered. Stages 2 and 4 of this book's model obviously relate to B (draw out stories and strengths; restory with exploration and decision making). The client moves from "I don't know what to do" to "At least I can do this" or "Now, I have an action plan to deal with this crisis."

On the other hand, sometimes action and coping may come before storytelling. For example, in hurricane, flood, or fire situations, the first client question may be "Where are my family members?" "How do I get my medications?" "Where am I going to sleep tonight?" These may need to be answered (if possible) even before a solid relationship bond is established.

Let the client be your partner and guide as you work through the ABC framework. Nonetheless, all dimensions need to be covered eventually.

C: COPING

This stage focuses on goal setting and generalizing ideas from the B stage to the real world outside the session. It corresponds with Stages 3 and 5 of the five-stage model. The main goal here is to help client generalize thoughts, feelings, and behavior to daily life. The focus is on "doing" and "action." The schemas have been reorganized, and the client is able to restory the situation and move on.

Table 14.1 summarizes the similarities and differences between the five-stage and ABC models of interviewing.

▲ **TABLE 14.1** Relationship Between the Five-Stage Interview and the ABC Crisis Model

Five-Stage Interview	ABC Crisis Intervention	Process Comments
Relationship (and preliminary goals)	A: Developing and maintaining contact	The listening skills are vitally important. Draw out story and strengths. Be ready to provide action and coping ideas for clients who need immediate answers.
Story and Strengths		
Goals (set more precisely)	B: Identifying the problem and providing therapeutic interaction	Set doable goals that can be achieved today or in the very near future. Continue to explore story, but never forget strengths that the client offers.
Restory		A story of possibility and hope may be generated.
Action	C: Coping	Develop a specific action plan and establish a contract for follow-up.

EXAMPLE INTERVIEW FOR THE ABC MODEL OF CRISIS COUNSELING

A: DEVELOPING AND MAINTAINING CONTACT

In crisis work, the relationship needs to be established quickly and genuinely. Warmth, genuineness, empathy, and acceptance are essential. The use of basic attending skills such as open questions, paraphrasing, encouraging, summarizing, and reflecting feelings are prominently utilized in the beginning phases in order to create safety and trust that will enable the client to accept validation and empowering statements without resistance.

The following interview presents a family breakup as the crisis situation.

Interviewer and Client Conversation	Process Comments
1. Kristi: Hi. What brings you here?	Open question
2. Sue: I've been so upset since my husband left me. I can't sleep or eat.	
3. Kristi: I can see by the tears in your eyes that you are in real pain.	Reflection of nonverbal feelings
4. Sue: Yes, it feels like I'll never get over it. It's even hard to focus on my kids and my job.	Crying openly now
5. Kristi: How many kids do you have?	Appropriate closed question to obtain facts
6. Sue: Just two. Jimmy is 5 and Heather is 9 now.	
7. Kristi: So, you've been having a hard time paying attention to your kids and to your job since your husband left. When did he leave?	Paraphrase that acknowledges feeling, followed by closed question to ascertain fact
8. Sue: Last week. I just don't understand why he left. I thought things were going fine. Sure, we fought sometimes because I'm tired a lot and maybe I don't pay enough attention to him. But I'm busy.	
9. Kristi: So you sound like you're confused about why he left, but you have an idea that it might have something to do with him feeling like you didn't pay enough attention to him?	Clarifying paraphrase that brings the two ideas the client expressed together. Here Kristi is identifying some key aspects of Sue's cognitive map or schema.
10. Sue: Well, he was always complaining that I spend all my time with the kids and that he feels left out a lot. He doesn't understand how hard it is to work full-time and take care of two kids. I'm always running around.	Sue elaborates in a bit more depth on her perceptions and thoughts.
11. Kristi: You sound a bit frustrated and discouraged.	Reflection of feelings

Interviewer and Client Conversation	Process Comments
12. Sue: Yes, it's unfair how much I do. All he has to do is work.	
13. Kristi: I hear more anger now.	Reflection of feelings
14. Sue: I am mad, but I don't want a divorce. I just want to communicate with him better.	The emotions underlying the cognitions are identified.
15. Kristi: So the way I hear you, your husband left you last week and since then you've been feeling sad and angry and are having difficulty focusing on work and your kids. You want to stay married and work on your problems, is that it?	Summarization. Here we see an outline of Sue's basic cognitive map.

A broad view of the major issues has been brought out. Kristi observes that this is a high-functioning and verbal client. Clients such as this can move rapidly. There are no safety or housing issues. This situation focuses on cognitive restructuring and coping skills. Special attention needs to be paid to how the children are doing.

B. IDENTIFYING THE NATURE OF THE CRISIS AND OFFERING THERAPEUTIC INTERACTION

Many of the components of the crisis have been provided by the client in the rapport-building phase. The emotional distress is sadness and anger, and the impairments in functioning mentioned so far are related to poor focus both with her children and at work. All this combines to produce sleeping and eating difficulties. The counselor can help the client become aware of her cognitions that have led to the feelings and her impairment in functioning.

Below you will see the counselor identify the specific stressors, the cognitions and emotions related to this stressor, and the resultant difficulty in personal functioning.

Interviewer and Client Conversation	Process Comments
1. Kristi: So the way I hear you, your husband left you last week [precipitating event] and since then you've been feeling sad and angry [emotional distress] and are having difficulty focusing on work and your kids [impairments in functioning]. You want to stay married and work on your problems, is that it?	Summarization. Here we see the cognitive behavior sequence of precipitated event, emotional result, and cognitive-behavior outcome.
2. Sue: I would want to work on it, but I don't think he does.	Sue examines her own cognitive map.
3. Kristi: What makes you think that?	Open question to explore cognitions and the client's "map" or schema about this issue

Interviewer and Client Conversation	Process Comments
4. Sue: Well, he said he doesn't love me anymore. It hurt so much.	
5. Kristi: What does that mean to you to make you feel so hurt?	Eliciting meaning
6. Sue: I'll probably just be alone forever now.	Mentions cognition and the deeper meaning of being alone
7. Kristi: In what way alone?	Eliciting meaning, what does the client mean by "alone"?
8. Sue: I'm afraid to be alone and start all over. Even though I knew our marriage was unhappy, I guess I would just tolerate it instead of dealing with my fears.	We are now into restorying as Sue explores the history of her marriage.
9. Kristi: What is most scary for you about that?	
10. Sue: I'm afraid to get close to someone else and feel hurt again. At least with my husband, I knew staying with him I would be miserable, but at least I could predict my life and I don't know what else to do.	Possible cognitive/emotional key. We see prediction and likely safety as a key element of her schema.
11. Kristi: It is often scary to start over. Many people prefer the certainty of misery rather than the misery of uncertainty. This scary feeling may at some point turn into excitement at the opportunity to have a more rewarding relationship. Even though you might feel like he has all the power because he left you, at this point you can focus on things you can do to move your life in the direction that you think is best.	Paraphrase. This support statement validates client's feelings and thoughts. A brief reframe of the scariness, possibly encouraging more self-confidence, is followed by an empowerment comment, emphasizing strengths.
12. Sue: Sure, but what can I do? He makes more money than me and I can't afford a lawyer like him. He'll probably take my kids away from me.	Sue explores the dangers she faces in the future.
13. Kristi: Well, I do know that in this state, all the money earned in a marriage is community property and that if the divorce does go through, you will be entitled to a fair share of the community holdings. As far as the kids, usually the courts focus on what is in the best interest of the children. The person making the most money does not automatically get everything.	Information giving is likely quite important in reassuring Sue about the future.

The above is only a taste of the issues that were discussed in this session. There is a need to explore Sue's negative self-concept in more depth and support her by drawing out wellness strengths and positive assets. This client has many strengths, and they need to be stressed more than we see in this brief example.

The children need attention. There are legal issues. How does she handle work, home, and children? Together all these issues and more are made more difficult by the present crisis of the husband's leaving.

The counselor has implanted more positive ideas via the reframing of the impossible situation into one with possibilities. As the interview progressed, this area was given considerable attention. Those facing immediate crisis have a feeling of hopelessness, and empowering them to regain control of their own thoughts, feelings, and behaviors is critical.

Two important issues in the treatment of this particular case are working through the immediate emotional crisis and planning for tomorrow and the coming week. The counselor needs to be aware that weekly sessions may not be enough. How soon should this client return? What are some longer-term goals that might be useful?

C: COPING

In this final stage of the ABC crisis intervention model, the focus is on what the client wants to do and what she will commit to do after leaving the session. The counselor needs to help the client develop a specific plan of action that best suits the client.

Rather than telling the client what to do, the counselor may ask the client what she would like to do in order to feel better and manage the crisis. As we work to establish mutually agreed-upon actions, the client can be encouraged to explore previous coping attempts, current coping attempts, and new coping behaviors. The counselor should pay attention to anything mentioned that hasn't been effective, but more attention needs to be paid during this stage to what works and what has helped. (This point is particularly characteristic of brief counseling.) This information can be used later when the counselor and client explore alternative coping behaviors.

Parenthetically, we should mention that Sue has been successful on the job, perhaps too successful as it has made time demands that affect her home life. Despite all the pressure, she has managed up to now, and that itself is a positive asset. Sue's family lives at a distance and cannot be much help, but she does have friends and close relationships at her church. By exploring these resources, the counselor has a positive base of strengths that are critical for coping with crisis.

Once the client agrees with one or more coping strategies, the counselor must get a commitment from the client that she will follow through with the plan. The counselor must set up a method of checking in with the client on this follow-through.

Interviewer and Client Conversation	Process Comments
1. Kristi: Well, I do know that in this state, all the money earned in a marriage is community property and that if the divorce does go through, you will be entitled to a fair share of the community holdings. As far as the kids, usually the courts focus on what is in the best interest of the children. The person making the most money does not automatically get everything.	Information giving that provides reassurance

Interviewer and Client Conversation	Process Comments
2. Sue: All this talk of the law is scary. I don't want to lose my kids. I couldn't live without them. I don't even know if I can live through all of this.	Sue's deeper fears appear.
3. Kristi: I'm a little concerned about what you've just said. Have you had any thoughts of suicide or have you tried to hurt yourself?	A brief suicide assessment was warranted. Always watch for depression in crisis cases.
4. Sue: No, not really. But it seems so overwhelming. Something has to be done because I can't continue like this.	Really depressed clients would not respond like this. Note that Sue is capable of thinking of doing something. Depression would be indicated very differently. For example, "No, I haven't really thought of suicide, but it seems hopeless now. I just sit and brood and cry. I can't sleep and I'm not doing well at work."
5. Kristi: Of course. And coming here was the first step in getting something productive done. I'm glad you haven't thought of hurting yourself. Can you make a commitment to me that you won't do anything to hurt yourself? If you have any feelings like that, you will talk to me first?	Getting a client contract not to hurt oneself has become an important part of suicide prevention. As you grow into the profession, you'll want to examine this issue in depth.
6. Sue: Yes. I don't really want to die. Somehow I'll get through this. The kids need me.	A life purpose—a sense of meaning—is critical. This is a very positive statement from Sue. We can see that she has resources to get through this. It is difficult to be fully depressed if one has discerned a purposeful life and found meaning.
7. Kristi: Yes they do. I'm happy to hear that. It sounds like they are really important to you. You seem like a good mother. You sought professional help when you saw signs in yourself that you were losing focus on them.	Validation and reframe through focusing on Sue's strengths. This is wellness and positive psychology in action.
8. Sue: Well, I would never want to do anything to hurt them. But what can I do to get better?	You'll love clients who start asking this question. Easy to work with!
9. Kristi: What comes to mind?	Open question often used in brief counseling. Keep this question in mind; it is originally drawn from psychoanalytic free association. The first thing that comes to mind is often an important clue, but also look for the "last thing" they could ever think of.

Interviewer and Client Conversation	Process Comments
10. Sue: Maybe I should talk to a lawyer and find out my options.	
11. Kristi: Sounds like a good idea. I know some lawyers will provide free or very low fee first-time consultations. What else might you do to feel better?	Validates the idea and adds information
12. Sue: I think I should try to talk to my husband and find out what he intends to do. But I'm scared. What if he files for divorce, what if he's with someone else, what if he tries to take the kids?	Sue generates another idea, but returns to her central anxieties,
13. Kristi: With all those fears, it does seem warranted to talk to him. Do you feel safe doing so? Is there someone who could help you or be with you?	Vital that you check safety concerns with women (or men) in crisis.
14. Sue: Yes, I'll be safe. He's never been violent with me. He may just lie to me or not answer my calls. I'll call today when I leave here. If he refuses to talk to me, I'm definitely going to call a lawyer just to get some information. If my husband talks to me and says he doesn't really want a divorce, I'm not sure what I'll do.	Reassuring to the counselor!
15. Kristi: Would you be interested in seeing a marriage counselor together if he wants to try to get the marriage back on track?	Here we see information giving in the form of a question.
16. Sue: Sure, if he's willing. I need someone to talk to. My friends are sick of me calling them up and crying every day.	It worked.
17. Kristi: Well, I can give you the name of several marriage counselors. Also, I do know of a few support groups for women going through separations and divorce. If you are interested, I can give you the phone numbers for these as well. Because everyone in these groups is going through the same thing, no one gets sick of hearing from you and they understand completely when you cry.	Uses client's information to come up with appropriate referrals. Be aware of all the resources available in your community that can help clients.
18. Sue: Thanks.	
19. Kristi: Please call me in 2 days and let me know what actions you have taken so far. We can schedule another appointment if need be then. Meanwhile, let's set up an appointment for next week.	Keep in contact in crisis, and make specific plans for the future.

As stressed in the Action stage of the five-stage interview (Chapter 7), follow-up and contact after the session are important. A beginning contract for action has been established. At the same time, we can see that there are many unsettled issues that call for more counseling sessions and perhaps even longer-term therapy, assuming financial resources are available.

This is a successful middle-class client and the type of situation that you are likely to encounter throughout your career, regardless of economic background. The resources of less advantaged clients will obviously be less than Sue's. This may require you to be more active and use more influencing skills in the crisis process.

What about situations in which immediate action to stabilize is required—floods, rape, suicide of a loved one? Again, safety considerations come first, then the basics of daily living, but the same ABC structure holds. Develop a relationship, draw out the basic story and explore alternatives, and work on coping and action for today and tomorrow. In these immediate and very serious types of crises, listening skills remain important—"listen with the ear of the heart." But you may have to give direction, advise the client on what to do, and pick up the cell phone and find some way that the client's basic needs can be met.

This is obviously only a beginning. I strongly recommend that all counselors and therapists take a course in crisis intervention and continuing education to keep up skills. Also, there are frequent opportunities for you to volunteer and be supervised in emergency situations. Seek a situation where you can obtain supervision. Working as a crisis counselor without training and/or supervision can be dangerous for the client.

▲ EXERCISE 14.1   **Practice With ABC Crisis Counseling**

Reading about crisis counseling is helpful, watching someone else do it is better, but the real learning happens when you practice skills yourself. We hope that you will take the time to try the following role-playing exercise, just in case you actually find yourself suddenly with people who really need your help.

Work with a partner, switching the roles of client and counselor. Plan for a minimum interview of 15 minutes, and record it if at all possible. More and more people have cell phones and cameras with video, and many also have access to video cameras. And don't forget that camera on more and more computers! There is nothing like seeing yourself as you are to bring the ideas of this text alive and make them truly useful.

Select a crisis for the role-play. Examples might be the sudden loss of a job, the death of a loved one, the discovery that one has a serious disease, a home invasion robbery, or the loss of one's home in a fire, earthquake, or flood. For the role-play, assume that this event has just happened today or yesterday.

Go through the specific stages of the ABC interview; then discuss the process, what you observed, and what you have learned.

A. Relationship, preliminary goals, and early storytelling. Establish a relationship, and be with the client as well as the needs. *Be with the client in the here and now.* This may mean using listening skills to draw out the story, even though the client presentation may be disorganized and not fully clear. It may mean reassurance that you are there to support. Rather than reflect feelings in depth, it is usually best to acknowledge emotions. Be prepared to offer advice or reassurance if the situation demands. In this process, note the schema maps of the presented problem.

B. Draw out the crisis story with more clarity and help define goals. Aim toward something concrete that the client can do today with your help. Reinforce hope for the future, with specifics. Point out client strengths from the past and present, and her or his basic strength to deal with the new challenges, but do not minimize the client's concerns. To clients in crisis, a blanket reassurance that all will be fine is almost sure to be rejected. Reality must be faced.

C. Coping. Work toward an action plan that has specific attainable goals for today. Make tentative longer-term plans. Help the client become aware of available resources and your commitment to be with the client over time.

 Complete Group Practice Exercise: Review and Discuss

14.3
BRIEF COUNSELING AND EXAMPLE INTERVIEW

by Allen E. Ivey, Robert Manthei, Sandra Rigazio-Digilio, and Mary Bradford Ivey

KEY CONCEPT QUESTION

▲ What are some basics, specifics of practice, and important ethical issues of brief counseling?

Brief counseling often sets goal setting as the first priority, and this is integrated with relationship building. But, as with the five stages, goals are often revised and clarified as the interview progresses. Drawing out client stories, issues, and concerns remains fundamental, but brief counseling focuses more on single issues and solving specific problems leading to action, much like crisis counseling. It is not oriented to personality organization and focuses on immediate change for the client.

Brief counselors seek to be brief. The brief approach may be a single interview, or it may extend to as many as five or ten sessions. Anticipate one to three interviews as typical. The key word is *brief*, emphasizing *solutions* rather than problems. Brief counselors believe that clients have their own answers and solutions available if we help them examine themselves, their skills, and their goals.

Brief counseling makes questioning the central skill. In the early stages of your practice with brief counseling, consider using the specific questions presented here and sharing them with your client. As you gain experience and confidence with this method, you may wish to continue sharing your interview plan—counseling and interviewing can be more powerful and real in an egalitarian co-constructed framework.

STAGE 1: RELATIONSHIP AND PRELIMINARY GOALS

RELATIONSHIP AND STRUCTURING THE SESSION

Use the relational and ethical strategies that we have stressed throughout this book. As you begin brief work, inform the client about the nature of the session. For example:

Many people can accomplish considerable progress in just a few sessions. What we are going to do here today is focus on solutions—the goals you want to achieve. Can you tell me what your goals are for today?

The words "for today" are important because they bring the client to the possible *here and now* rather than leading to a lengthy attempt to resolve everything at once. Some issues are too large to be resolved in a few sessions; your client may work on one primary issue now and leave the others for later. Returning to counseling is not failure; rather, it shows a willingness to work on the many complexities of daily life.

With children or adolescents, the wording may be better phrased as follows:

> Darryl, the teacher asked me to talk with you. Rather than talk about problems, I'd like to know how things might become better for you. Could you tell me one thing that you can do to feel better—happier about the rest of today? (Or "Before we begin, I want to know something that makes you happy. Tell me about what you like to do.")

Children and adolescents (as well as adults) may initially respond negatively to your questions and even say "Nothing." Remember the importance of rapport and listening—with many clients, a sense of humor helps! With experience, you will develop follow-up questions and help clients explore their issues in new ways. As these questions are also part of relationship building, keep in mind that what the client comes up with may be really a preliminary goal and later discussion can sharpen what the client wants to have happen.

IMPORTANT GOAL-ORIENTED QUESTIONS

Start solution-based thinking at the very beginning of the session. Even as you listen to the client's story and/or reasons for coming to the interview, you can ask the following:

▲ What is your goal here today?
▲ What do you want to happen today?
▲ What about that goal is important to you? (builds motivation to change)
▲ What has gotten better about your concern/issue/problem? What made that happen?
▲ What's keeping it from getting worse?
▲ Are there any exceptions in this problem? When is the problem not so much of a problem?
▲ What do you do right? What have you been doing to keep this issue from really dragging you down?

Your client will not always respond in depth to your questions. "What else?" prompts client thinking that may generate more complete answers and solutions. Don't always settle for the first answer. Eliciting meaning ("What does that mean to you?") often provides background and informs you and the client about motivation and values.

EXAMPLE INTERVIEW: BRIEF COUNSELING

This demonstration is a session conducted by Penny Ann John, a first-year graduate student at the University of Massachusetts, Amherst. We thank Penny for permitting us to share her work with you. As you will note, she worked with a verbal client volunteer with a fairly specific concern. As with all interviews,

Penny's work is not perfect, but it is a fine example of how the positive asset search and the focus on exceptions and solutions can make a difference in the life of volunteer and real clients. The interview has been edited for clarity, but it remains the work of Penny.

Particularly note how Penny uses the basic listening sequence and search for positive assets and wellness as a vital part of her example. Balancing the questioning style of brief work with listening skills generally strengthens the interview. As you read this session, think how you might have handled the interview in accordance with your own natural style of helping. It will not always be this easy and direct, but brief counseling often is this effective in a short time. For your first practice in brief counseling, we suggest that you find a classmate, friend, or family member. It will take some experience and practice to master these ideas with clients who have more complex issues or who may be resistant to the process. In your first efforts, making a list of useful questions and sharing these with your client will be helpful.

Interviewer and Client Conversation	Process Comments
1. Penny: Carter, we talked before about what we are going to do today, which is brief counseling. We are supposed to take an issue or concern for you and work through that and come up with some solutions. You said that you wanted to work on academic stress. Let us take a part of the larger issue—small parts of larger issues are often useful places to start. You've got a list of the questions, just as I have here. If you wish, add any questions I missed that you think are important.	When you work with a volunteer friend or classmate in a practice session, the relationship is already established and often the volunteer will know generally what to expect. Penny had told Carter about key ethical issues and confidentiality. Both Carter and Penny signed permissions forms allowing use of this interview in this book.
2. Carter: OK.	
3. Penny: So, to start, suppose you tell me what your goal is for today.	Note immediate focus on goal setting.
4. Carter: My goal for today is for us to brainstorm and come up with ideas to manage my stress because I am feeling really stressed out.	Of course, goal definition may not always be this quick and easy. Penny has a client who verbalizes well and who "buys in" to the brief model. We have here a good example of a preliminary goal, which needs to be made more specific as the interview moves on.

If the problem was clearly defined during the relationship stage and the goal is relatively clear, consider moving directly to mutually setting more precise goals (Stage 3). With Carter, the issue is to move beyond just exploration and define more precise goals, likely around stress management. In this case, Penny needs to draw out a bit more of the story. Here are some examples of goals that other clients have presented. Note that each is relatively clear, but would benefit from more exploration and clarity.

▲ I'd like to stop arguing so much with my partner.
▲ My son gets up and down during meals and is constantly leaving the table.
▲ I'd like to be able to speak up at meetings more effectively.
▲ Our lovemaking has become too routine. I want my partner to warm up to me.
▲ I want more challenge in my work.

All of the above require some awareness of times when the problem is *not* a problem. For example, "When are you able to avoid arguments with your partner? When has lovemaking been real to you?" These *exceptions to the problem* may serve as levers for positive change. For Carter, exceptions might be when she is comfortable with studying and times when she feels stressors are at a minimum. Asking her what she likes about her studies or a specific course that has been fun could also be useful.

On the other hand, if the concern is vaguely presented ("My relationship is falling apart"), if the client talks about confusing multiple issues, or if the client has difficulty identifying goals, more time needs to be spent in gathering data (Stage 2). Brief methods work best on only one problem at a time; other issues can be dealt with later.

STAGE 2: *STORY AND STRENGTHS*

The decision to omit the data-gathering portion of Stage 2 is not easy, and often you and the client will want to spend some time gathering data and examining key stories. This information should be aimed at returning to basic work on achieving goals.

THE MIRACLE QUESTION

This question, a favorite of brief counseling, was developed by de Shazer (1988, p. 5), a founder of brief counseling. Used in a variety of ways, it is designed to surprise the client and open her or him to new ways of thinking. Here you can obtain an ideal story that itself may represent a goal.

> Suppose when you go to sleep tonight, a miracle happens and the concerns that brought you in here today are resolved. But since you are asleep, you don't know the miracle has happened until you wake up tomorrow; what will be different tomorrow that will tell you that a miracle has happened?

Follow up the miracle question with "How will we know the issue has been resolved?" and "What are the first steps to keep the miracle going?" The miracle question can be asked in other ways. For example, "What would it be like if the problem went away?" "What would you like life to be like?"

Cultural/environmental/contextual issues such as gender, race/ethnicity, and spirituality factors may be part of the story process. The client, for example, may begin the session by stating the problem as depression over constant harassment. As you hear this story, you note that the client is focusing on self as if the problem is internal and provides no information regarding his circumstances. "Is the problem in you or is it in the system?" is the generic question here.

Of course you need to adapt this question by replacing "system" with a word specific to your client's situation (e.g., group, department, school). Also notice possible interactions in which both client and system affect the outcome.

This question is extremely important. It may not be as critical in Carter's situation, but we can help many clients if we use the cultural/environmental/contextual focus so that they begin to see their problems in a larger context. Then part of a healthy solution involves social justice—finding others who share these concerns and working toward community solutions. Pride in one's culture (African American, Asian American, gay, disabled, religion/spirituality, etc.) can be an important support in brief counseling.

Interviewer and Client Conversation	Process Comments
1. Penny: OK, sure. What brings this topic to your mind today versus talking about this another time?	Focus on *here and now* and the reason for wanting to discuss it *now*.
2. Carter: Well, I am a graduate student and it is that time of the semester. Everything is coming to what seems like crunch time, and that is when I feel the most stress. There are so many things on my plate.	Recognize this? Just before examination time, college counseling centers see a marked rise in students coming for help.
3. Penny: Right now it is stressful for you because it is coming toward the end of the semester . . . and you have a lot going on. OK? With all of this going on at the moment, what might be positive about your situation right now?	Reflection of feeling—"You feel X because Y"—followed by an open question oriented to strengths and solutions already existing in the client
4. Carter: Well, you know I have got to say, I have talked to a lot of people lately and they have a lot more to do . . . um hum . . . right. And I did not really realize that until I talked to them because I have been plugging right along and doing my papers so I don't have everything to do all at once. And that made me feel a lot better.	You may recognize this as well. Pressures do slowly mount before clients feel the full brunt of stress, which in turn affects their body through tension and often through sleeping issues. Carried far enough, the anxiety makes studying very difficult.
5. Penny: Great. So you seem pretty organized. You seem like you are getting things done but still have some work to do, but you have been doing things right along.	Positive feedback, emphasizing "what works" for Carter and her past successes. There is also a gentle reframe by stating "but still have some work to do." This phrasing suggests that Carter has the ability to resolve issues on her own.
6. Carter: Yeah, I really have. I am not sure why I feel so stressed because I know I will have the time to do it and I have been doing it so far, but I guess I still see the deadline at the end and it is getting closer, so it feels a little stressful.	Carter picks up on the feedback and positive reframe.
7. Penny: Have there been times in the past when you have felt this type of stress but have dealt with it in a positive way?	Brief question seeking exceptions to the problem and past successes

Interviewer and Client Conversation	Process Comments
8. Carter: Sure, last semester or even when I was working. There have been times when it seemed like I had a lot to do. I had a very busy job. There are things I like to do when I have time to do them. I like to go dancing. I like to be active. I like to be social and that always . . . it is a real release for me. It is like freedom. You know. (*Penny:* Right.) And then you can forget about it for a while. (*Penny:* Uh-huh.) Then it is really good and then I get rejuvenated and I can come back and do what I need to do. OK. It is just finding the time to do that.	Clients have past accomplishments and successes that can be drawn on to work on and solve their issues. An all too common issue with students is that they don't like to study and wish they were having fun and this awareness gets in the way of studying. Then they go out to enjoy themselves and sometimes it doesn't work because they feel guilty about not studying. The answer: "Relax, play when you play, but work when you work. Don't mix them up."
9. Penny: Oh, that is really terrific. So, in the past when you have been stressed, you have gone out dancing, you have done social things and you have done other things to keep your mind from it, and then you get more energized from it also.	Positive feedback, paraphrase of strength in dealing with stress
10. Carter: Yeah, it does really work.	It does work!
11. Penny: And then you are able to focus. Oh, great. What is different about the times when you don't feel stressed out?	Paraphrase, positive feedback in form of compliment, question searching for positive exceptions to the problem
12. Carter: One of two things. Either I am using the technique to not be stressed out by doing all the social things I need and all of the good things and fun things I enjoy, or there is less to do. There is not a crunch time or a deadline time. The summer. I guess when I feel the most organized, when I feel I have things under control, I feel less stressed.	We are finding that Carter has the answers in herself. Constantly search for how clients can solve their own issues.
13. Penny: OK, so when you feel organized and have things under control, you feel less stressed.	Brief summary or paraphrase/reflection of feeling

▲ EXERCISE 14.2 **Review for Focus**

Review the interview for focus. Note that virtually every one of Penny's comments focuses both on the client and on possible solutions to the problem. Her enthusiasm communicates hope for Carter. Your interest and belief in change will be heard by your clients.

STAGE 3: MORE PRECISE *GOAL* SETTING

The most important question at this stage is some variation of the following:

> We have heard your concern [summarize again, if necessary to keep interview on track and check accuracy]. . . . Now, what specifically do you want to happen? What are your goals here today? Be as precise as possible.

Interviewer and Client Conversation	Process Comments
1. Penny: OK, so when you feel organized and have things under control, you feel less stressed.	Repeat of Penny's last comment above
2. Carter: Right, and when I finish a project and when I see that it is completed and do one thing at a time, I feel less stressed. When I try to do three things and none of them are completed but I have done all this work, it is still stressful. I guess when I finish and look at it and say oh, it's done, I did this, whatever task it may be.	Note that Carter is defining her goal more precisely here.
3. Penny: You feel better when things are organized and you complete your papers. When you do a part of each of your projects but don't finish any complete class project, it doesn't feel so good because you don't feel like you have completed anything.	Paraphrase, reflection of feeling
4. Carter: Right, and it might be more work, but it doesn't look that way because I can't check it off the list. You know?	
5. Penny: Yeah. Say you woke up tomorrow and this stress was miraculously gone, what would it be like? What would it look like?	This is a good time for the miracle question, as we have an understanding of Carter's issues and her style. The miracle question often brings out new data, often unexpected, helping us find new solutions. We may find ourselves needing to totally redefine the problem or concern with data provided by the miracle question.
6. Carter: I would have everything done and I would be going on vacation.	Carter does find a miraculous answer that does define her long-term goals.
7. Penny: If you get everything done, then you could be on your vacation with your boyfriend, right?	Paraphrase with checkout
8. Carter: Yes, exactly.	

Sometimes the miracle question doesn't produce much in the way of useful data at first. Penny could have followed up with more specifics and asked, "What would you be doing differently?" In this case, we have a good idea of what Carter needs to do differently. Penny was on track, but could have asked for more concreteness. Another possibility: "Could you be more specific? What's the first thing you would notice that would be different if the stress were gone?" It takes time and practice to make the miracle question work.

Interviewer and Client Conversation	Process Comments
9. Penny: To recap, your goal has been to brainstorm and identify ways to deal with stress. We've identified some of the strengths in dealing with stress as your organization and your ability to do one thing at a time. And it helps, as you seem to be able to take time off and forget your studies for a while and enjoy yourself. Sounds like organization of your time these next few weeks will be important.	Summary building on the miracle answer, but with a focus on what has happened in the session to help Carter reach the desired result. The miracle question in this case provided a chance to summarize what would help Carter reach her goals.
10. Carter: Yes, that's it. I guess my goal is to cool down a bit as I know I can do it. Then my boyfriend and I can be off on vacation for a week.	A more precise goal has been established.

STAGES 4 AND 5: *RESTORYING* AND *ACTION* COMBINED IN BRIEF COUNSELING

When you have identified resources, found exceptions to the problem, and identified goals in the first three stages, you have already been brainstorming and exploring solutions. Constantly focus on the idea that something can be done. Your goal in this stage is to solidify and organize the solutions and move toward concrete action. Work on the clearly defined goals in specific manageable form. Every successful idea for solution needs to have a practical use outside the session. Here are some basic questions that focus on moving to the future.

What do you have to keep doing so that things continue to improve?

What will tell you that things are going well?

How can we take what we have learned today to daily life?

Think about change and how the client can "take it home." We need to restory and move to a new conversation about change and possibility. We need to transfer learning in brief counseling to the real world. We need to work on what clients have already done and their positives strengths and assets. The client often has the answer already if you can help bring it out and build on it.

Brief counseling represents a contract and commitment to clients. Do not leave them at this point. Stay with them until they accomplish their goals. Contract for specific follow-up in the next session or by phone. E-mail is not advised as you may open yourself too wide for constant contact. However, this general rule may be changing as we move "Facebookwise" to a society where privacy is considered less important.

Assign a task that the client can use to ensure transfer from the interview. Concrete, achievable tasks, set up in small increments, move the client toward significant change. To help the client take action, relapse prevention worksheets (Chapter 13, Box 13.2) and homework are especially appropriate.

Interviewer and Client Conversation	Process Comments
1. Penny: Let us talk about some of your strengths, your strengths in the way you can deal with this stress.	Directive, let's make positive assets and resources fully clear.
2. Carter: I don't know. I am pretty positive about things. I know that I will finish it and I will get everything done. I think that is the strength. It is not a question of if I will do everything; it is just as I am in the moment, things get hectic. I am not a defeatist. I know I will get my work done and I know I will graduate. I know I will and I know what I need to do . . . um hum . . . and I know the things I like to do if I could carve out the time to do them and make sure that I take that time for myself. Then it will be better. So, I think that is strength.	It is good for clients to say good things about themselves. Finding positives help them center themselves and move away from a total focus on difficulties and challenges.
3. Penny: Yeah. Some of your strengths are that you are positive, that you know what you need to do, you know how to do it, and you know how to get there. I also heard you say you were organized before. What are some of your strengths in other parts of your life?	Summary, open question that may lead to suggestions for dealing with stress
4. Carter: I think those strengths also follow through in other areas of my life. That I am a positive person, that I like to try new things, and am adventurous. I really like life and I think that has always helped me. Right. I think that is a strength and that I can do things.	Carter gets "into" her strengths. Positives really help the restorying/action process.
5. Penny: You like to try new things, are positive, and enjoy life. It is interesting. . . .	Paraphrase, reflection of feeling

Interviewer and Client Conversation	Process Comments
6. Carter: I just want to do everything very, very well, so sometimes that gets in the way. I want to do it perfectly, or as perfectly as I possibly can. Sometimes I feel like I am not doing my best and that bothers me. Even if it is stupid stuff. Even if I know I don't have to do the paper perfectly, I still try to. Right. Sometimes I just need to give myself a break.	"Sometimes I just need to give myself a break." This is an important part of resolving the studying issue. We can stop the cycle of guilt for not studying while refreshing ourselves in something that is relaxing and fun.
7. Penny: You have some really great strengths, and with those strengths you were able to obtain your goal today, which was to brainstorm solutions to reduce your stress. Your positive attitude and willingness to do things and being organized and a risk taker will help you work through this stress. On a scale of 1 to 10 where do you see yourself in regard to your stress level at the moment?	Positive feedback with another compliment and summarization, followed by scaling question. In many types of sessions, asking the clients to rate their issues on a scale helps you understand how deeply they experience their issues.

Scaling is a useful strategy often employed in brief and cognitive behavioral counseling, but also useful in other systems. Clients rate themselves on a scale of 1 to 10 on how they feel about issues of importance. This could be depression, the breakup of a relationship, or how likely the client is to drop out of school. As clients move along with your help, watch for their "numbers" to improve.

Interviewer and Client Conversation	Process Comments
8. Carter: It is not too bad. Let's say 10 is the most and 1 was the least. I am probably a 5. I don't think it is that bad; talking about it makes it a lot easier. Like I said, it is more when it is in the moment and I have had a crazy day. I was working all day, had my classes, I come home, and I have seven things to do and there are three messages on the answering machine and I think, I can't do everything. I probably could do most of them, and then I have to map it out and prioritize, but that is hard because it is so hard to say no. Especially when you want to do fun stuff. I guess it is a 5. Giving myself a break.	Carter very readily understands scaling and shows how it can be useful to her and to the counselor. Sometimes you will be surprised in interviews like this and get a low number (e.g., "3"). This is a clear sign that you likely need to search out other issues that may be affecting the client. This also would be a sign that more than brief counseling is needed.
9. Penny: It seems like you are handling your stress pretty well.	Positive feedback
10. Carter: Thanks, yeah, it is not too bad. It is not as bad as it seems in those moments. You know?	

Interviewer and Client Conversation	Process Comments
11. Penny: Um-hum. So there are times when you have a higher stress level and times when you have a lower stress level. When are times when you are a 1?	Encourage, paraphrase, question, search for exceptions. Another useful question could be "So, what do you have to do to reduce your stress from a 5 to a 4?" This question is an important part of the scaling process, particularly as it focuses on small change rather than total resolution.
12. Carter: When I can step away from responsibilities and play with my niece or go home and be with my family or out with my friends and dancing. I am a very active person, so I feel best when I am doing something outdoors or when I am moving or exercising.	Penny's question—focused on total removal of stress—works, but with many clients the smaller change from 5 to 4 would be more manageable.
13. Penny: So, you feel stress-free when you are moving, or exercising, or playing, or are social. On a scale of 1 to 10, what would be your ideal stress level at this time?	Paraphrase; the client's stress patterns are examined by scaling.
14. Carter: An ideal stress level would be a 3, because you can't always be playing. A certain amount of stress is good in order to be productive.	Carter's answer shows that asking for a "1" was perhaps too much, but nonetheless Penny's approach worked.
15. Penny: What do you want to happen precisely?	Move toward even more goal specificity with an open question.
16. Carter: I have 4 weeks left of school. I think I need to map out the next 4 weeks in how I can balance my productive work time and my social outlets. I have six papers left to write and 4 weeks, so that averages to about one and a half papers per week. Now that I put it in that perspective and I see it visually, it doesn't seem so bad after all. I feel a bit relieved already.	Carter really starts generating her own answers. A typical result of effective brief counseling.
17. Penny: Let's now generate a picture of what the next 4 weeks are going to look like for you.	Directive, concretizing the plan. Imaging that future is a useful strategy.
18. Carter: Well, I have four classes and have my assistantship, which is 10 hours per week. I also have these six papers and plenty of time to exercise and I can go out a few times over the next few weekends. My stress level has reduced tremendously already.	Carter takes over.

The interview continues. Penny and Carter work out a detailed action plan for the coming 4 weeks that includes plenty of time to meet the academic needs, but now includes a fun/relaxation plan. Endnote: Carter had a great time with her boyfriend!

WHEN BRIEF COUNSELING DOESN'T SEEM TO BE EFFECTIVE

Brief counseling, like any form of helping, has limits and is not appropriate for all people. If the client does not respond to brief counseling and you do not have other ideas, seek assistance from a supervisor, consultant, or more experienced colleague. Though this method can be helpful in resolving complicated issues, solution methods may not always be adequate in themselves. More complex matters may require experienced counselors or therapists who use other approaches. Seek supervision and referral as necessary.

Evidence that a brief counseling approach is not appropriate for a client includes the following:

1. The client presents with serious symptoms or problems (e.g., substance abuse, relational violence, child abuse, suicidal gestures, alcoholism, eating disorders) and does not readily respond to brief interventions.
2. The client is not able or willing to try the solutions generated in the session. For example, after several solutions have been successfully rehearsed in session, the client cannot or will not enact these solutions outside the counseling relationship.
3. During the final stage of counseling, the client may have difficulty making the solutions real in her or his life. Sometimes clients won't commit to try out ideas that previously looked quite promising.
4. The client's context may not be receptive to certain solutions. The interviewer may need to assess the potential impact of a solution on the client's relationships in a broader context—for example, a woman client's desire to take a much more assertive position in a marriage. If so, is there danger of abuse if the woman speaks up?

▲ EXERCISE 14.3   **Practice With Brief Counseling**

Again, there is a real need for some role-played practice. This should come fairly easily as the basic skills of brief counseling are similar to those you have learned in decisional counseling.

- ▲ Work with a partner, switching the roles of client and counselor. Plan for a minimum interview of 15 minutes.
- ▲ Select a concern for the role-play. This time the issues need to be very specific—for example, dealing with a specific conflict on the job, in the family, or with a partner. Aim for concreteness and clarity throughout the storytelling.
- ▲ If possible, record the session on audio or video, perhaps using your computer camera, cell phone, or personal camera to provide some instant feedback.
- ▲ Search through the discussion of brief counseling and select some specific questions listed or those used by Penny during the session. Have them available on your lap during this first practice session and be willing to feel awkward at times.

▲ Go through the following stages of brief counseling:

1. *Relationship and Preliminary Goals.* Establish a relationship, but with an early focus on what specific goals the client has. Use the questions provided in the text and the transcript.

2. *Story and Strengths.* As necessary, draw out the client story with more clarity, but ask questions focused on solutions. While it is important to hear the client's story, the main goal is to help the client resolve issues. Learning what worked and didn't work in the past can be helpful. The strengths of the client are critical in brief counseling; give them special attention. There are several useful questions in the text and transcript.

3. *More Precise Goals.* This may be quite brief if the goal is already precise. Otherwise, make sure that the goal is clear and doable.

4. *Restory and Action.* Once again, use the question list with a focus on developing a narrative or story that works for the client. There need to be specifics for action, and you may find yourself using more influencing skills. But effective questioning, using the model of the text and transcript, will often achieve the same result.

5. *Relapse Prevention.* This is your chance to practice relapse prevention, as discussed in Chapter 13. Commitment to real change is not easy, and the relapse prevention process has proven effective in enabling the client to develop and maintain change.

Complete Case Study: Minority Experience in Counseling Training
Complete Interactive Exercise 1: Microskills and Theoretical Approaches to the Interview
Complete Interactive Exercise 2: Basic Competence
Complete Group Practice Exercise: Review and Discuss

14.4
FURTHER APPLICATIONS OF MICROSKILLS:
Theories of Interviewing and Counseling

KEY CONCEPT QUESTIONS

▲ **What skill patterns are demonstrated in different theories of interviewing and counseling?**

▲ **What types of meaning or topic focus are stressed by varying theories?**

Equipped with the foundation skills of listening, observing, influencing, and structuring an interview, you are well prepared to enter the complex world of theory and practice. If you continue with the helping field, soon you will be encountering the 250 or more theories competing for your attention. However, those listed in Table 14.2 are the ones currently in most common use.

This is not the time to master these theories. Rather, we are just indicating how what you have learned so far relates to major dimensions of the helping field. They all use microskills; in addition, the five stages of the interview can help you become more competent in each system more quickly.

Each theory or system presented here is based on a narrative or story of what makes counseling and therapy work. The decisional narrative is focused on concrete decisions that people are constantly making; you may recall that all theories in some way are working with client decision making. Decisional counseling has been around for a long time. Ben Franklin reminded us that much of our life is concerned with making decisions. Decisional counseling (sometimes known as problem-solving counseling) may be the most widely practiced form of helping in current practice. Samples of decisional daily practice include social work, employment counseling, placement counseling, AIDS counseling, alcohol and drug counseling, school and college counseling, and the work of community volunteers and peer helpers.

Completing a full interview using only listening skills provides an introduction to Carl Rogers's (1961) person-centered theory. Rogers is our central theorist on the importance of the relationship and working alliance in the interview. You likely have discovered that many clients are self-directed and, with a good listener, can do much to resolve their issues on their own. Mastery of completing a full interview using only listening skills does not make you a person-centered counselor, but it is a very important start. Study Rogers's work in detail and complete an interview using *no questions* at all, a very different task from the one used in brief counseling or decisional work.

The emphasis in cognitive-behavioral ABC crisis intervention is on cognitive schemas/maps, making decisions to change thoughts and behaviors along with underlying emotions. In brief approaches, you found that your listening skills and the positive asset search enabled you to conduct a very different type of decision-making interview. Questions are a central skill of four models presented here. But these theories also need relationship and the working alliance, derived from Carl Rogers's person-centered theory. Without skills in relationship and listening, no counseling or therapy system will be effective.

If you have demonstrated competence in basic skills and strategies, you are well prepared to increase your mastery of the interviewing approaches presented in this book, plus a beginning understanding of person-centered helping. Later, you can more effectively become competent in other theoretical/practical strategies. Note the following key points as you think about these theoretical/practical methods.

▲ Different theoretical/practical systems use varying patterns of microskills. These can range from almost all questions in the brief approach to no questions at all with the person-centered orientation.

▲ Theoretical/practical systems tend to focus on different areas of meaning and tell different stories about the interviewing, counseling, and therapeutic process.

Table 14.2 illustrates in summary form how the microskills can be used in different approaches to the interview. The table shows that widely varying styles of helping can be understood and practiced via the microskills system.

▲ **TABLE 14.2** Microskills Patterns of Differing Approaches to the Interview

MICROSKILL LEAD	Decisional counseling	Person-centered	Cognitive behavioral ABC crisis counseling	Brief counseling	Frankl's logotherapy and meaning	Motivational interviewing	Interpersonal psychodynamic	Gestalt	Feminist therapy	Business problem solving	Medical diagnostic interview	Eclectic/ metatheoretical
BASIC LISTENING SKILLS												
Open question	●	○	●	●	●	●	◐	●	◐	◐	◐	◐
Closed question	◐	○	●	◐	●	◐	○	◐	◐	◐	◐	◐
Encourager	●	◐	◐	◐	●	●	◐	◐	◐	◐	◐	◐
Paraphrase	●	●	◐	◐	●	●	◐	○	◐	◐	◐	◐
Reflection of feeling	●	●	◐	◐	◐	●	◐	○	◐	◐	◐	◐
Summarization	◐	◐	◐	●	●	◐	◐	○	◐	◐	◐	◐
INFLUENCING SKILLS												
Reflection of meaning	◐	●	○	○	●	◐	◐	○	●	○	○	◐
Interpretation/reframe	◐	○	○	○	◐	●	●	●	◐	◐	◐	◐
Logical consequences	◐	○	◐	○	○	●	○	○	◐	●	◐	◐
Self-disclosure	◐	◐	○	◐	◐	◐	○	◐	◐	◐	○	◐
Feedback	◐	◐	◐	◐	◐	◐	○	◐	◐	◐	◐	◐
Information/ psychoeducation	◐	○	●	○	○	◐	○	◐	◐	●	●	◐
Directive	◐	○	●	○	◐	◐	○	●	◐	●	●	◐
CONFRONTATION (Combined Skill)	◐	◐	◐	◐	◐	●	◐	●	●	◐	◐	◐
FOCUS												
Client	●	●	●	●	●	●	●	●	◐	◐	◐	◐
Main theme/problem	●	○	●	●	●	◐	○	◐	◐	●	●	◐
Others	◐	○	◐	◐	◐	◐	◐	◐	◐	○	○	◐
Family	◐	○	◐	◐	◐	◐	◐	○	◐	○	○	◐
Mutuality	○	◐	○	◐	◐	○	○	○	◐	○	○	◐
Counselor/interviewer	○	◐	○	○	◐	○	○	○	◐	○	○	◐
Cultural/ environmental context	◐	○	◐	◐	◐	○	○	○	●	◐	○	◐
ISSUE OF MEANING (Topics, key words likely to be attended to and reinforced)	Decision making	Self-actualization	Change in thoughts, behaviors, and feelings	Solved key issue	Meaning, discernment	Change	Unconscious motivation	Here-and-now behavior	Problem as a "woman's issue"	Problem solving	Diagnosis of illness	Varies
AMOUNT OF INTERVIEWER TALK-TIME	Medium	Low	High	Medium	Medium	Medium	Low	High	Medium	High	High	Varies

LEGEND   ● Frequent use of skill   ◐ Common use of skill   ○ Occasional use of skill

CHAPTER SUMMARY

| Key Points of "Applications of Microskills" | CourseMate and DVD Activities to Build Interviewing Competence |

14.1 Intentionality, Microskills, and Applications with Multiple Approaches to the Interview

▲ Microskills and the five-stage structure of the interview can be used to understand various theories and how theory may be used in practical ways.

1. Group Practice Exercise: Reflect, Identify, Share, and Plan for Advancement
2. Weblink Exercise: Counseling Approaches. Counseling resource website with hot links to pertinent topics.
3. Weblink Exercise: Examples of Treatment Modalities.

14.2 Cognitive-Behavioral Crisis Counseling: The ABC Model and Example Interview

▲ The ABC model of crisis intervention is a cognitive-behavioral strategy that focuses on cognitive change. Theory is very much integrated into practice. Through client-selected goals for change, role-playing, and interviewer encouragement, the client learns to cope effectively with crises.

▲ The five-stage interview provides a solid outline for the practice of the ABC model. Listening skills remain important, but working collaboratively to modify negative cognitions becomes central.

1. Group Practice Exercise: Review and Discuss. Use this exercise to discuss the ABC model of crisis intervention.

14.3 Brief Counseling and Example Interview

▲ Brief counseling operates on the theory that clients have their own answers in their past success experiences and wellness strengths. Emphasis is on immediate goal setting and clarity. Questions are a major skill, although many practitioners consider that listening carefully to the client's full story is critical.

▲ Skilled questioning is the major skill of brief counseling. We recommend, at least in the early stages of your practice, that you have, and share with your client, a list of questions for each stage of the interview.

▲ Think about referral or another approach to helping if the client does not respond to brief methods.

1. Case Study: Minority Experience in Counseling Training. Supervisor/Employee Conflict in Japan—Will I Be Fired? A good case to review: What are the issues? How would you help? How would you use the basics of ABC crisis counseling and brief counseling to help the client?
2. Interactive Exercise 1: Microskills and Theoretical Approaches to the Interview. The two approaches in this chapter are best understood by conducting an interview.
3. Interactive Exercise 2: Basic Competence. Review and select the most appropriate response from the point of view of brief therapy.
4. Group Practice Exercise: Review and Discuss. Use this exercise to discuss brief counseling.

Key Points of "Applications of Microskills"

 CourseMate and DVD Activities to Build Interviewing Competence

14.4 Further Applications of Microskills: Theories of Interviewing and Counseling

▲ What you have learned so far relates to all theories and strategies, as they all use microskills. Also, the five stages of the interview will help you quickly become more competent in each system.

▲ If you have developed competence in the basic microskills and strategies, you are well prepared to understand person-centered helping and to become competent in other strategies.

▲ Different theories of counseling use varying patterns of microskills. Some use many questions (e.g., brief counseling approaches) and others use no questions at all (e.g., person-centered orientation).

▲ Theoretical/practical systems tend to focus on different areas of meaning and tell different stories about the interviewing, counseling, and therapeutic process.

 Assess your awareness, knowledge, and skills as you conclude the chapter:

1. Flashcards: Use the flashcards to check your understanding of key concepts and facilitate memorization of key information.

2. Self-Assessment Quiz: The quiz will help you assess your current knowledge and prepare for course examinations.

3. Portfolio of Competencies: Evaluate your present level of competence in attending and listening skills using the downloadable Self-Evaluation Checklist. Self-assessment of your attending skills competence demonstrates what you can do in the real world.

DETERMINING PERSONAL STYLE
Self-Evaluation and the Future

DETERMINING PERSONAL STYLE AND THEORY

SKILL INTEGRATION AND APPLICATIONS

INFLUENCING SKILLS AND STRATEGIES

FOCUSING

CONFRONTATION

THE FIVE-STAGE INTERVIEW STRUCTURE

REFLECTION OF FEELING

ENCOURAGING, PARAPHRASING, AND SUMMARIZING

OPEN AND CLOSED QUESTIONS

ATTENDING BEHAVIOR

ETHICS, MULTICULTURAL COMPETENCE, AND WELLNESS

When you learn, teach. When you get, give.

—*Maya Angelou*

How can determining your own style help you and your clients?

CHAPTER GOALS

It's nearly the end of this book, and it's time to spend some time reviewing and evaluating your progress, including your competency in the microskills and other major concepts discussed in this book. As a conclusion, we'll identify the many possible uses of microskills and the five-stage interview framework.

Awareness, knowledge, and skills developed through the concepts of this chapter will enable you to

▲ Think more about your personal style of interviewing and where it might lead you in the future.

▲ Review your competency in skills, strategies, and approaches to interviewing and counseling.

▲ Commit to a lifetime of movement, change, and constant growth as a helping professional.

Assess your awareness, knowledge, and skills as you begin the chapter:

1. Self-Assessment Quiz: The chapter quiz will help you determine your current level of knowledge. You can take it before and after reading the chapter.
2. Portfolio of Competencies: Before you read the chapter, fill out the downloadable Self-Evaluation Checklist to assess your existing knowledge and competence in attending skills. Then, at the end of the chapter, complete the checklist again to summarize your competencies after study and practice.

15.1

INTENTIONALITY AND FLEXIBILITY IN YOUR INTERVIEWING STYLE AND IN YOUR FUTURE CAREER

KEY CONCEPT QUESTION

▲ **What are the key issues of intentionality and flexibility as emphasized in this text?**

This chapter is focused on you and your review of your experience of *Essentials of Intentional Interviewing*. Having reached the apex of the microskills hierarchy, you can stand back and review where you are and where you want to go.

Determining Personal Style and Theory	*Predicted Result*
As you work with clients, identify your natural style, add to it, and think through your approach to interviewing and counseling. Examine your own preferred skill usage and what you do in the session. Integrate learning from theory and practice in interviewing, counseling, and psychotherapy into your own skill set.	You, as a developing interviewer or counselor, will identify and build on your natural style. You will commit to a lifelong process of constantly learning about theory and practice while evaluating and examining your behavior, thoughts, feelings, and deeply held meanings.

Chapter 13 gave you a chance to evaluate and classify your interviewing style. We encourage you to do this again and again and obtain feedback from others, including your clients. Whether it's a mock interview or one with a real client, creating a transcript and then evaluating your microskills is one of the best ways to learn. If you haven't already done so, we urge you to go back so that you can define your skills and areas for growth more precisely.

Some interviewers and counselors have developed individual styles of helping that require their clients to join them in their view of the world. Such individuals have found the one "true and correct" formula for the session; clients who have difficulty with that formula are often termed "resistant" and "not ready" for counseling. Such counselors and therapists can and do produce effective change, but they may be unable to serve very many types of clients. The complexity of humanity and the helping process is missing from their orientation. They do not recognize that some other approach may be more effective than their own fixed style.

Remember, all counseling is multicultural, and your awareness, knowledge, and skills of multicultural counseling competencies are critical to building rapport with your clients. Consequently, it is urgent that you remain aware that your client may not prefer the skills, strategies, and theories that you favor. Expand your skills and knowledge in areas where you are now less comfortable. With patience, study, and experience, you will increase your ability to work with more and more varied clients. The opportunity to learn from clients who are different from us is one of the special privileges of being an interviewer, a counselor, or a therapist.

Many of you will continue to study interviewing and counseling and will have long and useful careers in the helping fields. But it is equally likely that you may find yourself managing your own business, becoming a salesperson, or working as a physical therapist or nurse, a teacher, or any of a myriad of other careers. Regardless of where you go, communication skills will be basic to the job and to your advancement. Microskills, the first systematic communication skills program in the world, is now used throughout that world in training settings ranging from AIDS peer counselors in Africa, to executives in Sweden and Japan, to Aboriginal social workers in Australia. You will find many places to use your interviewing and communication skills. And don't forget the importance of listening carefully and fully to your loved ones and friends. Research shows that relationships "wear out" over time if not renewed. Intentionality and flexibility are critical wherever we go.

Box 15.1 presents an international perspective on using microskills throughout a professional career. Note how Dr. Brodhead has intentionally moved ahead through opportunities to apply her knowledge and experience in new settings. She is now in a field somewhat distant from her original work in teaching and counseling. Her experience provides an illustration that a career in counseling involves constant learning.

Developing your own personal approach to interviewing and counseling involves a multiplicity of factors. Be ready to learn new things and change throughout your career.

Complete Weblink Exercises:
▲ The Call to an Authentic Life—Markers on the Path to Personal Authenticity
▲ Synthesis of Scholarship in Multicultural Education

BOX 15.1   National and International Perspectives on Counseling Skills

Using Microskills Throughout My Professional Career
MARY RUE BRODHEAD, EXECUTIVE, CANADIAN FOOD INSPECTION AGENCY

My first encounter with the microskills program was through my master's program in teacher education. I wanted to learn about counseling so that I could better reach teacher trainees. What I learned very rapidly was that teachers often fail to listen to their students; in fact, one research study found that out of nearly 2,000 teacher comments, there was only one reflection of feeling and, of course, most comments focused on providing information. Often when teachers did listen, they weren't able to recognize the verbal and nonverbal cues that were reflecting the reality of their students.

After completing my degree, I entered teacher education and found that microskills led to more student-centered teaching. I also found that teachers who matched their students' cognitive/emotional style were more effective (see Ivey et al., 2005). If a student is concrete, the teacher needs to provide specific examples and use concrete questions. If the student is more reflective, then abstract formal-operational strategies can be used. Bringing in emotional involvement via microskill strategies enriches teaching.

Microskills in Counseling: A Multicultural Experience
I next lived on an island off Vancouver where my husband Dale and I worked with the Kwakiutl nation. I trained teachers, but I also counseled and led a group of school dropouts in a special program under the School Board. I actually taught them the same interviewing skills that you are learning in this book. Needless to say, the multicultural orientation helped sensitize me so that I adapted my counseling and teaching to better fit Kwakiutl style.

One of our first activities was a trip in the Chief's fishing boat to gather Christmas trees for the old people, a cherished tradition in the community. This was the first time that most of these young people had participated in the ritual. As they delivered the trees, the recipients, in an expression of gratitude, invited them in for something to eat and began to tell stories. These old people, thrilled to have an audience, would shower attention on the alienated youth who, in response, would listen with attention and respect. And so an upward cycle of communication began. Eventually we raised money to buy tape recorders and to capture these tales for the future, leading to further strengthening of the students' listening and questioning skills. Attending behavior increased respect and communication on both sides.

Microskills in the Workplace: Training Executives to Listen, Hear, and Understand
After three years in British Columbia, Dale and I returned to Ottawa, where I directed an employment equity program. The task centered on providing culturally sensitive counseling and opportunities to members of the Canadian four "equity groups" (visible minorities, Aboriginal peoples, persons with disabilities, and women in nontraditional occupations) to develop the skills needed for career development and success in the Federal Public Service.

As part of my work, I used microskills to train government officials to listen and really hear and understand the variety of perspectives, strengths, and styles of working found within members of our multicultural workforce. Here I learned the importance of language, and I became reasonably fluent in French, an essential skill in bilingual, multicultural Canada. I developed a multicultural counseling course used within 15 universities in Canada that included many ideas presented here.

Microskills in Management: Creating a Culture of Learning and Team Building
My current agency is responsible for animal health, plant protection, and food safety. It is one of those "science departments" involving agriculture, fisheries, scientific research, and many other issues. Given the diversity of my workforce, my first task has been to build a "culture of learning" in which

(Continued)

BOX 15.1 National and International Perspectives on Counseling Skills (Continued)

vast amounts of information and knowledge can be shared effectively and efficiently. Microskills are key elements in management training, and a good communication skills workshop can be vital in team building. And, of course, all my managers need to listen to and motivate those with whom they serve.

Looking back over my career, it is amazing to find that the basic microskills have been useful in my teaching, counseling, multicultural work, and governmental leadership positions. "Training as treatment" and "teaching competence" in microskills can help us all make a difference throughout our careers.

15.2
TIME TO STOP, REFLECT, ASSESS, AND TAKE STOCK

KEY CONCEPT QUESTIONS

▲ **How have you changed in your thoughts about interviewing and counseling?**

▲ **Where are you now that you are finishing this book and the course of study?**

▲ EXERCISE 15.1 **How Have You Changed?**

Think about yourself, your values, and your personal experiences throughout this course. What have you learned about yourself? What is your current perception of interviewing? How have you changed? What else would you like to change?

▲ EXERCISE 15.2 **Revisiting Your First Audio or Video Recording**

In Chapter 1 (pages 21–22) we suggested that you record an interview. This exercise documented the baseline of your natural style and skill level. Now is a good time to go back and examine your first interview. What do you notice and feel about your skills? Do you prefer some skills over others? What would you do differently now? If you were an observer, what would you suggest to the interviewer? What other skills would you use as a result of being in this course? Do the same with the interview completed in Chapter 13 (page 300). After reviewing your interviews, consider: What are your strengths and assets? How would you bring these to action?

You have been presented with more than 30 major concepts and skill categories Within those major divisions are more than 100 specific methods, theories, and strategies. Ideally, you will commit them all to memory and be able to draw on them immediately in practice to facilitate your clients' development and progress in the interview. As you grow as an interviewer, counselor, or therapist and your competence and skill mastery increase, you will find the ideas expressed here becoming increasingly clear and a natural part of your practice.

Retaining and mastering the concepts of the microskills hierarchy may be facilitated by what is termed **chunking.** We do not learn information just in bits and pieces; we organize it into patterns. The microskills hierarchy is a pattern that can be visualized and experienced. For example, at this moment you can probably immediately recall that attending behavior has certain major concepts "chunked" under it (culturally appropriate visuals, vocal tone, verbal following, and body language—"three V's + B"). You can probably also recall the basic listening sequence, the purpose of open questions, and perhaps which questions lead to which likely outcomes (for example, *how* questions lead to process and feelings).

Identify and classify the concept. The microskills language provides a vocabulary and communication tool with which to understand and analyze your interviewing and counseling behavior and that of others. If the concept or skill is present in an interview, can you label it? If you can identify concepts, you have most likely chunked most of the major skill points together in your mind. You may not immediately recall all the types of focus, but when you see an interview in progress you will probably recall which one is being used.

Demonstrate basic competence. Basic competence means you will be able to understand and practice the concept in an interview. Continued practice and experience become the foundation for later intentional mastery.

Demonstrate intentional competence. Skilled interviewers, counselors, and therapists can use the microskills to produce specific, concrete effects with their clients. Appendix I presents the full Ivey Taxonomy and reviews specific aspects of intentional prediction. If you reflect feelings, do clients actually talk more about their emotions? If you provide an interpretation/reframe, does your client see her or his situation from a new perspective? If you work through some variation of the positive asset search, does your client actually view the situation more hopefully? If you conduct a well-formed, five-stage interview, does your client's self-concept and/or developmental level change? Does your interview produce client changes?

▲ EXERCISE 15.3   **Determine Your Mastery of Skills**

Use the self-assessment in Table 15.1 to assess your competence and mastery of each concept.

BRAIN RESEARCH: IMPLICATIONS FOR THE FUTURE OF INTERVIEWING AND COUNSELING

When you interact with clients, your brain functioning as well as your client's can be modified. These changes can be measured through a variety of brain imaging techniques, especially functional magnetic imaging (fMRI).

> We are not prisoners of our genes or our environment. Poverty, alienation, drugs, hormonal imbalances, and depression don't dictate failure. Wealth, acceptance, vegetables, and exercise don't guarantee success. . . . Experiences, thoughts, actions, and emotions actually change the structure of our brains. . . . Indeed, once we understand how the brain develops, we can train our brains for health, vibrancy, and longevity. (Ratey, 2001, p. 17)

You likely have noticed stories on television and in the popular media on brain research. This research has now reached a state of precision where it has important

▲ **TABLE 15.1** Self-Assessment Summary

Take some time to review your own competence level in each of the major microskill areas and indicate your personal competencies on this chart. You may use this as a summary and as a plan for future growth. Can you identify each skill or concept and classify its place in the interview? Can you demonstrate basic competence by using the skill in the session? Most important, can you obtain specific and predictable results in client behavior as a result of your use of the skill or concept? Feel free to use the glossary to help refresh your memory.

Skill or Concept	*Identification and Classification*	*Basic Competence*	*Intentional Competence*	*Evidence of Achieving Competence Level*
1. Attending behavior				
2. Questioning				
3. Observation skills				
4. Encouraging				
5. Paraphrasing				
6. Summarizing				
7. Reflecting feelings				
8. Basic listening sequence				
9. Positive asset search				
10. Empathy				
11. Five stages of the interview				
12. Confrontation				
13. The Client Change Scale				
14. Focusing				
15. Eliciting and reflection of meaning				
16. Interpretation/reframe				
17. Logical consequences				
18. Self-disclosure				
19. Feedback				
20. Information/psychoeducation				
21. Directives				

▲ **TABLE 15.1** (Continued)

Skill or Concept	Identification and Classification	Basic Competence	Intentional Competence	Evidence of Achieving Competence Level
Key issues and practical applications of microskills and the five-stage interview				
22. Analysis of the interview (Chapter 13)				
23. Ethics				
24. Multicultural competence				
25. Social justice				
26. Wellness				
27. Community genogram				
28. Decisional counseling				
29. Person-centered counseling				
30. Cognitive-behavioral crisis counseling				
31. Brief counseling				
32. Defining personal style and self-assessment				

implications for you as an interviewer or counselor. Chapter 7 summarized research on empathy, and there we learned that the healthy person's brain resonates (empathizes) with the experience of the client. The empathic listening helper facilitates client physical and mental health.

As this book is being written, the National Institute of Mental Health is starting to turn away from subjective diagnosis via the Diagnostic and Statistical Manual and will now be putting funding into brain-based neuroscience and diagnosis. Neuroscience is so new and so important that we suggest you use your search engine to explore these new developments with key words such as *brain research, neuroscience*, and *National Institute of Mental Health Neuroscience Research*. If you keep exploring, you can find many training programs for understanding how the brain functions, and even how it relates to counseling. Those with newer cell phones can download an "app" that presents the brain in three-dimensional color.

Following are some key concepts for your future study and practice.

NEUROSCIENCE AND NEUROPLASTICITY: IMPLICATIONS FOR THE INTERVIEW

Neuroscience is the study of the brain and nervous system. The science's most important discovery for interviewing and counseling is *neuroplasticity*. Formerly, it was believed that we humans lost brain cells and neurons only as we age. Now we know that the brain develops new neurons and neural networks throughout the lifespan, including in old age. Each new situation provides an opportunity for learning and thus adding new networks. In short, new information on neuroplasticity has shown us that the brain can be completely rewired.

What does this mean for you and the helping process? When you interact with clients, your brain functioning as well as your client's can be measured through a variety of brain imaging techniques, especially functional magnetic imaging (fMRI). An example of neuroplasticity is that both you and your client may learn, change, and develop new neural connections as a result of your interaction. Neuropsychology's research on emotion validates past counseling and therapy theory and research from a new perspective.

SYSTEMATIC LEARNING AND THE BRAIN

Restak (2003) found that training volunteers in movement sequences produced sequential changes in activity patterns of the brain as the movements became more thoroughly learned and automatic. Systematic step-by-step learning, such as that emphasized in this book, is an efficient learning system used in ballet, music, golf, and many other settings. If there is sufficient practice, changes in the brain may be expected and increased ability in demonstrating these skills will appear in areas ranging from finger movements to dance—and from the golf swing to interviewing skills. Practice of specific skills facilitates a natural automatic style.

ATTENTION AS A CENTRAL PROCESS

Attention is not just a psychological concept—it is measurable through brain imaging. When a person attends to a stimulus (e.g., the client's story), many areas of the brain of both interviewer and client become involved (Posner, 2004). Two key factors in attention are arousal and focus. By giving attention to the client, the interviewer's and the client's thought processes have both been activated. Selective attention is a critical aspect of listening—"Focus is brought about by . . . a part of the thalamus, which operates rather like a spotlight, turning to shine on the stimulus" (Carter, 1999, p. 186). Just as attending behavior underlies all the microskills of this book, attention and selective attention provide the physiological foundation for personal growth. When we listen and attend selectively, our brain and the client's brain are reacting and changing.

BRAIN RESEARCH SUPPORTS THE WELLNESS APPROACH

The left hemisphere of the frontal cortex is associated more with positive emotions such as happiness and joy; the right hemisphere and amgydala (deep in the brain) are associated more often with negative feelings. In depression and deep sadness, brain scans reveal that the positive areas are less active (Davidson, Pizzagalli,

Nitschke, & Putnam, 2002). Exercise increases blood flow throughout the body and brain, supporting neuronal growth. Happiness involves physical pleasure, positive meanings, and the absence of negative emotions (Carter, 1999; Davidson et al., 2002). In effect, positive thoughts and action can help override fear, anger, and sadness.

What, specifically does this mean for your practice? First and foremost, always prescribe and recommend exercise for your clients. There is extensive research presenting this as the treatment of choice for all clients. A past president of the American Psychiatric Association has even said that failing to inform clients of the benefits of exercise is unethical. Exercise increases blood flow not only to the body, but also to the brain where new neurons are formed. Dr. John Ratey teaches a 10-week course on exercise at Harvard Medical School. His book *Spark* (Ratey, 2008) is one of the best places to begin your reading on neuroscience.

When clients focus solely on problems and negative emotions, we can help them through a wellness approach that strengthens and nourishes the individual. In this way we can help clients "build a tolerance for negative emotions and gradually acquire a knack for generating positive ones" (Damasio, 2003, p. 275). So the wellness approach is not just "window dressing." This strengthening can be both psychological (reminding clients about positive experiences and personal strengths) and physical (encouraging exercise and sports, nutrition, and adequate sleep). A base of strengths facilitates problem solving and working through the many complex issues we all face. A general theme running through neuroscience research in this area is that it takes four positives to counter one negative.

Broader areas of research are now entering the world of neuroscience. A study has found that a sense of meaning and purpose was the most important area for mental health among older African Americans (Alim et al., 2009). Eliciting and reflecting meaning, along with the discernment process are likely to be recognized more and more in the future as a central issue.

Stress destroys brain cells (Ratey, 2001). Human development depends on cultural/environmental/contextual influences. If clients grow up and live in stressful, impoverished situations, both mental and physical development will be impaired, often permanently. This becomes an issue for our advocacy and the need for our participation in social justice work. We need to balance individual interviewing and counseling with a commitment to social justice, action in the community, and efforts to prevent problems and issues before they occur.

LOOKING TO THE FUTURE

Throughout your career, you will want to explore how brain research relates to interviewing practice and everyday life. Brain research is not in opposition to the cognitive, emotional, and behavioral emphasis of interviewing and counseling. Rather, it will help us pinpoint types of interventions that are most helpful to the client. In fact, one of the clearest findings is that the brain needs environmental stimulation to grow and develop. You can offer a healthy atmosphere for client growth and development.

Complete Group Practice Exercise: Share Your Reflections and Your Plans to Continue Your Journey

BOX 15.2   Your Natural Style and Story of Interviewing and Counseling

Consider these issues as you continue the process of identifying your natural style and future theoretical/practical integration of skills and theory.

Goals	What do you want to happen for your clients as a result of their working with you? What would you *desire* for them? How are these goals similar to or different from those of decisional, brief, person-centered, and cognitive-behavioral crisis intervention? What else?
Skills and strategies	You have identified your competence levels in these areas. What do you see as your special strengths? What are some of your needs for further development in the future? What else?
Cultural intentionality	Consider the RESPECTFUL model. With what multicultural groups and special populations do you feel capable of working? What knowledge of diversity do you need to gain in the future? How aware are you of your own complex multicultural background? What else?
Theoretical/practical issues	What applications and what theoretical/practical story would you provide now that summarizes how you view the world of interviewing and counseling? Where would you like to focus your efforts and interests next? What else?

15.3
YOUR PERSONAL STYLE

KEY CONCEPT QUESTIONS

▲ **What is your personal authentic style and your story of interviewing and counseling?**

▲ **What is your planned personal journey in interviewing and counseling?**

There are two major factors to consider as you move toward identifying your own personal style. Unless a skill or practical application of a set of skills harmonizes with who you are, it will tend to be false and less effective; however, modifying your natural style and theoretical orientation will be necessary if you are to be helpful to many clients. Remember that you are unique, and so are those whom you would serve. We all have varying issues and problems. We all come from varying families, differing communities, and distinct views of gender, ethnic/racial, spiritual, and other multicultural issues. Please turn to Box 15.2 to review and consolidate what you consider to be your natural style. How have you developed and changed since you have been studying interviewing?

▲ EXERCISE 15.4   **Identify and Define Listening Skills**

Summarize your own story of interviewing and counseling. You may want to base it on your interview transcript completed in Chapter 13 and on your success in microskill practice exercises. Use the following questions to review your goals, your special skills, and your plans for the future.

What have you learned by completing this exercise? What can you say about your natural style of interviewing and counseling? Where would you like to go next?

Complete Interactive Exercise 1: Self-Awareness and Emotional Understanding Self-Assessment

Complete Interactive Exercise 2: Ethical Self-Assessment

Complete Interactive Exercise 3: Multicultural Self-Assessment

Complete Interactive Exercise 4: Comparing Your Pre- and Post-Self-Assessments

Complete Interactive Exercise 5: Assess Your Competencies

YOUR JOURNEY IN INTERVIEWING AND COUNSELING BEGINS NOW

You did it! You've reached the end of the book, and many of those concepts, theories, and skills that once seemed awkward and overwhelming are now more familiar and a natural part of your counseling style. We have come to the end of this phase of your interviewing and counseling journey. You have been introduced to the foundation skills and how they are applied in a variety of theoretical/practical interviews. The basic listening sequence has become an important part of your interviewing practice and way of thinking. You have a basic story of the interview to build on and work with in many types of situations.

The real journey begins now. Many of you will be moving on to individual theories of counseling, exploring issues of family counseling, becoming involved in the community, and learning the many aspects of professional practice. Others will be moving into careers in one of the many fields that use the microskills and the five stages. Among these are professional coaches, librarians, teachers, nurses, physicians, lawyers, business and salespeople, and teachers of peer helpers in a multitude of settings throughout the world. Still others of you may find find this presentation sufficient for your purposes; perhaps you will use the skills in other work settings or in developing more effective partner or family communication.

We have designed this book as a clear summary of the basics of effective interviewing and counseling; a naturally skilled person can use the information here for many effective and useful helping sessions. To assist you in your continued development, we list below a few key references. Enjoy delving more deeply into our exciting field.

You've reached the end of this book, and yet your journey into the field of interviewing and counseling is just beginning. Throughout this process you have learned about your many strengths and the positive values you bring to the interview. The awareness, knowledge, and skills gained from this book will enable you to help others find their own life directions, make critical decisions, and resolve issues, all the while becoming more sure of themselves and more empowered to help others. This field is full of interesting and involved people working in counseling, psychology, social work, business, medicine, and teaching, among other areas, each seeking to make a difference not only for their clients, but also for their communities and the world. At the core of all of these professions is listening; it works every time and everywhere. We hope you too will use your attending and influencing skills to make a difference in our world!

We have enjoyed sharing this time with you. Many of the ideas in this book come from interaction with students. We hope you will take a moment to provide us with your feedback and suggestions for the future. Please see the evaluation forms

at the end of this book or download them from the book's website (www.cengage
.com/counseling/ivey). This book and the materials available through the website will
be constantly updated with new ideas and information. You have joined a never-
ending time of growth and development. Welcome to the field of interviewing and
counseling!

"Trust yourself. You know more than you think you do" (Dr. Benjamin Spock).

Allen, Mary, Carlos, and Kathryn

SUGGESTED SUPPLEMENTARY READINGS

The literature of our field is extensive, and you will want to sample it on your own. We
would like to share some books that we find helpful as next steps to follow up ideas
presented here. All of these build on the concepts of this book, but we have recommended several books that take different perspectives from our own.

MICROSKILLS

Evans, D., Hearn, M., Uhlemann, M., & Ivey, A. (2011). *Essential interviewing*
(8th ed.). Belmont, CA: Brooks/Cole.

Microskills in a programmed text format.

Ivey, A., Gluckstern, N., & Ivey, M. (2006). *Basic attending skills* (3rd ed.).
Framingham, MA: Microtraining Associates.

Brief and perhaps the most suitable book for beginners and those who
would teach others microskills. Supporting videotapes are available (www
.emicrotraining.com).

Zalaquett, C., Ivey, A., Gluckstern, N., & Ivey, M. (2008). *Las habilidades aten-
cionales básicas: Pilares fundamentales de la comunicación efectiva.* [Book and
Training Videos]. Framingham, MA: Microtraining Associates and
Alexander Street Press.

An introduction to the microskills and the five steps of the interview in
Spanish. Una introducción a las microhabilidades y la entrevista de cinco
etapas en Español. [Libro y Videos de entrenamiento.]

www.emicrotraining.com

Visit this website for up-to-date information on interviewing and counseling, microskills, and multicultural counseling and therapy. There are
links to professional associations, ethical codes, and many multicultural and
professional sites. You will also find interviews with leaders of our field.

THEORIES OF INTERVIEWING AND COUNSELING WITH AN ORIENTATION TO DIVERSITY

Ivey, A., D'Andrea, M., & Ivey, M. (2012). *Theories of counseling and
psychotherapy: A multicultural perspective* (7th ed.). Thousand Oaks, CA:
Sage.

The major theories are reviewed, with special attention to issues. Includes
many applied exercises to take theory into practice.

Sue, D., & Sue, D. M. (2007). *Foundations of counseling and psychotherapy: Evidence-based practices for a diverse society.* New York: Wiley.

An excellent text, focusing on evidence-based approaches.

SUGGESTIONS FOR FOLLOW-UP ON SPECIFIC THEORIES

Most of the following books are classics. In many ways they are more comprehensive, thorough, and important than more recent writings. We commend them all to you.

Decisional Counseling

D'Zurilla, T., & Nezu, A. (2006). *Problem-solving therapy* (3rd ed.). New York: Springer.

An entire counseling model derived from decision making.

Parsons, F. (1967). *Choosing a vocation.* New York: Agathon. (Originally published 1909)

It is well worth a trip to your library to read Parsons's work. You will find that much of his thinking is still up to date and relevant.

Brief Counseling

Connell, B. (2005). *Solution-focused therapy.* Beverly Hills, CA: Sage.

The basics of brief counseling in brief form.

de Shazer, S., & Dolan, Y. (2007). *More than miracles: The state of the art of solution-focused brief therapy.* Binghamton, NY: Haworth.

Person-Centered Counseling and Humanistic Theory

Frankl, V. (1959). *Man's search for meaning.* New York: Pocket Books. (Originally published 1946)

This is one of the most memorable books you will ever read. It describes Frankl's survival in German concentration camps through finding personal meaning. You will find that referring your future clients to this book is an excellent counseling and therapeutic tool in itself.

Rogers, C. (1961). *On becoming a person.* Boston: Houghton Mifflin.

The classic book by the originator of person-centered counseling.

Cognitive-Behavioral Counseling

Alberti, R., & Emmons, M. (2008). *Your perfect right: A guide to assertiveness training* (9th ed.). San Luis Obispo, CA: Impact.

The classic book by the originators of assertiveness training.

Davis, M., Eshelman, E., & McKay, M. (2008). *The relaxation and stress reduction workbook* (6th ed.). Oakland, CA: New Harbinger.

One of many books on cognitive and behavioral counseling. This is clear, direct, and easily translatable into microskill approaches to the interview. If you are in practice and looking for concrete ideas, this is the book to buy.

Dobson, K. (Ed.). (2009). *Handbook of cognitive-behavioral therapies* (3rd ed.). New York: Guilford Press.

A comprehensive and specific presentation of key skills, strategies, and theories.

Multicultural Counseling and Therapy

Sue, D. W., Carter, R., Casas, M., Fouad, N., Ivey, A., Jensen, M., LaFromboise, T., Manese, J., Ponterotto, J., & Vazquez-Nuttall, E. (1998). *Multicultural counseling competencies.* Beverly Hills, CA: Sage.

The original and most comprehensive coverage of the necessary skills and competencies in the multicultural area.

Sue, D. W., Ivey, A., & Pedersen, P. (1999). *A theory of multicultural counseling and therapy.* Pacific Grove, CA: Brooks/Cole.

A general theory of multicultural counseling and therapy with many implications for practice.

Sue, D. W., & Sue, D. (2007). *Counseling the culturally diverse* (5th ed.). New York: Wiley.

The classic of the field. This book helped launch a movement.

Integrative/Eclectic Orientations

Ivey, A., Ivey, M., Myers, J., & Sweeney, T. (2005) *Developmental counseling and therapy: Promoting wellness over the lifespan.* Boston: Lahaska/Houghton-Mifflin.

Lazarus, A. (2006). *Comprehensive therapy: The multimodal way.* Baltimore: Johns Hopkins Press.

The basic book for multimodal therapy. You will find the BASIC-ID model useful in conceptualizing broad treatment plans.

CHAPTER SUMMARY

Key Points of "Determining Personal Style"	**CourseMate and DVD Activities to Build Interviewing Competence**

15.1 Intentionality and Flexibility in Your Interviewing Style and in Your Future Career

▲ Intentional competence is reflected in what your client does, not in your execution of the skills. If your effort is not successful, can you flex and intentionally use another skill to reach the desired result? Reformulate your approach to the client on the spot?

1. Website: The Call to an Authentic Life—Markers on the Path to Personal Authenticity.
2. Website: Synthesis of Scholarship in Multicultural Education—Monograph by Dr. Geneva Gay, University of Washington–Seattle, North Central Regional Educational Laboratory.

15.2 Time to Stop, Reflect, Assess, and Take Stock

▲ Assessing your positive assets, strengths, and resources is a good place to start as you increase clarity and awareness of your personal style.

▲ Expand your skills and knowledge in areas where you are now less comfortable.

1. Group Practice Exercise: Share Your Reflections and Your Plans to Continue Your Journey.

Key Points of "Determining Personal Style"

CourseMate and DVD Activities to Build Interviewing Competence

▲ Over time you will increase your abilities to work with more and more varied clients.

▲ Remember that all counseling is multicultural and your awareness, knowledge, and skills of multicultural counseling competencies will be essential to build rapport with your clients.

15.3 Your Personal Style

▲ Become aware of your preferred skills, strategies, and theories.

▲ Determine your natural style. Remember that you are unique and will work best with skills that are natural to you.

▲ Those whom you serve are also unique. Modifying your natural style and learning skills you are less comfortable with will empower you to help more clients.

▲ It is critical that you maintain a personal authenticity in interviewing and counseling practice and continue to study and learn over time.

1. Interactive Exercise 1: Self-Awareness and Emotional Understanding Self-Assessment.
2. Interactive Exercise 2: Ethical Self-Assessment.
3. Interactive Exercise 3: Multicultural Self-Assessment.
4. Interactive Exercise 4: Comparing Your Pre- and Post-Self-Assessments.
5. Interactive Exercise 5: Assess Your Competencies.

Assess your awareness, knowledge, and skills as you conclude the chapter:

1. Flashcards: Use the flashcards to check your understanding of key concepts and facilitate memorization of key information.

2. Self-Assessment Quiz: The quiz will help you assess your current knowledge and prepare for course examinations.

3. Portfolio of Competencies: Evaluate your present level of competence in attending and listening skills using the downloadable Self-Evaluation Checklist. Self-assessment of your attending skills competence demonstrates what you can do in the real world.

APPENDIX I
THE IVEY TAXONOMY
Definitions of the Microskills Hierarchy and Predicted Results From Using Skills

Ethics Observe and practice ethically and follow professional standards. Particularly important issues for beginning interviewers are *competence, informed consent, confidentiality, power,* and *social justice.*	*Predicted Result* Client trust and understanding of the interviewing process will increase. Clients will feel empowered in a more egalitarian session. When you work toward social justice, you contribute to problem-prevention in addition to healing work in the interview.
Multicultural Issues Base interviewer behavior on an ethical approach with an awareness of the many issues of diversity. Include the multiple dimensions from the RESPECTFUL model (Chapter 1).	*Predicted Result* Anticipate that both you and your clients will appreciate, gain respect, and learn from increasing knowledge in ethics and multi-cultural competence. You, the interviewer, will have a solid foundation for a lifetime of personal and professional growth.
Wellness Help clients discover and rediscover their strengths through wellness assessment. Find strengths and positive assets in the clients and in their support system. Identify multiple dimensions of wellness.	*Predicted Result* Clients who are aware of their strengths and resources can face their difficulties and discuss problem resolution from a positive foundation.
Attending Behavior Support your client with individually and culturally appropriate visuals, vocal quality, verbal tracking, and body language.	*Predicted Result* Clients will talk more freely and respond openly, particularly about topics to which attention is given. Depending on the individual client and culture, anticipate fewer breaks in eye contact, a smoother vocal tone, a more complete story (with fewer topic jumps), and a more comfortable body language.
Client Observation Skills Observe your own and the client's verbal and nonverbal behavior. Anticipate individual and multicultural differences in	*Predicted Result* Observations provide specific data validating or invalidating what is happening in the session. Also, they provide guidance for

nonverbal and verbal behavior. Carefully and selectively feed back observations to the client as topics for discussion.	the use of various microskills and strategies. The smoothly flowing interview will often demonstrate movement symmetry or complementarity. Movement dissynchrony provides a clear clue that you are not "in tune" with the client.
Open and Closed Questions Open questions often begin *who, what, when, where,* or *why.* Closed questions may start with *do, is,* or *are. Could, can,* or *would* questions are considered open, but have the advantage of giving more power to the client, who can more easily say what he or she wants to respond.	*Predicted Result* Clients give more detail and talk more in response to open questions. Closed questions provide specific information, but may close off client talk. Effective questions encourage more focused client conversations with more pertinent detail and less wandering. *Could, would,* and *can* questions are often the most open of all.
Encouraging Encourage with short responses that help the client keep talking. These responses may be verbal (repeating key words and short statements) or nonverbal (head nods and smiling).	*Predicted Result* Clients elaborate on the topic, particularly when encouragers and restatements are used in a questioning tone of voice.
Paraphrasing Shorten or clarify the essence of what has just been said, but be sure to use the client's main words when you paraphrase. Paraphrases are often fed back to the client in a questioning tone of voice.	*Predicted Result* Clients will feel heard. They tend to give more detail without repeating the exact same story. If a paraphrase is inaccurate, the client has an opportunity to correct the interviewer.
Summarizing Summarize client comments and integrate thoughts, emotions, and behaviors. Summarizing is similar to paraphrasing but used over a longer time span.	*Predicted Result* Clients will feel heard and often learn how the many parts of important stories are integrated. The summary tends to facilitate a more centered and focused discussion. The summary also provides a more coherent transition from one topic to the next or a way to begin and end a full session.
Reflection of Feelings Identify the key emotions of a client and feed them back to clarify affective experience. With some clients, the brief acknowledgment of feelings may be more appropriate. Often combined with paraphrasing and summarizing.	*Predicted Result* Clients experience and understand their emotional state more fully and talk in more depth about feelings. They may correct the interviewer's reflection with a more accurate descriptor.

Empathic Response	*Predicted Result*
Experiencing the client's world as if you were the client; understanding his or her key issues and feeding them back to clarify experience. This requires attending skills and using the important key words of the client, but distilling and shortening the main ideas. In additive empathy, the interviewer may add meaning and feelings beyond those originally expressed by the client. If done ineffectively, it may subtract from the client's experience.	Clients will feel understood and move on to explore their issues in more depth. Empathy is best assessed by clients' reaction to a statement.
Basic Listening Sequence	*Predicted Result*
Select and practice all elements of the basic listening sequence: using open and closed questions, encouraging, paraphrasing, reflecting feelings, and summarizing. These are supplemented by attending behavior and client observation skills.	Clients will discuss their stories, issues, or concerns, including the key facts, thoughts, feelings, and behaviors. Clients will feel that their stories have been heard.

THE FIVE STAGES/DIMENSIONS OF THE WELL-FORMED INTERVIEW

1. Initiate the Session	*Predicted Result*
Develop rapport and structuring. "Hello, what would you like to talk about today?"	The client feels at ease with an understanding of the key ethical issues and the purpose of the interview. The client may also know you more completely as a person and professional.
2. Gather Data	*Predicted Result*
Draw out client stories, concerns, problems, or issues. "What's your concern?" "What are your strengths and resources?"	The client shares thoughts, feelings, behaviors; tells the story in detail; presents strengths and resources.
3. Set Goals Mutually	*Predicted Result*
"What do you want to happen?"	The client will discuss directions in which he or she might want to go, new ways of thinking, desired feeling states, and behaviors that might be changed. The client might also seek to learn how to live more effectively with situations that cannot be changed at this point (rape, death, an accident, an illness). A more ideal story ending might be defined.

4. Explore and Create	*Predicted Result*
Explore alternatives, confront client incongruities and conflict, restory. "What are we going to do about it?" "Can we generate new ways of thinking, feeling, and behaving?"	The client may reexamine individual goals in new ways, solve problems from at least those alternatives, and start the move toward new stories and actions.
5. Conclude	*Predicted Result*
Plan for generalizing interview learning to "real life" and eventual termination of the interview or series of sessions. "Will you do it?"	The client demonstrates change in behavior, thoughts, and feelings in daily life outside of the interview.
Confrontation	*Predicted Result*
Supportively challenge the client: 1. Listen, observe, and note client conflict, mixed messages, and discrepancies in verbal and nonverbal behavior. 2. Point out internal and external discrepancies by feeding them back to the client, usually through the listening skills. 3. Evaluate how the client responds and whether it leads to client movement or change. If the client does not change, the interviewer flexes intentionally and tries another skill.	Clients will respond to the confrontation of discrepancies and conflict with new ideas, thoughts, feelings, and behaviors and these will be measurable on the 5-point Client Change Scale. Again, if no change occurs, *listen*. Then try an alternative style of confrontation.
Client Change Scale (CCS)	*Predicted Result*
The CCS helps you evaluate where the client is in the change process. Level 1. Denial. Level 2. Partial examination. Level 3. Acceptance and recognition, but no change. Level 4. Generation of a new solution. Level 5. Transcendence.	You will be able to determine the impact of your use of skills and the creation of the *New*. Suggest new ways that you might try to clarify and support the change process though more confrontation or the use of another skill that might facilitate growth and development.
Focusing	*Predicted Result*
Use selective attention and focus the interview on the client, problem/concern, significant others (partner/spouse, family, friends), a mutual "we" focus, the interviewer, or the cultural/environmental/context. You may also focus on what is going on in the *here and now* of the interview.	Clients will focus their conversation or story on the dimensions selected by the interviewer. As the interviewer brings in new focuses, the story is elaborated from multiple perspectives.

Reflection of Meaning Meanings are close to core experiencing. Encourage clients to explore their own meanings and values in more depth from their own perspective. Questions to elicit meaning are often a vital first step. A reflection of meaning looks very much like a paraphrase, but focuses beyond what the client says. Often the words *meaning, values, vision,* and *goals* occur in the discussion.	*Predicted Result* The client discusses stories, issues, and concerns in more depth with a special emphasis on deeper meanings, values, and understandings. Clients may be enabled to discern their life goals and vision for the future.
Interpretation/Reframing Provide the client with a new perspective, frame of reference, or way of thinking about issues. Interpretations/reframes may come from your observations; they may be based on varying theoretical orientations to the helping field; or they may link critical ideas together.	*Predicted Result* The client may find another perspective or meaning of a story, issue, or problem. Their new perspective may have been generated by a theory used by the interviewer, from linking ideas or information, or by simply looking at the situation afresh.
Self-Disclosure As the interviewer, share your own related past personal life experience, *here and now* observations or feelings toward the client, or opinions about the future. Self-disclosure often starts with an "I" statement. Here-and-now feelings toward the client can be powerful and should be used carefully.	*Predicted Result* The client is encouraged to self-disclose in more depth and may develop a more egalitarian interviewing relationship with the interviewer. The client may feel more comfortable in the relationship and find a new solution relating to the counselor's self-disclosure.
Feedback Present clients with clear information on how the interviewer believes they are thinking, feeling, or behaving and how significant others may view them or their performance.	*Predicted Result* Clients may improve or change their thoughts, feelings, and behaviors based on the interviewer's feedback.
Logical Consequences Explore specific alternatives with the client and the concrete positive and negative consequences that would logically follow from each one. "If you do this . . . , then. . . ."	*Predicted Result* Clients will change thoughts, feelings, and behaviors through better anticipation of the consequences of their actions. When you explore the positives and negatives of each possibility, clients will be more involved in the process of decision making.

Information and Psychoeducation	*Predicted Result*
Share specific information with the client—e.g., career information, choice of major, where to go for community assistance and services. Offer advice or opinions on how to resolve issues and provide useful suggestions for personal change. Teach clients specifics that may be useful: help them develop a wellness plan, teach them how to use microskills in interpersonal relationships, educate them on multicultural issues and discrimination.	If information and ideas are given sparingly and effectively, the client will use them to act in new, more positive ways. Psychoeducation that is provided in a timely way and involves the client in the process can be a powerful motivator for change.
Directives	*Predicted Result*
Direct clients to follow specific actions. Directives are important in broader strategies such as assertiveness or social skills training or specific exercises such as imagery, thought-stopping, journaling, or relaxation training. They are often important when assigning homework for the client.	Clients will make positive progress when they listen to and follow the directives and engage in new, more positive thinking, feeling, or behaving.
Skill Integration	*Predicted Result*
Integrate the microskills into a well-formed interview and generalize the skills to situations beyond the training session or classroom.	Developing interviewers and counselors will integrate skills as part of their natural style. Each of us will vary in our choices, but increasingly we will know what we are doing, how to flex when what we are doing is ineffective, and what to expect in the interview as a result of our efforts.
Determining Personal Style and Theory	*Predicted Result*
As you work with clients, identify your natural style, add to it, and think through your approach to interviewing and counseling. Examine your own preferred skill usage and what you do in the session. Integrate learning from theory and practice in interviewing, counseling, and psychotherapy into your own skill set.	You, as a developing interviewer or counselor, will identify and build on your natural style. You will commit to a lifelong process of constantly learning about theory and practice while evaluating and examining your behavior, thoughts, feelings, and deeply held meanings.

APPENDIX II
THE FAMILY GENOGRAM

The individual develops in a family within a culture. You and your clients will more easily understand the self-in-relation concept if you help them draw a family genogram. We suggest that you consider developing both family and community genograms with many of your clients (see Chapter 9 on focusing). If you keep the genograms displayed during the session, they will remind you and your clients of the cultural/environmental context in which we all live. Moreover, some clients find them comforting as the genograms bring their family history to the interview. In a sense, we are never alone; our family and community histories are always with us.

Much important information can be collected in a family genogram. Many of us have family stories that are passed down through the generations. These can be sources of strength (such as a favorite grandparent or ancestor who endured hardship successfully). These family stories are real sources of pride and can be central in the positive asset search. There is a tendency to look for problems in the family history and, of course, this is appropriate. But use this important strategy positively whenever possible. Be sure to search for positive family stories as well as problems. How can family strengths help your client?

Children often enjoy the family genogram, and a simple adaptation called the "family tree" makes it work for them. The children are encouraged to draw a tree and put their family members on the branches, wherever they wish. This strategy has the advantage of allowing children to present the family as they see it, permitting easy placement of extended family and important support figures as well as immediate family members. Many adolescents and adults may also respond better to this more individualized and less formal approach to the family.

Box 1: Drawing a Family Genogram illustrates the major "how's" of developing a family genogram. The classic source for family genogram information is McGoldrick and Gerson (1985). Specific symbols and conventions have been developed that are widely accepted and help professionals communicate information to each other. There is a convention of placing an "X" over departed family members. Once we were demonstrating the family genogram strategy with a client and she commented, "I don't want to cross out my family members—they are still here inside me all the time." We believe that it is important to be flexible and work with the clients' view of family and their choice of symbols. The family genogram is one of the most fascinating exercises that you can undertake. You and your clients can learn much about how family history affects the way individuals behave in the here and now.

We have found family genograms helpful and use them frequently; however, there are situations in which some clients find them less satisfying than the community genogram. There is a Western, linear perspective to the family genogram that does not fit all individuals and cultural backgrounds. It is important to adapt the family genogram to meet individual and cultural differences. You will find *Ethnicity and Family Therapy* a most valuable and enjoyable tool to expand your awareness of racial/ethnic issues (McGoldrick, Giordano, & Garcia-Preto, 2005). The family

BOX 1   Drawing a Family Genogram

This brief overview will not make you an expert in developing or working with genograms, but it will provide a useful beginning with a helpful assessment and treatment technique. First go through this exercise using your own family; then you may want to interview another individual for practice.

Basic Relationship Symbols

Close	═══════
Enmeshed	≡≡≡≡≡
Estranged	—⁄⁄—
Distant	- - - - -
Conflictual	ΛΛΛΛ
Separated	—⁄—

1. List the names of family members for at least three generations (four is preferred) with ages and dates of birth and death. List occupations, significant illnesses, and cause of death, as appropriate. Note any issues with alcoholism or drugs.

2. List important cultural/environmental/contextual issues. These may include ethnic identity, religion, economic, and social class considerations. In addition, pay special attention to significant life events such as trauma or environmental issues (e.g., divorce, economic depression, major illness).

3. Basic relationship symbols for a genogram are shown on the left, and an example of a genogram is shown below.

4. As you develop the genogram with a client, use the basic listening sequence to draw out information, thoughts, and feelings. You will find that considerable insight into one's personal life issues may be generated in this way.

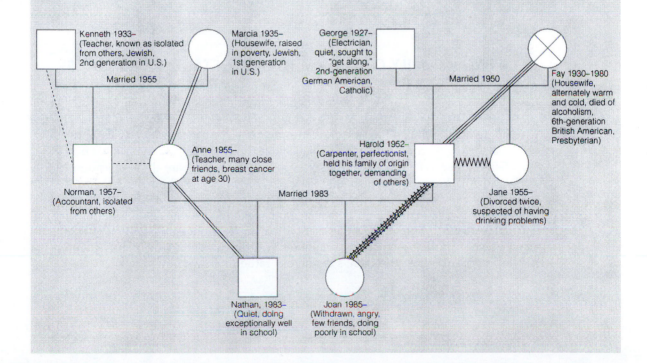

genogram is most effective with a client who has a nuclear family and actually can trace the family over time. We developed the community genogram because some of our clients were uncomfortable with the family genogram. Clients who have been adopted sometimes find the genogram inappropriate. Single-parent families may also feel "different," particularly when important caregivers such as extended family and close community friends are not included. We have talked with gay and lesbian clients who have very differing views of the nature of their family.

▲ EXERCISE 1

Developing a Family Genogram

Develop a family genogram with a volunteer client or classmate. After the two of you have created the genogram, ask the client the following questions and note the impact of each question. Change the wording and the sequence to fit the needs and interests of the volunteer.

- ▲ What does this genogram mean to you? (individual focus)
- ▲ As you view your family genogram, what main theme, problem, or set of issues stands out? (main theme, problem focus)
- ▲ Who are some significant others, such as friends, neighbors, teachers, or even enemies, who may have affected your own development and your family's? (others focus)
- ▲ How would other members of your family interpret this genogram? (family, others focus)
- ▲ What impact do your ethnicity, race, religion, and other cultural/environmental/ contextual factors have on your own development and your family's? (C/E/C focus)
- ▲ As an interviewer working with you on this genogram, I have learned _____ (state your own observations). How do you react to my observations? (interviewer focus)

USING A FAMILY GENOGRAM TO UNDERSTAND FAMILY ISSUES

Developing a genogram with your clients and learning some of the main facts of family developmental history will often help you understand the context of individual issues. For example, as you look at the family genogram in Box 1, what might be going on at home that results in Joan's problems at school? Why is Nathan doing so well? How might intergenerational alcoholism problems play themselves out in this family tree? What other patterns do you observe? What are the implications of the ethnic background of this family? The person with a Jewish and Anglo background represents a bicultural history. Change the ethnic background and consider how this would impact counseling. Four-generation genograms can complicate and enrich your observations. (Note: The clients here have defined their ethnic identities as shown. Different clients will use different wording to define their ethnic identities. It is important to use the client's definitions rather than your own.)

GLOSSARY

ABC model of crisis intervention is a three-stage system for conducting brief mental health interviews with clients whose level of functioning has significantly diminished following a psychosocial stressor. It is a problem-focused approach and is most effectively applied immediately or within 4–6 weeks of the stressor.

Acknowledging feelings is a brief acknowledgment of feeling that is just as helpful as a full reflection of feeling. In acknowledging feelings, you state the feeling briefly ("You seem to be sad about that," "It makes you happy") and move on with the rest of the conversation.

Active listening is a communication process that requires active participation, decision making, and responding on our part.

Additive empathy means that the interviewer response adds a link to something the client has said earlier or a new idea or frame of reference that helps the client see a new perspective. Wellness and the positive asset search can be vital parts of additive empathy.

Advocacy is speaking out for your clients; working in the school, community, or larger setting to help clients; and also working for social change.

Attending behavior is defined at supporting your client with individually and culturally appropriate visuals, vocal quality, and body language.

Basic empathy means that the interviewer response is roughly interchangeable with those of the client. The interviewer is able to say back accurately what the client has said.

Basic listening sequence (BLS) is used to draw out clients' stories by using open and closed questions, encouraging, paraphrasing, reflecting feelings, and summarizing. Your goal is to draw out the key facts, thoughts, feelings, and behaviors of the client's story.

Being-in-relation is similar to self-in-relation, but points out that all of us are beings who grow in the here-and-now moment as we interact with others.

Body language is nonverbal communication via the movements, gestures, or postures of the body. It is essential to build trust, show interest, express warmth, and demonstrate attentiveness and authenticity.

Bombardment/grilling means using too many questions. This may give too much control to the interviewer and tend to put many clients on the defensive.

Boundaries of competence are limits within which counselors practice, based on their education, training, supervised experience, state and national professional credentials, and appropriate professional experience. Counselors are expected to gain knowledge, personal awareness, sensitivity, and skills pertinent to working with a diverse client population.

Brief counseling may be a single interview, or it may extend to as many as 5 or 10 sessions. The key word is *brief*, emphasizing solutions rather than problems. Brief counselors believe that clients have their own answers and solutions available if we help them examine themselves and their goals.

Checklist is a list of specific items that need to be considered in the interview, particulary the first interview. Even the most experienced interviewer may forget to cover important basics during a session.

Checkout, or **perception check**, is a brief question at the end of the communication, asking the client for feedback on whether what you said was relatively correct and useful. Periodically checking with your client (a) helps to communicate that you are listening and encourages her or him to continue and (b) allows the client to correct any wrong assumptions you may have. You can also paraphrase or reflect feelings and use a checkout by ending your comment/sentence with a raised tone, indicating a questioning voice to check for accuracy.

Chunking refers to learning information not just in bits and pieces, but by organizing it into patterns.

Client Change Scale (CCS) is a five-level model that provides you with a systematic way to evaluate the effectiveness of interventions and to track client change in the here and now of the interview. At Level 1, clients may deny issues, at Level 2 deal with them partially, and at Level 3 acknowledge fairly accurately what is going on. Levels 4 and 5 indicate significant changes.

Clinical mental health counselors are skilled in both diagnostics and therapeutic treatment, in addition to having excellent interviewing and counseling skills. Clinical counselors need additional training in all these areas and in working with more severe client cases.

Clinical social workers focus on psychotherapy, but they are also involved in many interviewing and counseling activities. They have extensive training in the art of therapy, somewhat similar to mental health counselors and psychiatrists. In addition, they are fully trained in the broad field of social work and thus are particularly aware of contextual issues such as socioeconomics, the impact of community, and social justice concerns.

Closed questions enable you to obtain specific information and tend to be answered in very few words. Closed questions often begin with *is, are*, or *do*.

Cognitive and emotional balance sheet is an extension of logical consequences, but with each alternative written down in a list of gains and losses.

Cognitive balance adds focus concepts to the balance sheet, helping clients see the broader implications of their actions and decisions. All this is done on a balance sheet developed jointly by the helper and the client.

Community genogram is a visual map of the client in relation to the environment showing both stressors and assets within the person's life. It is a useful way to understand your client's history and identify strengths and resources. It can be used by itself or to supplement the family genogram.

Confidentiality is respecting your client's right to privacy. Trust is built on your ability to keep client information confidential.

Confrontation is not a direct, harsh challenge but is a more gentle skill that involves listening to clients carefully and respectfully and helping them examine themselves or their situations more fully. Confrontation is not "going against" the client; it is "going with" the client, seeking clarification and the possibility of a new resolution of difficulties. Think of confrontation as a supportive challenge.

Contextual interviewing helps counselors to fully discover client uniqueness by understanding the broader context of the client (friends, family, community).

Corrective feedback is a delicate balance between negative feedback and positive suggestions for the future. When clients need to seriously examine themselves, corrective feedback may need to focus on things that clients are doing wrong or behavior that may hurt them in the future.

Counseling is a more intensive and personal process than interviewing. It is less about information gathering and more about listening to and understanding clients' personal issues, challenges, and opportunities and helping them develop strategies for further growth and change.

Creative brainstorming encourages clients to "loosen up" and let any thought come to their mind. You can help through your own creativity, but focus on helping clients generate their own solutions.

Cultural intentionality is acting with a sense of capability and deciding from among a range of alternative actions. The intentional individual can generate alternatives in a given situation and approach a problem from different vantage points, using a variety of skills and personal qualities and adapting styles to suit different individuals and cultures.

Decisional counseling may be described as a practical model that recognizes decision making and the microskills as a foundation for most—perhaps all—systems of counseling. Another term for decisional counseling is problem-solving counseling.

Determining personal style and theory means that as you work with clients, identify your natural style, add to it, and think through your approach to interviewing and counseling. Examine your own preferred skill usage and what you do in the session. Integrate into your own skill set your learning from theory and practice in interviewing, counseling, and psychotherapy.

Directives direct clients to follow specific actions. Directives are important in broader strategies such as assertiveness or social skills training or specific exercises such as imagery, thought-stopping, journaling, or relaxation training. They are often important when assigning homework for the client.

Disclosure of what is going to happen in the session is an important part of the interview. Acquaint the client with what you are doing and its potential benefits. The general rule is to avoid surprises. Disclosure tends to build comfort and trust even when the next step of the interview may not be comfortable.

Dual relationships occur when a helper has more than one relationship with a client. Another way to think of this is as a conflict of interest.

Eliciting meaning often precedes reflection. Clients do not often volunteer meaning issues, even if they are central to their concerns; through careful attending and questions, you can bring out client stories and meaning.

Emotional balancing is giving special attention to how the client might feel after each possible decision.

Empathy is defined as experiencing the world as if you were the client, but with awareness that the client remains separate from you. Part of empathy is communicating that you understand.

Encouragers are verbal and nonverbal expressions the counselor or therapist can use to prompt clients to continue talking. Encouragers are minimal verbal utterances ("ummm" and "uh-huh"), head nods, open-handed gestures, and positive facial expressions that encourage the client to keep talking. Silence, accompanied by appropriate nonverbal communication, can be another type of encourager.

Encouraging is comprised of short responses that help clients keep talking. They may be verbal (repeating key words and short statements) or nonverbal (head nods and smiling).

Ethics means observing, practicing, and following professional standards. Particularly important issues for beginning interviewers are competence, informed consent, confidentiality, power, and social justice.

External conflict refers to conflict with others (friends, family, coworkers, or employers) or with a difficult situation, such as coping with failure, living effectively with success, having difficulty achieving an important goal financially or socially, or struggling to navigate one's cultural or ethnic identity.

Family genogram is a visual map of the client in relation to his or her family and brings out additional details of family history. It can be used by itself or to supplement the community genogram.

Feedback presents clients with clear information on how the interviewer believes they are thinking, feeling, or behaving and how significant others may view them or their performance.

HIPAA privacy refers to the Health Insurance Portability and Accountability Act (HIPAA) requirements regarding the protection and confidential handling of protected health information.

Incongruity occurs when clients have difficulty making important decisions, they feel confusion or sadness, or they have mixed feelings and thoughts about themselves or about their personal or cultural background.

Influencing is part of all interviewing and counseling. Through selective attention and the topics you choose consciously or unconsciously to emphasize (or ignore), you influence what the client says. It is impossible not to influence what happens in the interview. Confrontation, focusing, reflection of meaning, and interpretation/reframing have been identified as strategies of interpersonal influence, but listening skills can be as or more influential. Help clients set their own direction as much as possible. *Influencing* can be defined as "a power affecting a person, thing, or course of events, especially one that operates without any direct or apparent effort" (*Free Dictionary*, 2010).

Information includes advice, opinions, and suggestions; it can be an important part of interviewing and counseling.

Informed consent is the consent of a client to participate in the interview or therapy after being informed of its benefits, limitations, potential risks, and other pertinent information.

Intentional prediction is working intentionally in the interview so that you can anticipate predictable client responses. And even if the expected does not happen, you can intentionally flex and come up with a helpful alternative comment.

Intentional wellness plan includes both an assessment and a plan. An effective wellness plan can improve the quality of life and work of your clients.

Intentionality is acting with a sense of capability and deciding from among a range of alternative actions.

Internal conflict is a synonym for *incongruity*, but focuses on conflict inside the person. It may or may not involve external conflict as well.

Interpretation is an explanation of the meaning or significance of something. Closely allied to reframing, it is an important influencing skill to help clients discover new ways of thinking. It often comes from a specific theoretical framework.

Interviewing is the basic process used for gathering data, helping clients resolve their issues, and providing information and advice to clients.

Ivey Taxonomy provides definitions of the microskills and predicted results from using each skill.

Language of emotion is the vocabulary that people use to name and understand emotions.

Linking is the bringing together of two or more ideas to provide the client with a new insight.

Listening is the core of developing a relationship and making real contact with clients.

Logical consequences is a strategy in which you explore with the client specific alternatives and the concrete positive and negative consequences that would logically follow from each one. "If you do this . . . , then. . . ."

Meaning is defined as "what a word means" or its purpose and significance (Encarta, 2009).

Mentoring is the process by which a more experienced person passes on knowledge and helps a beginning professional open doors to opportunities.

Microaggressions are small insults and slights that accumulate and magnify over time. They are usually associated with discriminatory remarks and over time can become traumatic and injurious to physical and mental health.

Microskills are communication skill units that help you to interact more intentionally with a client.

Movement complementarity is paired movements that may not be identical but are still harmonious.

Movement synchrony refers to the situation in which people who are communicating well "mirror" each other's body language.

Multicultural competence is your awareness of yourself as a cultural being and your knowledge of, and skills to work effectively with, people different from you.

Multicultural issues call for interviewer behavior based on an awareness of the many dimensions of diversity identified in the RESPECTFUL model.

Multiculturalism includes race/ethnicity, gender, sexual orientation, language, spiritual orientation, age, physical ability/disability, socioeconomic status, geographical location, and other factors.

Multiple questions are another form of bombardment. Throwing out too many questions at once may confuse clients; however, it may enable clients to select which question they prefer to answer.

Negative consequences are undesirable, painful, or aversive results of our decisions or actions. Potential negative consequences could include the harm caused by smoking while pregnant, the disruption of long-term friendships, or the problem of moving children to a new school.

Negative feedback may be necessary when the client has not been willing to hear corrective feedback. However, it is to be used rarely and with care and empathy. Otherwise, the client is almost certain to reject what you say.

Nonattention is lack of, reduction, or removal of attending behavior; usually leads to ineffective communication.

Open questions are those that can't be answered in a few words. They encourage others to talk and provide you with maximum information. Typically, open questions begin with *what, how, why,* or *could*.

Oppression is the continued mistreatment and exploitation of a group of people by another person, organization, or culture.

Paraphrasing helps to shorten and clarify the essence of what has just been said, but be sure to use the client's main words when you paraphrase. Paraphrases are often fed back to the client in a questioning tone of voice.

Person-as-community points out that our family and community history live within each of us.

Person-centered theory was founded by Carl Rogers, who revolutionized the helping field by clearly identifying the importance of listening and empathy as the foundation for human change. A major assumption of person-centered theory is that the client is competent and self-actualizing. Interviewer skills are used to help the client uncover internal strength and resilience.

Political correctness (also **politically correct, P.C.,** or **PC**) is a term used to describe language that is calculated to provide a minimum of offense, particularly to the racial, cultural, or other identity groups being described. It has developed negative interpretations in parts of society. The issue is providing RESPECT for the client and using the language the client prefers. Avoid the politics and pay attention to the client before you.

Positive consequences are desirable and enjoyable results of our decisions or actions. Potential positive consequences could include a pay raise, a better school system, or a healthier environment. It is part of the logical consequences skill.

Positive feedback is concrete, strengths-based feedback that helps clients restory their problems and concerns. Whenever possible, find things right about your client.

Positive psychology brings together a long tradition of emphasis on positives within counseling, human services, psychology, and social work.

Power is part of the act of helping. Simply by being a client, one begins counseling with perceived lesser power than the interviewer. This power relationship may remain, or you can work toward a more equal relationship with the client.

Praise and supportive statements convey your positive thoughts about the client, even when you have to give troubling feedback. ("You can do it and I'll be there to help.")

Privilege is power given to people through cultural assumptions and stereotypes. There are many types of privilege including, among others, White privilege, social class privilege, economic privilege, religious privilege, and male privilege.

Problem solving consists of defining the problem, generating or brainstorming alternatives, and then deciding among the alternatives. Benjamin Franklin developed the first working model.

Psychiatrists are medical professionals who have a traditional M.D. degree plus additional study, internships, and residencies in hospitals. While once they typically engaged in psychoanalytically oriented therapy, they now tend to emphasize diagnosis and treatment with medication.

Psychoeducation is similar to information giving, but the counselor takes on more of a teaching role and provides systematic ways for clients to increase their life skills. Psychoeducation has become prominent in cognitive-behavioral counseling in recent years, as a way to prepare the client to meet challenging situations.

Psychologists evaluate, diagnose, and treat a wide variety of human issues; their role may range from working with the severely distressed to providing typical counseling services, sometimes with a major emphasis on education and prevention.

Psychotherapy focuses on deep-seated personality or behavioral difficulties.

Questions are an essential component in many theories and styles of helping and are used for gathering needed information. Open questions tend to draw out the most information, while closed questions can be useful for specifics.

Questions as statements are a way in which some interviewers use questions to sell their own points of view. If you are going to make a statement, do not frame it as a question.

Referral includes identifying the needs of the client that cannot be met by the counselor or agency and helping the client find appropriate help for those needs. When referral is necessary, provide sufficient support during the transfer process.

Reflection of feeling identifies a client's key emotions and feeds them back to clarify affective experience. With some clients, the brief acknowledgment of feeling may be more appropriate. Often combined with paraphrasing and summarizing.

Reflection of meaning focuses beyond what the client says. Often the words *meaning, values*, and *goals* will appear in the discussion. Clients are encouraged to explore their own meanings in more depth from their own perspective. It provides more clarity on values and deeper life meanings.

Reframing provide the client with a new perspective, frame of reference, or way of thinking about issues. This is closely related to interpretation, but reframing more often comes from the interviewer's frame of reference. Careful work with clients often enables them to do the reframing themselves.

Relationship—story and strengths—goals—restory—action is a five-stage model of the interview; it includes a basic decision-making model and adds relationship building, issues of confidentiality, and ethical concerns.

RESPECTFUL model (D'Andrea & Daniels, 2001) points out that all of us are multicultural beings and identifies ten dimensions of multiculturalism to consider.

Restatement is another type of extended encourager in which the counselor or interviewer repeats short statements, two or more words exactly as used by the client.

Selective attention is what we choose to listen to, consciously or unconsciously. Watch your patterns. Therapists and clients can miss issues because of selective attention.

Self-disclosure means that you as interviewer share your own related past life experience, here-and-now observations or feelings toward the client, or opinions about the future. Self-disclosure often starts with an "I" statement. Here-and-now feelings toward the client can be powerful and should be used carefully.

Self-in-relation defines the "individual self" as existing in relationship to others. It is an expansion of the former view of an "independent self-constructing self."

Social justice is the movement toward a more equitable society. It includes interviewers' efforts to effect social change, particularly on behalf of vulnerable and oppressed individuals and groups, as well as to help individual clients understand that their difficulties may not be caused by their behavior, but by an intolerant or oppressive system. Social justice information and exploration can also be part of an effective interview, particularly with clients who have experienced oppression.

Stuckness is an inelegant but highly descriptive term coined by the Gestalt theorist Fritz Perls to describe the opposite of intentionality. Other words that represent the same condition include *immobility, blocks, repetition compulsion, inability to achieve goals, lack of understanding, limited behavioral repertoire, limited life script, impasse*, and *lack of motivation*. Stuckness may also be defined as an inability to resolve conflict, reconcile discrepancies, and deal with incongruity.

Subtractive empathy refers to the interviewer response to the client that is less than what the client has said and perhaps even distorts or misunderstands the client. In this case, the listening skills are used inappropriately and take the client off track.

Summarizing uses client comments and integrates thoughts, emotions, and behaviors. Summarizing is similar to paraphrasing but used over a longer time span. It typically, but not always, includes a summary of emotions as well.

Supervision is an educational and monitoring role in which the supervisor simultaneously teaches and supports beginning professionals.

Treatment planning refers to specific written goals and objectives, which are increasingly standard and often required by agencies and insurance companies.

These goals are best negotiated with the client and represent that important third stage of the interview. Treatment planning also stresses the critical importance of action to ensure that these goals are actually implemented.

Verbal tracking is effectively following the client's story. Don't change the subject; stay with the client's topic.

Verbal underlining is giving louder volume and increased vocal emphasis to certain words and phrases.

Visual/eye contact refers to looking at people when you speak to them. Cultural and individual variations are important to note.

Vocal qualities refer to your vocal tone and speech rate, which should be appropriate to the client and the situation.

Wellness helps clients discover and rediscover their strengths through wellness assessment. Find strengths and positive assets in the client and in the support system; identify multiple dimensions of wellness.

***Why* questions** can put interviewees on the defensive and cause discomfort. As children, most of us experienced some form of "Why did you do that?" Any question that evokes a sense of being attacked can create client discomfort and defensiveness. On the other hand, "why" is often basic to helping a client explore issues of meaning and life vision. The person who has a why will find a how.

REFERENCES

Adams, D. (2005, April 25). Cultural competency now law in New Jersey. Available at http://www.ama-assn.org/amednews/2005/04/25/prl20425.htm

Alberti, R., & Emmons, M. (2001). *Your perfect right: A guide to assertiveness training* (8th ed.). San Luis Obispo, CA: Impact. (Original work published 1970)

Alim, T., Feder, A., Graves, R., Wang, Y., Weaver, J., Westphal, M., et al. (2009). Trauma, resilience, and recovery in a high-risk African-American population. *American Journal of Psychiatry, 165,* 1566–1575.

American Counseling Association. (2005). *ACA code of ethics.* Alexandria, VA: Author.

American Psychological Association. (2002). *Ethical principles of psychologists and code of conduct.* Washington, DC: Author.

Asbell, B., & Wynn, K. (1991). *Touching.* New York: Random House.

Barrett, M., & Berman, J. (2001). Is psychotherapy more effective when therapists disclose information about themselves? *Journal of Consulting and Clinical Psychology, 69,* 597–603.

Blair, R. (2001). Neurocognitive models of aggression, the antisocial personality disorders, and psychopathy. *Journal of Neurology, Neurosurgery, and Psychiatry, 71,* 727–731.

Blanchard, K., & Johnson, S. (1981). *The one-minute manager.* San Diego, CA: Blanchard-Johnson.

Boyle, P., Buchman, A., Barnes, L., & Bennett, D. (2010) Effect of a purpose in life on risk of incident Alzheimer disease and mild cognitive impairment in community-dwelling older persons. *Archives of General Psychiatry, 67,* 304–310.

Brammer, L., & MacDonald, G. (2002). *The helping relationship* (8th ed.). Boston: Allyn & Bacon.

Camus, A. (1955). *The myth of Sisyphus.* New York: Vintage Books.

Canadian Counselling Association. (2007). *Code of ethics.* Ottawa, Ontario: Author.

Carkhuff, R. (2000). *The art of helping in the 21st century.* Amherst, MA: HRD Press.

Carter, R. (1999). *Mapping the mind.* Berkeley: University of California Press.

Center for Credentialing and Education. (n.d.). *Distance Credentialed Counselor* (DCC). Downloaded from http://www.cce-global.org/home

Chang, E. C., D'Zurilla, T. J., & Sanna, L. J. (Eds.). (2004). *Social problem solving: Theory, research, and training.* Washington, DC: American Psychological Association.

Corey, G., Corey, M. S., & Callanan, P. (2010). *Issues and ethics in the helping professions* (8th ed.). Pacific Grove, CA: Brooks/Cole.

Council for Accreditation of Counseling & Related Educational Programs (CACREP). (2009). *2009 CACREP Standards.* Downloaded from http://www.cacrep.org/doc/2009%20Standards%20with%20cover.pdf

Damasio, A. (2003). *Looking for Spinoza: Joy, sorrow, and the feeling brain.* New York: Harvest.

D'Andrea, M., & Daniels, J. (2001). RESPECTFUL counseling: An integrative model for counselors. In D. Pope-Davis & H. Coleman (Eds.), *The interface of class, culture and gender in counseling* (pp. 417–466). Thousand Oaks, CA: Sage.

Daniels, T. (2010). A review of research on microcounseling: 1967–present. In A. Ivey, M. Ivey, & C. Zalaquett, *Intentional interviewing and counseling: Your interactive resource* (CD-ROM) (7th ed.). Belmont, CA: Brooks/Cole.

Daniels, T., & Ivey, A. E. (2006). *Microcounseling: Making skills work in a multicultural world.* Springfield, IL: Thomas.

Davidson, R., Pizzagalli, D., Nitschke, J., & Putnam, K. (2002). Depression: Perspectives from affective neuroscience. *Annual Review of Psychology, 53,* 545–574.

Decety, J., & Jackson, P. (2004). The functional architecture of human empathy. *Behavioral and Cognitive Neuroscience Reviews, 3,* 71–100.

de Shazer, S. (1988). *Clues: Investigating solutions to brief therapy.* New York: Norton.

de Shazer, S. (1993). Creative misunderstanding: There is no escape from language. In S. Gilligan & R. Price (Eds.), *Therapeutic conversations.* New York: Norton.

de Waal, E. (1997). *Living with contradiction: An introduction to Benedictine spirituality.* Harrisburg, PA: Morehouse.

Duncan, B. L., Miller, S. D., & Sparks, J. A. (2004). *The heroic client: A revolutionary way to improve effectiveness through client-directed outcome-informed therapy.* New York: Jossey-Bass/Wiley.

D'Zurilla, T. (1999). *Problem-solving therapy.* New York: Springer.

Egan, G. (2007). *The skilled helper* (8th ed.). Belmont, CA: Wadsworth.

Ekman, P. (2007). *Emotions revealed*. New York: Holt Paperbacks.

Ericsson, A. K., Charness, N., Feltovich, P., & Hoffman, R. R. (2006). *Cambridge handbook on expertise and expert performance*. Cambridge, UK: Cambridge University Press.

Fall, K., Fang, F., Mucci, L. A., Ye, W., Andrén, O., Johansson, J. E., et al. (2009). Immediate risk for cardiovascular events and suicide following a prostate cancer diagnosis: Prospective cohort study. *PLoS Medicine, 6*(12), e1000197.

Fang, F., Keating, N. L., Mucci, L. A., Adami, H. O., Stampfer, M. J., Valdimarsdóttir, U., & Fall, K. (2010). Immediate risk of suicide and cardiovascular death after a prostate cancer diagnosis: Cohort study in the United States. *Journal of the National Cancer Institute, 102*(5), 307–314.

Farnham, S., Gill, J., McLean, R., & Ward, S. (1991). *Listening hearts*. Harrisburg, PA: Morehouse.

Frankl, V. (1959). *Man's search for meaning*. New York: Simon & Schuster.

Frankl, V. (1978). *The unheard cry for meaning*. New York: Touchstone.

Free dictionary. (2011). Farlex, Inc. Available at http://www.thefreedictionary.com

Fukuyama, M. (1990, March). *Multicultural and spiritual issues in counseling*. Workshop presentation for the American Counseling Association Convention, Cincinnati.

Gergen, K., & Gergen, M. (2005, February). The power of positive emotions. *The Positive Aging Newsletter*, www.healthandage.com

Hall, E. (1959). *The silent language*. New York: Doubleday.

Hall, R. (2007). Racism as a health risk for African Americans. *Journal of African American Studies, 11*, 204–213.

Hargie, O., Dickson, D., & Tourish, D. (2004). *Communication skills for effective management*. Basingstoke, UK: Palgrave.

Hölzel, B., Carmody, J., Vangel, M., Congleton, C., Yerramsetti, S. M., Gard, T., & Lazar, S. W. (2011). Mindfulness practice leads to increases in regional brain gray matter density. *Psychiatry Research: Neuroimaging, 191*, 36–43.

Ishiyama, I. (2006). *Anti-discrimination Response Training (A.R.T.) program*. Framingham, MA: Microtraining Associates.

Ivey, A. E. (1996a). The community genogram: Identifying strengths. Downloaded from www.emicrotraining.com/resources/Community_genogram.pdf

Ivey, A. E. (1996b). Developing a community genogram: Identifying personal and multicultural strengths. Downloaded from www.emicrotraining.com/resources/CommunityGenogramInstructions.pdf

Ivey, A. E. (2000). *Developmental therapy: Theory into practice*. Framingham, MA: Microtraining Associates. (Originally published 1986)

Ivey, A. E. (2010). Decisional counseling. In A. E. Ivey, M. B. Ivey, & C. P. Zalaquett, *Intentional interviewing and counseling* (7th ed.). Belmont, CA: Brooks/Cole.

Ivey, A., D'Andrea, M., & Ivey, M. (2012). *Theories of counseling and psychotherapy: A multicultural perspective* (7th ed.). Thousand Oaks, CA: Sage.

Ivey, A. E., & Gluckstern, N. (1974). *Basic attending skills*. North Amherst, MA: Microtraining Associates.

Ivey, A. E., Gluckstern, N., & Ivey M. B. (2006). *Basic attending skills* (3rd ed.) [Book and accompanying videos]. Framingham, MA: Microtraining Associates.

Ivey, A. E., & Ivey, M. B. (1987). Decisional counseling. In A. E. Ivey, M. B. Ivey, & L. Simek-Downing, *Counseling and psychotherapy* (pp. 25–48). Englewood Cliffs, NJ: Prentice-Hall.

Ivey, A. E., & Ivey, M. B. (2007). *Intentional interviewing and counseling: Your interactive resource* [CD-ROM]. Belmont, CA: Brooks/Cole.

Ivey, A. E., Ivey, M. B., Myers, J., & Sweeney, T. (2005). *Developmental counseling and therapy: Promoting wellness over the lifespan*. Boston: Lahaska/Houghton-Mifflin.

Ivey, A. E., Ivey, M. B., & Zalaquett, C. P. (2010). *Intentional interviewing and counseling: Facilitating client development in a multicultural society* (7th ed.). Belmont, CA: Brooks/Cole.

Ivey, A. E., & Matthews, W. (1984). A meta-model for structuring the clinical interview. *Journal of Counseling and Development, 63*, 237–243.

Ivey, A. E., Pedersen P., & Ivey, M. B. (2001). *Intentional group counseling: A microskills approach*. Belmont, CA: Brooks/Cole.

Janis, I., & Mann, L. (1977). *Decision making: A psychological analysis of conflict, choice, and commitment*. New York: Free Press.

Jordan, J., Hartling, L., & Walker, M. (Eds.). (2004). *The complexity of connection: Writings from the Stone Center's Jean Baker Miller Training Institute*. New York: Guilford Press.

Kanel, K. (2008). *Crisis counseling: The ABC model* [DVD]. Framingham, MA: Microtraining Associates.

Kanel, K. (2011). *A guide to crisis intervention* (4th ed.). Belmont, CA: Brooks/Cole.

Keller, S. M., Zoeliner, L. A., & Feeny, N. C. (2010). Understanding factors associated with early therapeutic alliance in PTSD treatment: Adherence, childhood sexual abuse history, and social support. *Journal of Consulting and Clinical Psychology, 78*(6), 974–979.

The King Center. (2010). *Kingian principles of nonviolence.* Downloaded from http://www.thekingcenter.org/ProgServices/Default.aspx

Kolb, B., & Whishaw, I. (2003). *Fundamentals of human neuropsychology* (5th ed.). New York: Worth.

Kübler-Ross, E. (1969). *On death and dying.* New York: Macmillan.

Lane, P., & McWhirter, J. (1992). A peer mediation model: Conflict resolution for elementary and middle school children. *Elementary School Guidance and Counseling, 27,* 15–23.

Likhtik, E., Popa, D., Apergis-Schoute, J., Fidacaro, G. A., & Paré, D. (2008). Amygdala intercalated neurons are required for expression of fear extinction. *Nature, 454,* 642–645.

Marci, C. D., Ham, J., Moran, E., & Orr, S. P. (2007). Physiologic correlates of perceived therapist empathy and social-emotional process during psychotherapy. *Journal of Nervous and Mental Disease, 195*(2), 103–111.

Marlatt, G. A., & Donovan, D. M. (Eds.). (2007). *Relapse Prevention: Maintenance strategies in the treatment of addictive behaviors.* New York: Guilford Press.

Marshall, J. (2010, February 22). *Discovery News.* Retrieved May 10, 2010, from http://news.discovery.com/human/poverty-children-income-adults.html

McGoldrick, M., & Gerson, R. (1985). *Genograms in family assessment.* New York: W. W. Norton.

McGoldrick, M., Giordano, J., & Garcia-Preto, N. (2005). *Ethnicity and family therapy* (3rd ed.). New York: Norton.

McIntosh, P. (1988). *White privilege and male privilege: A personal account of coming to see correspondences through work in women's studies.* Wellesley, MA: Wellesley College Center for Women.

Meara, N., Pepinsky, H., Shannon, J., & Murray, W. (1981). Semantic communication and expectation for counseling across three theoretical orientations. *Journal of Counseling Psychology, 28,* 110–118.

Meara, N., Shannon, J., & Pepinsky, H. (1979). Comparisons of stylistic complexity of the language of counselor and client across three theoretical orientations. *Journal of Counseling Psychology, 26,* 181–189.

Myers, J. E., & Sweeney, T. J. (2004). The indivisible self: An evidence-based model of wellness. *Journal of Individual Psychology, 60,* 234–244.

Myers, J. E., & Sweeney, T. J. (Eds.). (2005). *Counseling for wellness: Theory, research, and practice.* Alexandria, VA: American Counseling Association.

National Association of Social Workers. (1999). *Code of ethics.* Washington, DC: Author.

National Organization for Human Services. (1996). *Ethical standards for human service professionals.* Available at http://www.nationalhumanservices.org/index.php?option=com_content&view=article&id=43&Itemid=90

Obonnaya, O. (1994). Person as community: An African understanding of the person as intrapsychic community. *Journal of Black Psychology, 20,* 75–87.

Office of the Surgeon General. (1999). *Mental health, culture, race, and ethnicity.* Washington, DC: Department of Health and Human Services.

Parsons, F. (1967). *Choosing a vocation.* New York: Agathon. (Originally published 1909)

Posner M. (Ed.). (2004). *Cognitive neuropsychology of attention.* New York: Guilford Press.

Power, S., & Lopez, R. (1985). Perceptual, motor, and verbal skills of monolingual and bilingual Hispanic children: A discrimination analysis. *Perceptual and Motor Skills, 60,* 1001–1109.

Ratey, J. (2001). *A user's guide to the brain.* New York: Vintage.

Ratey, J. (2008). *Spark: The revolutionary science of exercise.* New York: Hatchette/Little Brown.

Ratts, M., Toporek, R., & Lewis, J. (2010). *ACA advocacy competencies: A social justice framework for counselors.* Washington, DC: American Counseling Association.

Restak, R. (2003). *The new brain.* New York: Rodale.

Rigazio-DiGilio, S. A., Ivey, A. E., Grady, L. T., & Kunkler-Peck, K. P. (2005). *Community genograms: Using individual, family, and cultural narratives with clients.* New York: Teachers College Press.

Rogers, C. (1957). The necessary and sufficient conditions of therapeutic personality change. *Journal of Consulting Psychology, 21,* 95–103.

Rogers, C. (1961). *On becoming a person.* Boston: Houghton Mifflin.

Roysicar, G., Arredondo, P., Fuertes, J., Ponterotto, J., & Toperek, R. (2003). *Multicultural competencies, 2003.* Washington, DC: Association for Multicultural Counseling and Development.

Schwartz, J., & Begley, S. (2002). *The mind and the brain: Neuroplasticity and the power of mental force.* New York: Regan.

Seligman, M. (2009). *Authentic happiness.* New York: Free Press.

Shenk, D. (2010). *The genius in all of us: Why everything you've been told about genetics, talent, and IQ is wrong.* New York: Doubleday.

Shostrum, E. (1966). *Three approaches to psychotherapy* [Film]. Santa Ana, CA: Psychological Films.

Siegel, D. (2007). *The mindful brain.* New York: Norton.

Singer, T., Seymour, B., O'Dougherty, J., Kaube, H., Dolan, R., & Frith, C. (2004). Empathy for pain involves the affective but not sensory components of pain. *Science, 303,* 1157–1161.

Sklare, G. (2004). *Brief counseling that works: A solution-focused approach for school counselors and administrators.* Beverly Hills, CA: Corwin.

Smith, B., Tooley, E., Montague, E., Robinson, A., Cosper, J., & Mullins, P. (2009, August 8). Teaching resilience, sense of purpose in schools can prevent depression and improve grades. *Science Centric* [Online publication]. Available at http://www.sciencecentric.com/news/09080806-teaching-resilience-sense-purpose-schools-can-prevent-depression-improve-grades.html

Sue, D. W. (2010). *Microaggressions and marginality: Manifestation, dynamics, and impact.* New York: John Wiley.

Sue, D. W., & Sue, D. (2007). *Counseling the culturally diverse: Theory and practice* (5th ed.). New York: Wiley.

Sweeney, T. (1998). *Adlerian counseling: A practitioner approach.* Philadelphia: Taylor & Francis.

Tyler, L. (1961). *The work of the counselor* (2nd ed.). East Norwalk, CT: Appleton & Lange.

University of Massachusetts Memorial Medical Center, Behavioral Medicine Clinic. (2004). *Treatment plan.* Unpublished document, Griswold Mental Health Clinic, Palmer, MA.

Zalaquett, C. P., Ivey, A. E., Gluckstern-Packard, N., & Ivey, M. B. (2008). *Las habilidades atencionales básicas: Pilares fundamentales de la comunicación efectiva.* [Manuals and videos]. Framingham, MA: Microtraining Associates.

Zalaquett, C. P., & Lopez, A. D. (2006). Learning from the stories of successful undergraduate Latina/Latino students: The importance of mentoring. *Mentoring & Tutoring, 14*(3), 337–353.

Zhan-Waxler, C., Radke-Yarrow, M., Wagner, E., & Chapman, J. (1992). Development of concern for others. *Developmental Psychology, 28,* 128–136.

Zur, B. (2005). *The HIPAA compliance kit.* Sonoma, CA: O. Z. Publications.

NAME INDEX

SUBJECT INDEX